The Evolution of Sex Determination

# The Evolution of Sex Determination

**Leo W. Beukeboom**

**Nicolas Perrin**

OXFORD
UNIVERSITY PRESS

**OXFORD**
UNIVERSITY PRESS

Great Clarendon Street, Oxford, OX2 6DP,
United Kingdom

Oxford University Press is a department of the University of Oxford.
It furthers the University's objective of excellence in research, scholarship,
and education by publishing worldwide. Oxford is a registered trade mark of
Oxford University Press in the UK and in certain other countries



First published 2014
First published in paperback 2015



Published in the United States of America by Oxford University Press
198 Madison Avenue, New York, NY 10016, United States of America

British Library Cataloguing in Publication Data
Data available

Library of Congress Cataloging in Publication Data
Data available

ISBN 978–0–19–875367–4

*To Linda and Sjors, for the time they had to be without me*
*To my parents, for the provided opportunities they never had themselves*

*To Camille, Felix, and little princess Lissah*
*As you see, dad can be very serious when it comes to sex*

# Preface

The mechanisms that determine sex have been intriguing biologists for a long time. One reason may lie in the widespread occurrence and fundamental importance of sex throughout all forms of life. Another may stem from the profound sympathy we share with our fellow scientists for categories and numbers: clear-cut male and female phenotypes render studies on sex ratios and sex determination both tractable and appealing. Still another might simply be that we, biologists, share with our fellow humans an obsession for sex, hard-wired in our brains.

In 1983, James Bull supplied evolutionary biologists with a state-of-the-art synthesis on the 'Evolution of Sex Determining Mechanisms'. From his own words, writing this book was motivated by the author's astonishment in face of the bewildering variety of sex-determination systems. He intended his review to provide both an account of this variety, and a discussion of the processes that might have generated it. Accordingly, the first and largest part comprised a review of the main systems described so far, their occurrence throughout the animal kingdom, and a discussion of the possible transitions and evolutionary paths among them. The second part was more specifically devoted to the evolution of sex chromosomes, including a discussion of the possible role of sex-antagonistic selection in driving their dynamics. Bull's thorough synthesis successfully integrated current knowledge on the patterns of sex-determination systems and the processes driving their evolution; his book was very influential and is still much quoted.

More than 30 years have now passed, and, surprisingly, no full account of the field has been published since this landmark publication. The present book intends to fill this gap. In some way, it is an updating of Bull's work. We will re-appreciate and further expand several important insights from his book; we share, among others, the central idea that seemingly discrete sex-determination systems are connected by an underlying continuum. The proximate and ultimate causes driving transitions along this continuum are certainly key issues to address; this is the topic of our final chapter (Chapter 7), which expands in particular on the prominent role of genetic conflicts in this context. Along the same line, our Chapter 4 (quantitative genetics of sex determination) opposes the classical dichotomous view of environmental versus genotypic sex determination as distinct and mutually exclusive processes. In line with Bull's (1983) concept of 'sex tendency' (itself an extension of Bridges' (1925) genic balance theory), we rather see sex as a threshold trait, underlain by continuously distributed liability factors, under both genetic and environmental influence. Our Chapter 5 on the evolution of sex chromosomes updates and expands the second part of Bull's book. We further detail the selective forces acting on non-recombining genomic regions and document recent advances in the genomic aspects of Y- and W-chromosome degeneration. We also discuss recent insights in the distinct evolutionary fates of recombining sex chromosomes (X, Z, and pseudo-autosomal regions), as well as on other sex-limited chromosomes (U and V) evolving under haplo-genotypic sex determination.

Such common themes are to be expected: we evidently share Bull's astonishment of the bewildering diversity and impressive dynamics of sex-determination systems. Accounting for such dynamics and diversity naturally emerges as a central goal of any book with an evolutionary perspective on the topic. However, whereas many crucial questions raised in his time remain open issues today, much progress has also been made in domains that were largely unexplored (or even unforeseen) some 30 years ago. Hence, there are necessarily differences in scope.

One such difference concerns the phylogenetic coverage of our book. Bull's review was essentially limited to animal sex determination. Angiosperms were briefly mentioned, mostly through a list of families in which male or female heterogamety had been documented, and the suggestion that, due to the haploid phase of their life cycles, plants might provide good models to test theories on the evolution of sex chromosomes. In the 1980s, the eukaryote radiation outside animals, land plants and higher fungi, was largely *terra incognita* (the 'protist' black box). Since then, the full extension and richness of living forms has been progressively unveiled; land plants, animals, and fungi now appear as the emerging tips of an iceberg of eukaryotic variety. Importantly, meiotic sex turns out to be one of the unifying features of this radiation: the same genes controlling meiosis have been found in all main eukaryotic lineages. Details of sexual cycles have been described in increasing numbers of lower fungi, chlorophytes, rhodophytes, phaeophytes, diatoms, oomycetes, alveolates, amoeobozoa, and many other basally diverged lineages. Deciphering the structure and function of mating types has provided new perspectives on the origin of the male and female sexes. Our Chapter 2 (including both a printed part and electronic supplements) accounts for this newly discovered world. We cover the diversity of eukaryotes and propose a typology for sexual cycles; we delineate the similarities and differences between sexes and other mating categories, discuss ultimate causes for their emergence, and suggest possible evolutionary scenarios to connect the dots: how did, for instance, the mating types found in haplontic unicellular lineages evolve towards the diplo-genotypic sex determination found in multicellular plants or animals?

Other obvious developments concern the molecular mechanisms of sex determination. Bull's (1983) fourth chapter mostly focused on a list of few genes known at that time to affect sex determination in three laboratory models (mice, *Drosophila*, and *Caenorhabditis*), with a description of the phenotypes of some mutations. Since then, biology has undergone its molecular and genomic revolutions. Large arrays of new laboratory tools now allow functional analysis of the gene cascades and feedback loops that, from the initial triggers to the downstream effectors, determine the sexual fate of cell lineages. The genetic pathways of sex determination have been dissected in much detail, not only in classical laboratory models, but also in increasing numbers of non-model animals, plants, algae, and fungi. We now know a lot about the molecular mechanisms that determine differentiation of floral meristems and self-incompatibility systems in angiosperms, or mating-type signalling and zygote formation in fungi or algae. Our Chapter 3 summarizes recent advances on the molecular mechanisms that orchestrate sexual cycles; besides their intrinsic interest, such mechanisms also matter because they provide crucial insights in transition processes.

Our book also extends and refines the concept of sex determination. For reasons presented in Chapter 1, we think that the field of sex determination is not limited to the initial triggers that decide sexual fates in gonochoric animals or dioecious angiosperms, but also includes the range of developmental processes of the male and female functions (often referred to as 'primary sex differentiation', or 'sex development' in Bull's words). In line with this extensive definition of the field, we also discuss the processes of sex differentiation in sequential and simultaneous hermaphrodites.

Finally, we further expand the scope of the topics covered, by indulging in a few incursions upstream and downstream of sex determination itself, into domains that were explicitly avoided in Bull's book. Placing sex determination in a broader context, our Chapter 1 will briefly present current thinking about what sex is and why it evolved, and how sexes relate to mating types and other mating categories such as self-incompatibility systems or induction types. Our Chapter 6 will outline the consequences of specific sex-determination systems for the evolution of life histories, genetic systems, sex allocation, sexual selection, and speciation. We found it useful to delineate how these aspects feed back on each other in an evolutionary context.

Our book is thus meant as a contemporary overview and a comprehensive treatise of patterns and processes of sex determination for the beginning of this century. It makes also clear that the field is now expanding at an unprecedented pace. With the ongoing popularization of genomic tools, and the rapid development of next-generation sequencing techniques, we expect a flurry of important new discoveries in the coming years, which promise to drastically affect our vision of the field. It is highly unlikely that another 30 years will pass until the next synthesis on sex-determination systems becomes necessary.

**Lausanne and Groningen**
*May 2014*

# Acknowledgements

Many people and institutions have been indispensable for the writing of this book and greatly contributed to its coming into being. LWB was able to spend a sabbatical at the Department of Ecology and Evolution, University of Lausanne, thanks to grants from the Herbette Foundation and the Swiss National Science Foundation (grant IZK0Z3-142108). We took this opportunity to organize a workshop on sex-determination systems (La Sage, 6–9 June 2012), which provided some foundations for this book; this was helped by grants from the Herbette Foundation and the CUSO (Conférence Universitaire de Suisse Occidentale). France Pham provided much appreciated administrative support and Josiane Bonetti hard-to-access literature. We are also very grateful to Emma Plender-Hartman for efficiently composing the reference list and help with literature collection and indexing. Jeannette Steenmeijer spontaneously contacted us, not knowing we were writing a book; her enthusiastically and skilfully designed illustrations have significantly shaped the layout. We thank Ian Sherman and Lucy Nash of Oxford University Press for their guidance and diligent insistence on meeting the deadlines, and their editorial team for production of the book, in particular, Vicki Harley and Smita Gupta for copy-editing the manuscript.

We are indebted to a large number of people for scientific input. Daniel Bopp, Sander van Doorn, Mark Kirkpatrick, and Derek Roff critically read draft versions of several chapters and provided constructive comments. Tanja Schwander and Heath Blackmon shared information on insect sex determination. We take full responsibility for any mistake that might remain. We thank all participants of the sex-determination workshop in La Sage for inspiring presentations and vivid discussions, in particular Judith Mank, Doris Bachtrog, Katie Peichel, Nicole Valenzuela, Ikuo Miura, Eric Haag, Francesc Piferrer, John Pannell, and Tobias Uller. NP benefited from discussion with participants of the NESCent Tree of Sex working group; he offers special thanks to Mark Kirkpatrick and Sally Otto for inspiring discussions on the dynamics of recombining sex chromosomes. LWB acknowledges the many enlightening discussions with participants of the bi-annual meetings of European sex-determination researchers (Nöthiger meeting), in particular with long-term colleagues Martin Beye, Daniel Bopp, Giuseppe Saccone, Erwin Schmidt, and Walter Traut.

Over the many years we have also greatly benefited from discussions with our local colleagues and students. LWB particularly acknowledges stimulating interactions with Jetske de Boer, Sander van Doorn, Bart Pannebakker, Ido Pen, Tanja Schwander, Corneel Vermeulen, Franjo Weissing, Bregje Wertheim, and Louis van de Zande, as well as with past and present graduate students of sex determination, Barbara Feldmeyer, Elzemiek Geuverink, Bernd Grillenberger, Magdalena Kozielska, Wen-Juan Ma, Irene Mateo-Leach, and Eveline Verhulst. He also thanks all Evolutionary Genetics group members for their understanding and patience during his absence. Louis van de Zande deserves extra credit for efficiently taking over the group's leadership during his sabbatical leave.

NP wants to collectively thank all his colleagues from the Department of Ecology and Evolution for the friendly and stimulating atmosphere they contribute to create. He benefited in particular from helpful discussions on sex-determination and genetic systems with Laurent Keller, Michel Chapuisat, John Pannell, Jerôme Goudet, Laurent Lehmann, Séverine Vuilleumier, and Tanja Schwander. His reflection was stimulated by many interactions with past and present PhD students and post docs, among which Julie Jaquiéry, Christine Grossen, Olivier Blaser, Caroline Betto-Colliard, Christophe Dufresnes, Nicolas Rodrigues, Elisa Cavoto, Hélène Jourdan-Pineau, Thomas Broquet, Matthias Stöck, Alan Brelsford, and Samuel Neuenschwander.

Last but not least, we are very grateful to our families for bearing with us during this endeavour. We were a significant amount of time absent from our families, either mentally or physically, or both. We therefore contribute this book to our partners and children.

# Contents

CHAPTER 1

# What are sexes, and why are there sexes?

Sex occupies a central place in the human mind. The distinction between males and females is deeply hardwired in our brains; both the biological reality of **sexes** and their social construction (**genders**) accompany our daily life. Maleness and femaleness are fundamental symbolic elements in every human culture: gender is key to interpreting and organizing the world; according to Taoism, for instance, all phenomena in the universe stem from the interaction between the female (yin) and male (yang) principles. The number of genders is not necessarily limited to two, and neither is the notion of gender limited to social partners: many cultures and languages also assign genders to objects and words, which may be masculine, feminine, or neuter. The origin of sexes and genders, and their exact nature, has given rise to a diversity of myths and speculations. On biblical accounts, separate human genders originated when Eve was extracted from Adam's side (suggesting that the first human being was **hermaphrodite**; Mittwoch 2005). In the *Symposium*, a philosophical text of Plato (fourth century BC) dedicated to the nature of sex and love, Aristophanes is reported to say that in primal times people had doubled bodies, with faces and limbs turned away from one another. There were three genders: all male, all female, and 'androgynous', which were half male, half female. To limit their power, Zeus decided to cut these original people in half. Ever since, people run around looking for their lost half in order to recover their primitive nature: people descending from androgynes are therefore heterosexuals, while women separated from women, and men from other men, run after their own kind.

There seem to be good biological reasons to this overwhelming fascination for sex and sexes: they are required for human reproduction. That it takes a male and a female to make a baby has certainly been realized for long time. However, contrasting with the prominent role of females, who endure the full load of pregnancy and lactation, the exact contribution of males to the reproductive process has been matter of long debates and alternative interpretations. The Greek philosopher Pythagoras (570–495 BC) originated the 'spermist' view, according to which fathers contribute the essential characteristics to offspring, while mothers only contribute the material substrate. In 1676, the Dutch draper Antonie van Leeuwenhoek (1632–1723), used his homemade 270-fold magnifying lens to investigate 'what, without sinfully defiling myself, remains as a residue after conjugal coitus', and established that this mysterious substance—semen—was inhabited by a multitude of tiny eel-like animals. He suggested these 'spermatic animalcules' to play a vital role by providing its substance to the embryo, while the egg would provide nourishment. In 1694, Nicolaas Hartsoeker (1656–1725), assistant of van Leeuwenhoek and co-discoverer of spermatozoids, illustrated his spermist view with a picture of a tiny preformed human ('homunculus') within a human sperm cell. In the eighteenth century, the Italian priest Lazzaro Spallanzani (1729–1799) demonstrated experimentally that male sperm is necessary for fertilization of female ova: by fitting trousers to male frogs (in order to prevent semen from mixing with eggs) he found that eggs were not fertilized 'for want of being bedewed with semen'.

*The Evolution of Sex Determination*. Leo W. Beukeboom and Nicolas Perrin.

However, he considered the proteinaceous and fatty substances forming the bulk of the semen to play the essential role, and assimilated Leeuwenhoek's animalcules to parasites, akin to protozoa. This opinion was still shared in the nineteenth century, notably by the Germano-Russian naturalist Karl Ernst von Baer (1792–1876, otherwise discoverer of the mammalian ovum), who accordingly coined the name 'spermatozoa' in 1827. The dispute about the exact nature of spermatozoa was settled in 1841 by the Swiss embryologist Albert von Kölliker (1817–1905), who, after examining fertilization processes in a diversity of marine animals under the microscope, concluded that spermatozoa were not independent animals, but the products of cells in the testis, and that contact with the egg was needed for successful reproduction. The fusion between egg and sperm nuclei was actually observed in 1876 by the German zoologist Oskar Hertwig (1849–1922).

Similarly, the way sexes are determined has been interpreted in many ways across time and cultures (Mittwoch 2005). Parmenides (sixth century BC) proposed that the side of the womb in which the embryo develops determines its sex; according to Anaxagoras (fifth century BC), however, the side of the father's testis was the deciding factor. Aristotle (fourth century BC) criticized both theories, citing evidence that in animals, embryos of both sexes can be found at the same side of the uterus, and that men with a single testis can have children of both sexes. Following Empedocles' ideas about the four elements, Aristotle proposed that males are characterized by an abundance of fire (the superior element), and for this reason are hot and dry, while females, with an abundance of water, are rather cold and wet. Accordingly, he suggested the sex of a child to be determined by the heat of the male partner during sexual intercourse. In his book, "L'art de faire des garçons" ('The art of having boys'), the French anatomist Michel Procope-Couteau (1684–1753) resurrected the views of Parmenides and Anaxagoras, and suggested the best way to control a child's sex would be to remove the testis or ovary connected with the unwanted sex; though a less drastic mean for ladies would be to lie on the correct side, and let gravity do the rest (Mittwoch 1985). Environmental or **epigenetic** theories of sex determination have actually prevailed until recently. At the end of the nineteenth century, food was thought to be the most decisive factor, and sex to be determined by the mother's nutrition during the first three months of pregnancy; a poor diet was thought to result in males, and a better diet in females.

These epigenetic views of **sex determination** were progressively abandoned following the discovery of sex chromosomes (Chapter 5). Henking (1891) first observed that one element in meiosis of male firebugs *Pyrrhocoris apterus* (with an XX-XO system) was only transmitted to half of the sperm. He referred to this element as 'X' for unknown. McClung (1902) postulated this to be a chromosome, responsible for the inheritance of sex, and called it the **'X' chromosome**. The involvement of sex chromosomes in sex determination was further supported by the work of Bridges (1916, 1925) who analysed *Drosophila* individuals with aberrant constitutions of sex chromosomes (Box 3.18). At about the same time (early twentieth century) biologists realized that human chromosomal constitution also differs between males and females; out of our 23 pairs of chromosomes, one is heteromorphic in males (with a small **Y chromosome** pairing with its large X **homologue**), whereas this pair is **homomorphic** (XX) in females. As progressively realized later in the twentieth century, sex differences in humans are underlain by genetic differences: the small Y chromosome contains a few genes (including *Sry*) that are required for the determination and differentiation of maleness. A child's sex depends on whether the paternal sperm cell that fertilized the maternal ovum contributed an X or a Y chromosome. Since these early days, morphologically and genetically distinct sex chromosomes have been described from many taxa, and much has been learned about the origin and subsequent evolution of sex chromosomes (Chapter 5).

Given that sex seems such a crucial feature, essential for reproduction, one might expect sexes to play a key role in the life history of all living organisms, and all aspects of sex determination and differentiation to follow common pathways across lineages. This notion, however, has been largely dispelled by modern evolutionary biology. The broadest generalization that can be made is that most eukaryotes indulge in some form of sex. Besides this, nearly all of the attributes that we intuitively tend to assign to sex are rather lineage specific. Evolutionary perspectives on sex and sexes are actually delivering many counterintuitive insights. To mention just a few:

- Sex is fundamentally not a reproductive process; it just turns out to be associated with reproduction in many multicellular lineages. In fact, we do not know for sure why sex exists: costs are much more obvious than are benefits (Section 1.1.2).
- Sex does not imply sexual intercourse: external fertilization, following broadcast spawning of **gametes**, is common in many organisms. Moreover, it does not necessarily take a male and a female to make a baby: in many insects for instance (including ants and bees), males have no fathers (Box 3.21).
- Sex differentiation does not require genetic differences: in many sexually dimorphic species, sex is determined by environmental or social cues (Section 2.1.6).
- Males and females are latecomers in the long history of sex; instead of two differentiated sexes, many organisms have morphologically similar 'mating types', which in fungi may count by dozens or hundreds (Sections 1.2 and 3.2).

Among the many surprises and oddities that biological investigations on sexes are revealing, one is particularly intriguing: even in higher eukaryotes with well-differentiated male and female roles, the initial steps of sex-determination pathways are bewilderingly diverse and extraordinarily labile. Evolutionary biologists seek to understand this conundrum: what might drive the surprising dynamics of such a fundamental developmental process that, at the end, always leads to the same and simple outcome: the production of males and females? This issue will be addressed in the final chapter of this book (Chapter 7). Before reaching it, we will have to examine the diversity of sex-determination systems across eukaryote lineages (Chapter 2), detail the molecular mechanisms and quantitative genetics of sex determination (Chapters 3 and 4), investigate the evolutionary fate of sex chromosomes (Chapter 5), and outline the evolutionary correlates of sex-determination systems (Chapter 6).

Chapter 1 introduces a few important definitions and concepts. We will cover the diversity of mating categories (including mating types, sexes, genders, and **self-incompatibility (SI) systems**) and discuss why they exist; we will introduce **sex-ratio selection** and **sexual selection**, which are the main evolutionary forces moulding the determination and differentiation of sexes; and we will end up with a definition of what 'sex determination' is. Beforehand, however, one cannot escape the more fundamental questions of what sex is and why it evolved.

## 1.1 Meiotic sex

### 1.1.1 What is sex?

There are at least two biological definitions of sex. One considers as sex any genetic exchange between individuals. Another defines sex by the occurrence of meiosis; this is the one we will adopt here. Although these two definitions are largely overlapping, they present important differences: from our definition, virus transmission is not sex; reciprocally, **selfing** is a form of sex, as are diverse forms of **parthenogenesis**, as soon as they involve meiosis.

Meiotic sex is a complex two-step process, initiated by **syngamy**, that is, the fusion of two **haploid** cells to form a **diploid zygote**, and ending with the reduction to haploidy through meiosis, a specialized mechanism that ensures **recombination** between the two interacting genomes (Box 1.1). The haploid genomes resulting from recombination

**Box 1.1 Meiosis**

Meiosis bears a number of similarities with mitosis; both involve cell division and genetic recombination, for which they use many of the same mechanisms. However, the two processes also differ in several important aspects: mitotic recombination occurs between sister chromatids, is mostly of non-**crossover** (NCO) type, and results in two genetically identical diploid cells, while meiotic recombination occurs between **homologous** chromosomes, involves both crossovers (CO) and non-crossovers, and results in four genetically unique haploid cells (Fig. 1.1).

After a chromosome duplication has produced two pairs of sister chromatids attached at their **centromere** (step 3 in Fig. 1.1), the maternal and paternal homologues pair

*continued*

**Box 1.1** *Continued*

with each other; recombination is initiated by a **double-strand break** induced by the meiotic enzyme SPO11 (step 4a), followed by 'resection', during which DNA from the 5′ ends of the break is digested by exonucleases, creating 3′ overhangs of single-stranded DNA (step 4b). Specific recombinases (RAD51 and DMC1, the latter specific to meiosis) then attach to the single-stranded DNA, forming a nucleoprotein filament that searches for a similar DNA sequence from the duplex homologue, and invades it (step 4c). Upon strand invasion, a displacement loop (D-loop) is formed between the invading overhang and its homologue, allowing a DNA polymerase to extend the end of the invading 3′ strand, using the homologue as a template (step 4d). This changes the D-loop to a cross-shaped structure known as a **Holliday junction**. Following this stage, two distinct pathways, either double Holliday junction (step 4e) or synthesis-dependent strand annealing, lead to either CO or NCO recombination events, respectively (Andersen & Sekelsky 2010). CO recombination involves physical exchange of large segments of DNA between homologues, while NCO only involves a transfer of information between homologues (to repair the double-strand break), but no physical exchange. Even during meiosis, the majority of recombination events appear to be of NCO type (on average, there are less than two CO events per chromosome and per meiosis). The two sets of homologues then segregate in two daughter cells, each containing a haploid number of chromosomes composed of two chromatids (step 5). A second round of cell division then occurs, separating the sister chromatids to produce four haploid cells (step 6).

differ from those that entered fusion. The sexual cycle (Fig. 1.1) can thus be envisioned as a back and forth alternation of generations between a haploid and a diploid stage (Chapter 2).

Meiotic sex occurs in most lineages of eukaryotes (Chapter 2); it is widespread and conspicuous among large multicellular forms, but often difficult to assess in unicellular organisms, where it may be facultative, and only expressed in specific and cryptic conditions (Burt et al. 1996; Lee et al. 2010b). For these reasons, evidence for sex is sometimes indirect, provided by signs of genetic recombination or the presence of conserved and seemingly functional meiotic genes. Meiosis itself is both very complex and remarkably conserved, and as such unlikely to have arisen twice (Dacks & Roger 1999). Meiotic sex or meiosis-related genes are found in basal eukaryotes, such as *Trichomonas*, *Giardia* (Excavata), and *Ostreococcus* (Chlorophyta; e.g. Ramesh et al. 2005; Derelle et al. 2006; Malik et al. 2008; Schurko et al. 2009; Halary et al. 2011), supporting the idea that sex has a single evolutionary origin in basal eukaryotes, originating more than 1.5 billion years ago (Javaux et al. 2001). Sex has been lost in a few lineages (e.g. in about 0.1% of animals; Vrijenhoek 1998). Lineages that undergo only asexual reproduction (or obligate selfing) seem to show high extinction rates: they tend to occupy mostly the tips of phylogenetic branches and, thus, are recent and presumably short lived (Simon et al. 2003; Igic et al. 2008; Goldberg et al. 2010; Beck et al. 2011). Although amoebae have been assumed to be anciently asexual, Lahr et al. (2011) showed the majority of amoeboid lineages to be anciently sexual; most asexual groups have probably arisen recently and independently. There is, however, indisputable evidence for a few old asexual lineages (Judson & Normark 1996): in bdelloid rotifers, for instance, the differentiation between homologous alleles is so large that it can only be explained by a long divergence without recombination (the 'Meselson effect'; Kuhn et al. 2001; Enjalbert et al. 2002; Roose-Amsaleg et al. 2002; Schurko et al. 2009).

Several forms of asymmetric genetic exchange exist in prokaryotes, predating meiotic sex. Bernstein & Bernstein (2010) have pointed out fundamental similarities between meiotic sex and bacterial **transformation**. The latter consists in a stable and inheritable genetic change induced by the uptake of naked DNA from the environment. For transformation to happen, bacteria must be in a state of 'competence', which occurs as a time-limited response to environmental conditions (such as starvation and density). With rare exceptions, the DNA integrated into the host chromosome is derived from conspecific individuals, and is thus homologous to the resident chromosome. Transformation might primarily constitute an adaptation for repairing DNA (Michod et al. 2008). A key similarity between bacterial transformation and meiotic sex is that DNA originating from two individuals joins up so that homologous sequences are aligned with each other, followed by exchange of genetic

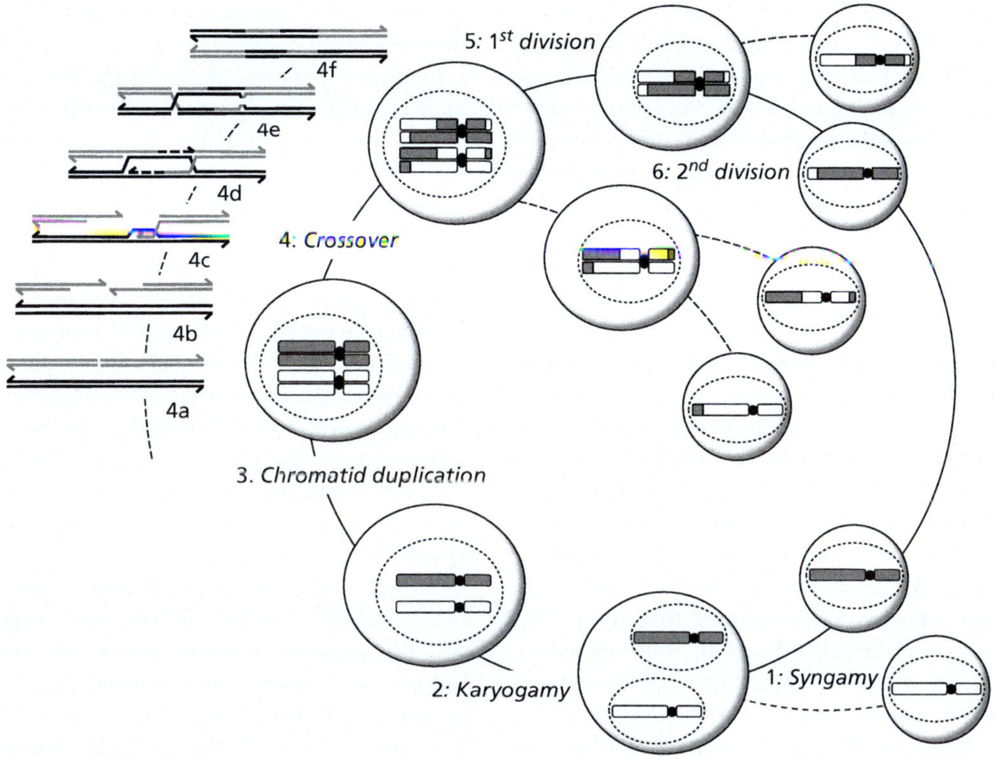


**Figure 1.1** Stages of meiosis. Arranged clockwise: 1. Fusion of two haploid gametes to form a di-haploid zygote (syngamy). 2. Fusion of the two haploid nuclei to form a diploid zygote (karyogamy). 3. Duplication of the two chromosome sets into sister chromatids, attached by their centromere. 4. Crossover recombination between the paternal and maternal homologous chromosomes; the outer arc details the establishment and resolution of a double Holliday junction: 4a double strand break, initiated by the enzyme SPO11; 4b resection; 4c strand invasion; 4d second end capture and synthesis; 4e ligation and formation of a double Holliday junction; 4f resolution of the double Holliday junction, resulting here in a crossover because different strands are cut at each junction. 5. First cell division, with segregation of the paternal and maternal centromeres. 6. Second cell division, with segregation of sister chromatids, forming four haploid cells.

information. This recombination process is catalysed in both prokaryotes and eukaryotes by evolutionarily related enzymes with similar functions. In particular, RAD51 and DMC1, involved in the search for homology and strand-pairing stages during eukaryote meiosis (Box 1.1), are **orthologues** of the recombinase RecA used for DNA repair and maintenance in bacteria. In both cases, the RecA or RecA-like protein assembles on single-stranded DNA to form a presynaptic filament, which then attaches to a duplex DNA molecule and searches for homology in its target. When the presynaptic molecule locates a homologous sequence in the duplex molecule, it forms a DNA joint, which is then processed further to complete homologous recombination repair (Bernstein et al. 2011).

### 1.1.2 What use is sex?

Meiotic sex is often associated with reproduction, and sometimes strictly so: as humans, we reproduce only sexually, as do all other mammals. Intrinsically, however, sex is not a reproductive process. Most lineages across the eukaryote phylogeny have maintained a form of asexual reproduction (occurring during either the haploid phase, the diploid phase, or both), and sex is often facultative (e.g. in cyclical parthenogens). In many groups the two processes are completely dissociated: dinoflagellates, for instance, only reproduce asexually, via haploid mitoses (Section S2.2.3b); during the sexual cycle, two haploid cells fuse to form a zygote, but a single haploid nucleus is left after meiosis (the three other products are discarded), so that sex actually halves the number of individuals. Similar patterns occur in the **diplontic** cycles of unicellular ciliates or diatoms.

What then is the aim of sex, if it is not reproduction? Given its early origin in eukaryotes and nearly universal distribution, it might seem that sex should provide large and obvious evolutionary advantages. However, it turns out that costs are much easier to identify than are benefits.

### 1.1.2a The costs of sex

Such costs are actually manifold. First are the intrinsic physiological costs of meiosis, which are essentially opportunity costs. Meiosis takes much longer to achieve than mitosis in unicellular organisms (Lewis 1983; Otto 2009; Lehtonen et al. 2012). During the time needed to conjugate and reorganize nuclei, a yeast cell could go through eight mitoses, resulting in $2^8 = 256$ daughter cells (Burt 2000). This may seem an extremely heavy cost (literally, a 256-fold cost of sex), largely alleviated, however, by expressing sex when environmental conditions are least favourable for growth. The fact that sex (like transformation in bacteria) is often expressed in stressful condition (e.g. under nitrogen depletion in *Chlamydomonas*) does not necessarily imply that it is adaptive in such conditions, but rather that it is expressed when opportunity costs are minimal. As a result of this association, zygotes often specialize into resting stages, both in unicellular and in multicellular eukaryotes (as in *Chlamydomonas* and *Daphnia*, respectively).

Additional costs accrue when sex involves a genetic exchange between different individuals (i.e. **outcrossing**, as opposed to diverse forms of **inbreeding** such as selfing, **automixis** or **same-clone mating**). First, outcrossing may induce an important **recombination load**, by breaking down beneficial combinations of genes built up by selection. Second, the mating process itself is costly. The building of reproductive male and female organs (e.g. flowers in angiosperms), and secondary sexual traits in many **gonochoric** animals, imposes a heavy burden on individual energy budgets. Finding a partner may be a difficult task when population density is low. Mating also facilitates the spread of sexually transmitted diseases, and may induce predation risks (the more so when it involves conspicuous behavioural and morphological displays).

Finally, in many species with separate sexes, males contribute only genes to the offspring. There is thus a cost of producing males (who tend to be very inefficient in converting resources into offspring) relative to asexual reproduction, where only females are produced. This is known as the 'two-fold cost of sex' (sometimes misinterpreted as a cost of 'genome dilution'; see Lehtonen et al. 2012 for a clarification). This two-fold cost may also be alleviated by producing males when the opportunity costs of sex are lowest (as do some cyclical parthenogens such as *Daphnia*, aphids or monogonont rotifers) or via the evolution of post-zygotic paternal care: what matters is not maleness per se, but how much males contribute to the reproductive effort.

It should be remembered, however, that all costs other than the intrinsic physiological costs of meiosis do not stem from sex per se, but only from the way sex is performed in some organisms. Sex originated in **isogamous** ancestors, long before the appearance of males and females, and thus did not incur the two-fold cost. It also first took place among similar genomes, and thus did not generate any recombination load: the first sexual eukaryote must have been automictic (thereby with little mating costs either). Finally, as already mentioned, expressing sex when opportunity for growth is lowest largely alleviates the intrinsic costs of meiosis. Hence, the origin of sex might have been an almost cost-free transition. However, features associated with its subsequent evolution (in particular **anisogamy** and gonochorism) do involve unavoidable costs. Many lineages evolved adaptations to alleviate these costs, but very few have renounced sex, pointing to strong evolutionary benefits.

### 1.1.2b The benefits of sex

A distinction must be made between the origin of sex, and its maintenance and subsequent evolution. The latter is likely driven—at least partially—by indirect benefits, among which is the building of genetic diversity for selection to work on. As already noticed, however, the first sexual eukaryote had to be automictic, so that genetic mixing was purposeless. In addition, indirect benefits are expressed in the long term, which might not suffice to counteract immediate costs; direct benefits seem thus required to account for its origin.

Meiotic sex might have first evolved as a repair mechanism (Bernstein et al. 2011). Environmental stress often leads to the production of reactive oxygen species, known to cause DNA damage. When a single strand is damaged, the information lost can be recovered by removing the damaged sequence and fill the resulting gap with a copy of the intact strand. However, reactive oxygen species also cause double-strand damage, including double-strand breaks; genetic information is thereby lost from both strands. Such damaging changes cannot be repaired in unicellular haploid organisms; if non-lethal, they will be transmitted to daughter cells and progressively accumulate through generations. In contrast, syngamy with a similar cell will produce a diploid zygote with two copies of each sequence, allowing repair of the damaged section via homologous recombination. Recombinational repair is effective against many different types of genomic damage, and particularly efficient at overcoming double-strand breaks. Thus, the most primitive form of sex may have involved haploid unicells copying intact strands from similar unicells to repair their own damaged DNA.

To what extent repair still plays a role in maintaining meiotic sex is a matter of debate (Bernstein et al. 2011). Explanations for the adaptive function of meiosis that focus exclusively on crossovers fail to account for the majority of recombination events, which are of the NCO type and therefore do not affect linkage disequilibria (Box 1.1); NCO recombination might still act as a repair mechanism. In addition, sex seems more ubiquitous than the benefits stemming from genetic variance would suggest : many sexual eukaryotes outcross only rarely or not at all; **mating-type switch** in yeast (Section 3.2.3a), **homothallism** in ascomycetes (Section 3.2.3b), and **autogamy** in ciliates (Section 3.4.2a) have evolved as ways to undergo meiotic sex with genetically identical individuals. In such cases, meiosis generates little or no genetic variation. According to this interpretation, DNA repair might provide sufficient short-term advantages to maintain meiosis, while the build-up of genetic variation provides long-term benefits (Bernstein et al. 2011).

Opposing this interpretation, apoptosis and cell turnover in multicellular organisms might seem a more efficient way of eliminating damaged cells than meiosis-induced repair. Furthermore, diploid organisms already carry two copies of each gene, which might be used as templates for repair without the need to undergo meiotic sex. Bdelloid rotifers, which suffer from extensive DNA damage during desiccation and rehydration processes (Gladyshev & Meselson 2008) do not undergo meiotic sex, but developed alternative ways to repair double-strand breaks via **homologous recombination** between non-sister co-linear chromosome pairs, maintained as templates (Welch et al. 2008).

Alternatively, homologous recombination might protect genomes against **selfish elements**. Filamentous ascomycetes activate a repeat-induced point mutation mechanism during their sexual cycle as a way to control **transposable elements**: duplicated sequences are detected and inactivated via C:G to T:A mutations (Galagan & Selker 2004). In *Aspergillus*, DNA and RNA viruses are only transmitted via asexual **spores** (conidiospores); **ascospores**, produced sexually by either outcrossing or same-clone mating, are free of these parasites (Coenen et al. 1997). Sex might also serve as a rejuvenating process via developmental resetting; Gorelick & Carpinone (2009) argued that multicellular organisms require meiosis in order to reset epigenetic signalling and maintain **ploidy** level throughout generations. In diatoms, the sexual cycle is required to restore cell size after a series of mitotic divisions (Section S2.2.2); in ciliates, sex allows avoidance of senescence by restoring telomere length after a series of fissions (Sections S2.2.3 and 3.4.2a). Finally, sex in many organisms is strictly associated with the formation of resistant propagules, required to survive stressful conditions; in *Chlamydomonas*, for instance, the same molecular mechanisms coordinate zygote formation, cell wall building, and stress resistance (Section 3.2.2e). Although some of the above associations might reflect fundamental evolutionary benefits of sex, interpretation of data is obviously complicated by adaptations that are not intrinsic to sex, but secondarily associated with it. In several instances, these secondary associations may suffice to prevent the loss of sex in the short term (e.g. Rispe et al. 1998; De Vienne et al. 2013).

Indirect benefits stemming from CO recombination and segregation are generally thought to play a crucial role in the long-term maintenance of sex, in line with Weismann's (1889) original suggestion that sex creates genetic variance for selection to work on (Burt 2000). However, although this idea is simple and appealing, its formalization has met unexpected difficulties. First, sex does not necessarily generate variance, it can also reduce it; second, variation generated through sex often reduces offspring average fitness (Otto & Lenormand 2002; Otto 2009). In infinite populations at equilibrium, sex only causes fitness improvements under a very narrow range of parameter values, where the fitness surface is negatively curved on a multiplicative scale, but positively curved on an additive scale (Barton 1995; Otto 2003). For most of the parameter space, sex breaks up successful gene combinations brought by selection, and is thus harmful in the short term. Basically the same conclusions hold in non-equilibrium populations under **directional selection** (mutation–selection balance): stringent conditions on the curvature of fitness surfaces are required for sex to be favoured, which makes genetic mixing unlikely as a general explanation for the widespread usage of sex (De Visser & Elena 2007; Otto 2009).

Temporal changes in the patterns of selection (e.g. stemming from host-parasite co-evolution; the **Red Queen hypothesis**; Bell 1982) may play a role: when conditions change rapidly over time, genetic associations built up by past selection can become detrimental. Modelling this process, however, shows that fluctuations must occur very rapidly, and selection be strong; in haploid models, for instance, **epistasis** must change sign on a timescale of two to five generations to select for high rates of recombination (Barton 1995; Otto 2009). From the work of Otto & Nuismer (2004), species interactions typically select against sex; although the Red Queen may favour sex under certain circumstances, it alone does not account for its ubiquity. Spatial variation might also contribute (because genetic associations created by migration can be locally detrimental) but cannot either provide a general explanation.

The conditions for the maintenance of sex become much less stringent in finite or structured populations. Without sex, as already pointed out by Fisher (1930) and Muller (1932), favourable mutations appearing in different individuals will compete against each other, and only one will reach fixation. Sex, in contrast, allows their combination to produce high-fitness genotypes, and thereby considerably speeds up adaptation. Infinite populations do not require recombination to achieve this result, because all possible combinations of mutations necessarily appear. Similarly, sex may act to bring together deleterious mutations and create severely unfit individuals, which are then eliminated from the population (i.e. sex aids in purifying selection). Formalization, indeed, shows that an interaction between **genetic drift** and selection offers the best conditions for recombination to evolve, as a way to generate new genetic combinations (Otto & Lenormand 2002; Otto 2009). Selection at one locus increases the amount of genetic drift at all other loci on the genome; such interactions between linked genes, known as **Hill–Robertson interference**, strongly reduce the ability of a population to respond to selection. In particular, drift creates negative disequilibria in which beneficial alleles are associated with deleterious ones; deleterious mutations that appear in a highly fit background increase in frequency through **hitchhiking**; reciprocally, favourable mutations may be eliminated because they are linked to unfavourable ones (**background selection**). Such interactions are obviously magnified in the absence of recombination, as the degeneration of sex chromosomes perfectly illustrates (Section 5.2). Recombination is also the only way to counteract the elimination, by drift, of the class of individuals with the lowest deleterious load (**Muller's ratchet**; Box 5.2).

Sex in diploid eukaryotes is often associated with outcrossing (i.e. genetic exchange between unrelated individuals), as a way to avoid the **inbreeding depression** stemming from the **homozygosity** of recessive deleterious mutations. Inbreeding avoidance is often enforced in diplontic or **sub-diplontic** lineages by specific adaptations such as SI systems in angiosperms and animals, **induction types** in oomycetes, or mating types in ciliates (Sections 2.1.7 and 3.4). Outbreeding enhancement (as any form of **heterosis**) arising from the masking of deleterious recessive mutations is clearly not a benefit of sex, just a way to counteract some of its short-term costs in diploid organisms. Inbreeding depression is not a problem in **haplontic** cycles (where sex first appeared), because deleterious mutations are always exposed to selection in the haploid stage. This genetic complementation might have played a role in the transitions from haplontic to diplontic cycles (Section 2.1.1a), but cannot be considered in any sense a benefit of sex. On the other hand, however, outbreeding also contributes to a genetic mixing between unrelated lineages that might help to build up genetic variance, thereby providing indirect benefits to sex.

Empirical evidence supports the view that sex may provide advantages over asexuality in the rather short term and the more so in more complex or variable environments. In particular, sex and crossovers improve the efficacy of selection in small populations, due to Hill–Robertson interferences (Burt 2000); small populations of *Drosophila* respond less to directional selection on bristle numbers in the absence of autosomal crossovers (McPhee & Robertson 1970; see Malmberg 1977 and Greig et al. 1998 for similar results on bacteriophages and yeasts respectively). Goddard et al. (2005) genetically manipulated yeasts to produce strains that differed only in their capacity for sexual reproduction; as expected, sex increased the rate of adaptation to a new harsh environment, but had no measurable fitness effect in a benign environment, where selection was mild. Interestingly, the recombination rate in *Drosophila* has been shown to increase as a correlated response to directional selection on other characters, such as positive or negative geotaxis (Korol & Iliadi 1994), resistance to DDT (Flexon & Rodell 1982), or to temperature fluctuations (Zhuchenko et al. 1985; Gorodetskii et al. 1990). Accordingly, domesticated mammals have more crossovers than do their wild counterparts (Burt & Bell 1987).

The genetic variance generated by sex also allows organisms to cope better with parasites: whenever host–parasite interactions are genotype-specific (e.g. in *Daphnia magna* and its parasitic bacterium *Pasteuria ramosa*; Luijckx et al. 2013), sex brings benefits by creating host genotypes to which the prevailing parasite is not adapted. In *Potamopyrgus* snails, sex is maintained in lakes where parasites are present, while asexual triploids dominate when parasites are absent (Jokela et al. 2009); self-fertilizing strains of the nematode *Caenorhabditis elegans* are more sensitive than outcrossing lineages to the bacterial pathogen *Serratia marcescens* (Morran et al. 2011). Outcrossing in *Aspergillus* was found to limit the spread of viruses significantly more than selfing, presumably due to the formation of new somatic incompatibility groups by sexual recombination (Coenen et al. 1997). **Purifying selection** seems also more efficient in the presence of sex (Rice 1994; Zeyl & Bell 1997). Phylogenetic analyses have provided evidence that selection is less effective at eliminating deleterious **non-synonymous mutations** in asexual lineages than in related sexual lineages of *Daphnia pulex* (Paland & Lynch 2006) and the freshwater snail *Campeloma* (Johnson & Howard 2007), as expected from Hill–Robertson interference among loci. More generally, the sporadic distribution of obligate asexuality is largely consistent with the idea that the loss of sex leads to a loss of adaptive potential.

By and large, it is fair to say that the 'evolutionary enigma' of sex (also qualified by Williams 1980 as a 'major unresolved mystery') is far from being fully resolved; however, a form of consensus is emerging that sex has no single evolutionary cause, and that a proper integration of the various above-mentioned causes into a pluralist approach has the potential to explain much of its occurrence (West et al. 1999).

## 1.2 Mating types, sexes, and genders

Surprisingly, members of sexual lineages are usually not free to recombine their genome with any of their conspecifics. Depending on lineages, the range of permissible partners is controlled by categories referred to as '**mating types**', 'sexes', or 'genders'. These categories certainly present similarities, but should still be kept distinct. Here below we provide a few definitions to clarify these important concepts.

i) We will refer to **'primary mating types'** to designate mating categories under **haplo-genotypic sex determination** with **isogamy** (Section 2.1.4). In such systems, fusion occurs between similar-sized gametes, but only if they belong to complementary mating types. The number of mating types is theoretically unlimited, and may reach hundreds in some basidiomycetes (Section 3.2.2c).

ii) 'Male' and 'female' functions will refer to the production of small and large gametes respectively (sperm and ova). As further developed in Section 2.1.2, anisogamy and **oogamy** independently evolved in a diversity of lineages from an ancestral state of isogamy. At this haploid and unicellular level, there are two sexes, and only two (Box 2.1). Male and female gametes likely evolved from primary mating types in most lineages (e.g. Volvocales; Nozaki et al. 2006; Nozaki 2008), but this is not an absolute rule: in ascomycetes, sexes emerged independently of pre-existing primary mating types (Section 2.1.7).

iii) Multicellular individuals differing in their allocation to the male and female functions will be referred to as 'sexes' or 'genders'. We thereby conform to widespread usage, despite some risks of confusion: the word 'sexes' is also used to characterize male and female gametes (as just mentioned), and 'genders' refers in the anthropological literature to the social construction of sexes in humans. Individual **gametophytes** or **sporophytes** specializing strictly in one or the other sexual function will be referred to as males or females respectively, while those producing both male and female gametes (either simultaneously or sequentially) will be referred to as hermaphrodites. At this level, there may be more than two genders or sexes: papaya, for instance, has three genders: males, females, and hermaphrodites (Box 3.11). In fact, the whole continuum from pure maleness to pure femaleness is possible. When there are two genders, these are not necessarily males and females: *C. elegans*, for instance, has males and hermaphrodites (Box 3.15), a system known as androdioecy. Angiosperms exhibit a wide variety of genders and sexual systems (Meagher 1988; Fig. 1.2). The vast majority of flowering plants (~96%) are **co-sexual**, bearing

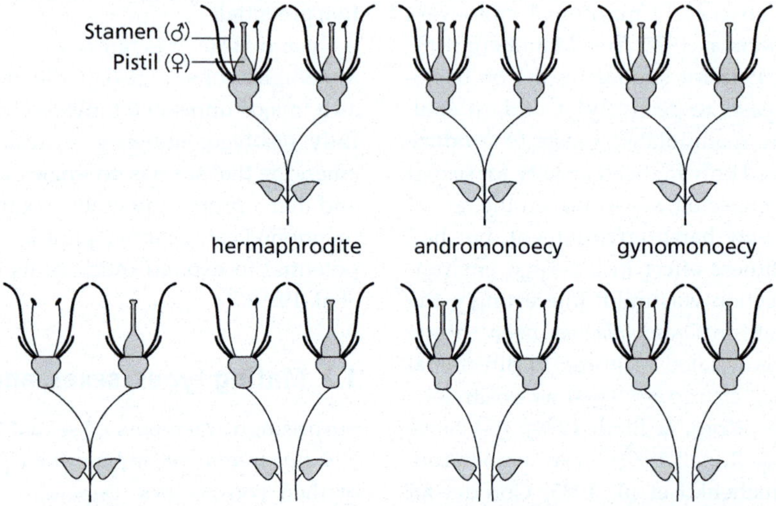


**Figure 1.2** The diversity of genders and sexual systems in flowering plants. Hermaphroditism is the ancestral state, where the male (stamen) and female (pistil) functions co-occur in the same 'perfect' flower. Andromonoecy refers to individual plants with both hermaphroditic and male flowers, gynomonoecy to individual plants with both hermaphroditic and female flowers. Monoecy arises when separate male and female flowers occur on the same plant, dioecy when male and female flowers occur on separate plants. Androdioecy refers to the coexistence of hermaphroditic and male individuals in the same population, and gynodioecy to the coexistence of hermaphroditic and female individuals.

the female and male functions on the same individual, either combined within the same 'perfect' flower (hermaphroditism) or separated into different flowers (**monoecy**). In some cases hermaphroditic flowers occur on the same individual with female-only flowers (**gynomonoecy**) or with male-only flowers (**andromonoecy**). A minority of plants are **dioecious**, having male and female flowers on different individuals. Populations are said **gynodioecious** when hermaphrodites coexist with **unisexual** female individuals, and **androdioecious** when hermaphrodites coexist with unisexual male individuals.

iv) Diploid hermaphrodites may also display mating categories (such as SI systems in angiosperms and ascidians, or **induction types** in oomycetes) evolved as a way to prevent or limit inbreeding (Sections 2.1.7 and 3.4). Such categories are sometimes also referred to as mating types (e.g. in ciliates), which is unfortunate because it only brings confusion. We will specifically refer to these as '**secondary mating types**', because such categories are secondarily superimposed to pre-existing sexual functions (male and female gametes).

### 1.2.1 What uses are primary mating types?

Having mating types presents the potentially important cost of limiting the range of possible partners, which may be alleviated by multiplying mating-type numbers, as many basidiomycetes do (Section 1.2.2); however, a majority of **heterothallic** lineages only present two mating types, which seems the worst number because it prevents mating with half of the potential partners. Billiard et al. (2011) discuss the potential benefits of mating types that might have driven their evolution despite this important cost.

#### 1.2.1a Inbreeding avoidance

Inbreeding avoidance can be rapidly discarded as a cause of primary mating types. First, in haplontic or **haplo-diplontic** lineages, most deleterious mutations are exposed to selection and thus eliminated in the haploid phase; there is no inbreeding load such as in diplontic or sub-diplontic lineages. Second, any meiosis necessarily produces complementary mating types in equal proportions, so that mating types cannot prevent selfing (defined as a fusion between gametes produced by the same diploid individual). Having two mating types (as do a majority of heterothallic lineages) allows exactly the same proportion of matings under selfing as under outcrossing (one half). Third, many heterothallic lineages have evolved ways of enforcing selfing despite maintaining mating types. In *Podospora anserina* and other filamentous ascomycetes meiosis is followed by a post-meiotic mitosis, hence producing four pairs of identical haploid nuclei. Each of the four ascospores resulting from a single meiosis then receives two nuclei, one of each mating type. As a consequence, cultures initiated by a single dispersing ascospore are self-fertile **heterokaryons** ('**pseudo-homothallic**'; Coppin et al. 1997). In *Microbotryum violaceum* and other bipolar basidiomycetes, the sporophyte produces diploid dispersing **teliospores**; meiosis only occurs after the spore has infected a new host, making sure that complementary mating types are found among the haploid yeasts resulting from a single dispersing spore. Selfing is actually the dominant reproductive mode in *M. violaceum* (Giraud et al. 2008).

#### 1.2.1b Sex-advantage enhancement

As already pointed out, the processes of syngamy and meiosis bear physiological and opportunity costs. If the benefits of sex only stem from a reshuffling of genomes, sex should only take place between genetically different individuals. In this context, mating types might have been selected to prevent same-clone mating. Opposing this argument, however, many ascomycetes have evolved homothallism via gene capture (Section 3.2.3b), thereby enforcing mating between genetically identical individuals: two mating-type alleles coexist in the genome, but are differentially expressed in each partner. Yeasts, in particular, have evolved highly sophisticated mechanisms of mating-type switching (Section 3.2.3a), aimed at allowing mating between genetically identical individuals, despite the need to present complementary mating types for cell fusion.

### 1.2.1c Control over organelle transmission

Hurst & Hamilton (1992; Hurst 1995; Hurst et al. 1996) proposed that mating types are required to control potential conflicts among cytoplasmic organelles. **Heteroplasmy**, which occurs when both gametes contribute mitochondria or chloroplasts to the zygote, opens the way to a deadly competition between organelles. These will be selected for better transmission to the next generation, through either active elimination of the competitor, or accelerated division (at the cost of efficiency), both resulting in decreased host fitness, and thereby eliciting a response from nuclear genes to enforce **uniparental** inheritance. According to this hypothesis, mating types (as well as sexes and genders) have evolved as a way to prevent costly competition, by fostering uniparental transmission. Thus, whenever mating involves cytoplasmic fusion, mating types or sexes should: (i) determine who transmits which organelle and (ii) signal who is a potential mating partner or not (in order to prevent mating between gametes which both suppress or both transmit their organelles).

There is ample support for this prediction across a diversity of phyla. Organelle transmission strongly correlates with sexes in plants, animals, and other anisogamic groups. Organelles are usually transmitted maternally (i.e. through the large gamete), but counterexamples exist (e.g. mitochondria are transmitted paternally in some gymnosperms and chytridiomycetes; Xu 2005). Biparental inheritance occurs in many mussels, but the paternal **haplotype** is then sequestered in the male **germ line**, which prevents any conflict (Breton et al. 2007). In isogamous groups also, mating types often correlate with organelle transmission. The locus that determines mating type (MAT) has been found to harbour genes that actively control organelle transmission in several lineages including algae, slime moulds, and fungi (Section 3.2.1).

However, this model also fails to account for a series of important cases. First are situations where transmission, though uniparental, is random with respect to mating types (such as in the slime mould *Didymium iridis*; Silliker et al. 2002; Scheer & Silliker 2006). Second, in some organisms (such as the yeast *Saccharomyces cerevisiae*), both mating types transmit mitochondria: potential conflicts are solved by mitochondrial fusion and mtDNA recombination (Takano et al. 2010); in the brown alga *Ectocarpus siliculosus*, both mating types transmit chloroplasts, which then segregate in different parts of the sporophyte after zygotic division (Peters et al. 2004); in the isogamous diatom *Pseudo-nitzschia*, both gametes contribute two plastids each, two of which are then randomly eliminated from the zygote, independent of parental origin (Levialdi Ghiron et al. 2008). Third are lineages (such as the acellular slime moulds *Physarum polycephalum*) where mitochondrial transmission is controlled by one locus, while gametic compatibility and cell fusion are controlled by other independent **MAT loci** (Moriyama & Kawano 2010; Section S2.2.7b). Finally, many lineages conserve mating types, despite absence of cytoplasmic fusion; in filamentous ascomycetes, for instance, organelle transmission is controlled by sexes (inheritance being maternal); the male parent only provides a nucleus, but has to transmit it to a partner with a complementary mating type. That mating types exist independently of sexes show that they must fulfil another (and possibly more fundamental and primary) function. As pointed out by Billiard et al. (2011), correlation is not indicative of causes or effects; organelle inheritance might have been superimposed on pre-existing mechanisms (first mating types, then sexes) that evolved for other reasons.

### 1.2.1d Developmental switch

Haag (2007) suggested that mating types evolved to control developmental switches between the haploid and diploid phases of the sexual cycle (see Perrin 2012 for further developments of this idea). Genomes must fulfil a diversity of alternative programmes along the sexual cycle: as a haploid gametophyte, an individual may either grow vegetatively (through haploid mitoses), or initiate **gametogenesis** and mating; as a diploid sporophyte, similarly, it may either grow vegetatively (through diploid mitoses) or initiate meiosis and **sporulation**. Only diploid sporophytes should switch on the meiotic programme, while only haploid gametophytes should switch on gametogenesis and mating (and only when other gametophytes are ready to do the same in the neighbourhood). Accordingly, mating types provide a ploidy-assessment function, mediated by complementary **transcription factors** encoded by MAT loci, with differential activity either as **homodimers** in haploid cells or as **heterodimers** in the diploid zygote. In addition, MAT genes also encode a mate-assessment function, mediated by complementary **pheromones** and receptors, which allows the mating programme to be switched on when conspecific haploid cells from a complementary mating type are ready to do the same in the immediate environment (Section 3.2.1). This developmental-switch model seemingly accounts for much of the structure and function of MAT loci (Section 3.2).

### 1.2.2 Frequency-dependent selection on mating types

The constraints imposed by mating categories induce a negative **frequency-dependent selection**, which has two main and far-reaching consequences. First, any new category arising by mutation, and able to mate with all other pre-existing categories, will spread; rare mutants are favoured because they benefit from a higher mating success. This process is expected to generate a high diversity of mating categories. Second, all coexisting categories are expected to reach the same frequency (with equal fitness) at equilibrium. This reasoning applies to both primary and secondary mating types (e.g. the SI systems of angiosperms).

Selection to increase the number of mating types should be strongest if mate availability is restricted by low dispersal ability or high habitat structure (Vekemans et al. 1998; Castric & Vekemans 2004; Busch & Schoen 2008). However, adding a third mating type to a dual system might not be an easy task, as illustrated, for example, by cellular slime moulds (Section 3.2.2d) and basidiomycetes (Box 3.4). The new mating type has to display hormones that activate all pre-existing receptors but not self, and receptors that react to all pre-existing pheromones but not self. Such evolutionary constraints (need to distinguish self from non-self) might, however, be alleviated if the two interacting molecules (pheromone and receptor) are produced by the **alternative splicing** of a single gene, as occurs in *Euplotes* ciliates (Vallesi et al. 2005). In primary mating types, similarly, the new transcription factors should function as active heterodimers with all pre-existing ones, but not self. As pointed out by Haag (2007), the biochemical problem of making specific, functional heterodimers becomes increasingly difficult with increasing numbers of mating types. An additional difficulty is that of controlling organelle transmission, whenever mating types were co-opted for this function, as illustrated by the plasmodial slime mould *P. polycephalum* (Section 3.2.2d).

Diverse mechanisms can be used to increase the numbers of mating types. This may be simply achieved by increasing the number of alleles at one locus, as exemplified by the primary mating types of cellular and acellular slime moulds (Section 3.2.2d) or the secondary mating types of ciliates with **synclonal** mating-type determination (Box 3.24). The ten mating types of *Euplotes patella* are determined by four co-dominant alleles at the MAT locus (i.e. four mating types are homozygous at this locus and six are heterozygous; Akada 1985). An alternative strategy was adopted by ciliates with cytoplasmic or **caryonidal** mating-type determination (such as *Tetrahymena*), which capitalize on differential epigenetics of the **macronucleus** to generate different mating types in genetically identical **clones** (Preer 2000; Box 3.24). Basidiomycetes succeeded in multiplying mating types by relying on functional redundancy (Casselton & Olesnicky 1998; Section 3.2.2c). **Karyogamy** and meiosis in this group are controlled at locus A (active heterodimers must be formed by combining an HD1 transcription factor from one partner with an HD2 factor from the other). The potential for compatibility was multiplied by several gene duplications, each gene copy encoding one HD1 and one HD2 factor. It is enough for the interacting gametophytes to be compatible at one copy only to produce active HD1/HD2 heterodimers. This allows *Coprinus cinereus*, for instance, to display as many as 160 mating types at its A locus (Kües & Casselton 1993), and *Schizophyllum commune*, up to 288 (Raper et al. 1958). This is in sharp contrast with yeasts, filamentous ascomycetes, or bipolar basidiomycetes, which mostly present two mating types only (as do some ciliates). However, multiplying the numbers of mating types might be less (or even not) required in lineages that evolved competences for same-clone mating (as allowed by mating-type switches in yeasts; Section 3.2.3a), same-spore mating (as in pseudo-homothallic filamentous ascomycetes or bipolar hemibasidiomycetes), or autogamy (which evolved in ciliates as an alternative to mating-type multiplication; Phadke & Zufall 2009; Box 3.24).

### 1.2.3 Sexes and sex-ratio selection

Males and females are latecomers in the long history of sex. Anisogamy independently emerged from an ancestral isogamous state in a diversity of lineages, among which oomycetes, diatoms, brown, red, and green algae (including the ancestors of land plants), fungi, animals, and still many others (Section 2.1.2). Anisogamy evolved under **disruptive selection** on gamete size (Box 2.1): under the opposing selective pressures to simultaneously maximize the number of gametes, their encounter rate, and the mass (and ensuing survival) of resulting zygotes, the evolutionarily stable strategy, for a large range of parameter values, involves one small and mobile gamete interacting with a large and sessile partner; intermediate gametes would do worse than small ones in terms of mobility and numbers, and worse than large ones in terms of provisioning. This explains why anisogamy evolved repeatedly, and why sexes (at the gametic level) are two and only two. Theory also shows that the emergence of sexes is greatly facilitated when evolving from pre-existing mating types, and predicts that genes controlling gamete size should be physically or functionally linked to those determining mating type (Charlesworth 1978a), as was recently established in Volvocales (Ferris et al. 2010; Charlesworth & Charlesworth 2010; Umen 2011).

Owing to anisogamy, frequency-dependent selection on sexes (i.e. **sex-ratio selection**) differs from that on mating types on several accounts. First, there is no space for additional sexes at the gametic level (i.e. a kind of gamete able to fuse with equal success with both sperm and ovum). Second, because the costs of producing sperm and ova differ greatly, the equilibrium expectation is not that of an equal number of male and female gametes, but of an equal investment in energy towards the male and female function. In the case of multicellular organisms with separate genders, this translates into an equal investment into sons and daughters, resulting in 1:1 **sex ratios** when the costs of producing males and females are identical. This is referred to as a **'Fisherian' sex ratio**, following Fisher's (1930) reasoning that, in outcrossed populations of sexually reproducing individuals, 'the collective investment by the parents in each sex ought to be equal'. If one sex is in excess, parents that have a bias towards producing the rare sex will have greater fitness.

The strength and efficacy of sex-ratio selection are particularly well illustrated by the swift response to sex biases induced by meiotic-drive sex chromosomes or cytoplasmic sex-ratio distorters (mitochondria or other **endosymbionts**; Section 6.3.2; Boxes 6.1 and 6.2). Following the experimental introduction of a driving X chromosome in populations of *Drosophila mediopunctata*, the proportion of males increased from 16% to 32% in about 50 generations, by accumulation of autosomal alleles redirecting the parental reproductive effort toward the rare sex (Carvalho et al. 1998). In Polynesian populations of the butterfly *Hypolimnas bolina* infected by maternally inherited *Wolbachia* that selectively kill male embryos, the proportion of males grew from 1% to 50% within 10 years, due to the spread of suppressor genes in the host (Hornett et al. 2006; Charlat et al. 2007).

Sex-ratio selection relates in important ways to the evolution of sex determination (Section 6.3). Fisher's prediction refers to equilibrium populations, but makes no predictions regarding the individual parental strategies or the mechanisms involved (Williams 1979). In infinite populations, the sex ratio is neutral at the family level, as long as it is 1:1 at the population level; half of mothers could specialize into male production, and the other half into female production (a situation referred to as **monogeny**). However, this variance in sex production among families entails a cost as soon as populations become limited; stochastic fluctuations in sex ratios exert strong selection against sex specialists. **Stabilizing selection** for balanced sex ratios is expected to reduce the genetic variation in parental strategies and increase **canalization** of the sex-determination mechanism. This certainly favours **heterogametic** sex-determination systems (XY or ZW), which are more likely to enforce Fisherian sex ratios in small populations (assuming **Mendelian segregation** of the sex chromosomes).

Evolutionarily stable sex ratios may also differ from Fisherian expectations. Fisher was explicit about the predictions of his model, less about its assumptions. Subsequent work has clarified several important underlying assumptions, among which are: separate sexes, biparentalism, parental control over sex ratios, Mendelian segregation of sex-ratio alleles, random mating in an infinite population, and no environmentally induced sex differences in fitness (Bull & Charnov 1988). Deviations from these assumptions have the potential to favour non-Fisherian sex ratios (reviewed in Charnov 1982; Karlin & Lessard 1986; Hardy 2002; West 2009). In his famous paper on 'Extraordinary sex ratios', Hamilton (1967) realized that, in structured populations consisting of local mating groups, competition among brothers for mates may select for female-biased sex ratios. This 'local-mate competition' is common in many parasitoid wasps, fig wasps, and mites; a spectacular example is provided by the parasitic mite *Acarophenax tribolii*, where competition occurs among brothers to inseminate their sisters within the maternal body (males then die before they are born). Reciprocally, competition between sisters over breeding opportunities is expected to induce selection for male-biased sex ratios (Clark 1978; Frank 1986). On balance, the optimal sex ratio (from the maternal perspective) tends to be biased towards the sex least likely to compete with relatives (Wild & Taylor 2004). Selection for non-Fisherian sex ratios is expected to disfavour simple male or female heterogametic systems. **Local mate competition**, in particular, might have contributed to the evolution of **haplodiploidy** and maternal control over sex ratios (Box 7.2). Similarly, environmental sex determination (**ESD**) might bring strong benefits over genetic sex determination (**GSD**) whenever environment affects offspring fitness in a sex-dependent way (Charnov & Bull 1977; Section 2.1.6c).

Fisher's sex-ratio principle assumes parental control over sex ratios; as long as all other underlying conditions are met, equal sex ratios are expected regardless of whether the control is parental or zygotic (the only difference being that the rate of evolution towards the equilibrium is half as fast under maternal control, because the trait is expressed in only one sex; Bulmer & Bull 1982). However, parental and offspring optima may differ when some conditions underlying Fisher's principle are not met. Under local mate competition, for instance, any offspring able to develop as the rare sex (male in this case) will gain strong reproductive benefits (Section 6.3.1). In addition, different genetic elements within an individual genome may have different optima regarding sex ratios; sex chromosomes and cytoplasmic elements are differentially transmitted by males and females, and thereby selected to overproduce the transmitting sex (Section 6.3.2). Such genomic conflicts are bound to impact the evolution of sex-determination systems.

In fact, sex-ratio selection is arguably the most important pressure influencing the evolution of sex-determination mechanisms. Shifts in optimal sex ratios, or restoration of Fisherian sex ratios following biases induced by environmental changes or genomic conflicts, are among the key ultimate causes driving the dynamics of sex-determination systems (Section 7.2.3).

## 1.3 Sex roles, sexual conflicts, and sexual selection

Another fundamental consequence of anisogamy is that it establishes distinct roles for sexes, thereby posing the basis for a more elaborate sex differentiation, driven by **sexual selection** and **sexual conflicts** (Section 6.4). A first and simple type of conflicts arises just from the differences in the molecular and physiological pathways required to build up the male or female function, be it only to produce male versus female gametes: alleles coding for large gametes are favoured in a female context, and disfavoured in a male context. A second and more complex type of conflicts arises from the fact that the male function allows production of many more gametes than the female function (meaning much more than required to fertilize all available female gametes). This leads to a much higher **potential reproductive success** for individual males than females (as well as a larger variance in effective male reproductive success, because averages cannot differ between sexes). This fundamental asymmetry offers opportunities for males to enhance their fitness by seeking more mating opportunities; females, which cannot increase their fitness by mating more, are rather selected to choose mating partners that provide the best direct or indirect benefits to their offspring (via paternal care or **good genes**, respectively; Box 6.3). It is worth noting, however, that such 'sex roles' are sometimes reversed: depending on circumstances, the **operational sex ratio** (i.e. the proportions of each sex that are effectively 'on the market', available for mating) may be female biased, so that males become the limiting sex.

Such ideas originally stem from Bateman's (1948 experiments showing that the number of offspring sired by male *Drosophila melanogaster* increased with their number of mates, whereas the fecundity of females did not increase with their number of matings. Bateman concluded that, because a single egg is more costly to produce than a single sperm cell, female fecundity is limited mainly by her ability to produce eggs, while male reproductive success is limited by the number of females he can inseminate. Bateman's ideas and their importance to evolutionary theory have been greatly expanded by Trivers (1972) and much refined since (reviewed in Arnqvist & Rowe 2005). As further developed by Parker (1979), a gene that helps males obtain more matings will increase in frequency even if the male phenotype thereby produced is harmful to females (e.g. through harassment or direct injury).

Sexual conflicts can have different genetic underpinnings. **Interlocus sexual conflicts** occur when the sexes employ different traits (controlled by different genetic loci) when struggling over the outcome of an interaction. This induces conflicts between loci that enhance male fitness at female's expense and loci that favour female's defence at the expense of male reproductive success; such conflicts may result in spectacular arms races, including spiny genitalia or toxic ejaculates. **Intralocus sexual conflicts** occur when the male and female traits are controlled by the same loci, but optima differ between sexes. Such **sexually antagonistic** (SA) genes are widespread, starting from the genes controlling gamete size in incipient anisogamous species. Intralocus sexual conflicts can be solved by physically linking the genes involved in sex determination and sex differentiation into a single, non-recombining sex locus (**supergene**); this occurs in some fishes, where colouration genes are tightly linked to the sex-determination locus (Section 5.1.1). Alternatively, sexual conflicts can be solved by the differential expression of genes in a male or a female context (as necessarily occurs in all species with ESD). The crucial point made here is that the fundamental asymmetry induced by anisogamy sets the stage for sexual selection and sex antagonisms, which are key players in the evolution of sex-determination systems (Sections 6.4, 7.2.2a, 7.2.3c).

## 1.4 Sex determination versus sex differentiation

A distinction is often made in the biological literature between sex determination and sex differentiation. The former is meant to designate the developmental step at which an individual fate is irremediably directed towards either the male or the female condition, whereas the latter describes subsequent steps in developmental pathways, during which the male and female phenotypes are progressively built up, according to the initial decision.

The scope of sex determination is sometimes restricted to the action of the initial trigger (e.g. *Sry* activation in **therians**), whereas the subsequent development of **gonads** into testes or ovaries (or of floral **meristems** into **carpels** and **stamens**) is referred to as 'primary sex differentiation'. This strict definition of sex determination and delimitation of its occurrence underlies the classical dichotomous view of sex determination, according to which sex is determined either by genes or by environment. Accordingly, differences in the timing of action of the initial trigger has been proposed as a criterion to distinguish GSD from ESD: in the former, sex is determined by genes at conception, while in the latter it is determined by environment during a sensitive period of embryonic development (e.g. Valenzuela et al. 2003).

As further discussed in Chapter 4, we think there are good reasons to reject this dichotomous view, which fails to account for the many complex systems of sex determination where both genes and environment contribute to the final outcome; this view also fails to account for the fundamental similarity in the downstream molecular and physiological mechanisms that lead to the production of males and females, whatever the initial trigger

(Chapter 3). On the same reasoning, there are good reasons to reject a fundamental distinction between sex determination and 'primary sex differentiation' (i.e. gonad development), if we want to understand the evolution of sex-determination systems. In all vertebrates, the same genes are involved in the cascade leading to gonadal development, sometimes with different timing of action and localization in the pathway. *Dmrt1*, for instance, takes a role of downstream effector in most lineages, but has moved up the hierarchy in some of them to occupy the top of the cascade; it does not make much sense to consider *Dmrt1* as either a sex-determination or a sex-differentiation gene depending on lineage.

In line with Uller & Helanterä (2011), we define sex determination as the whole process that leads to the development of differentiated reproductive organs (e.g. either testes or ovaries in animals). In vertebrates, the process starts when the precursor cells from the genital ridge develop into either **Sertoli** or **granulosa** cells; the timing and molecular processes are similar between ESD and GSD species. Along the same line, we think the concept of sex determination should not be limited to organisms with separate sexes, but also applies to hermaphrodites. Too strict a definition of sex determination fails to account for **sequential hermaphroditism** or mating-type switching, where a redifferentiation of sexes or mating types occurs, triggered by epigenetic cues (environmental or social information). Other epigenetic cues (positional information) are similarly used in **simultaneous hermaphrodites** to trigger the development of distinct cell lineages along either the male or the female pathways, based on the same gene networks and molecular cascades as in species with separate sexes. This is also a form of sex determination.

Sex is never determined at conception. Sex determination is a complex and dynamic process, starting with one or a series of initial cues (genetic, epigenetic, or often a mixture of both), and ending with the commitment of undifferentiated gonads into either testes or ovaries (or of undifferentiated meristems into either stamens or carpels). This commitment may be temporary and reversible (as in sequential hermaphrodites), and the initial decision potentially affected by mutations or environmental effects at any step along the sex-determination cascade (Chapter 3).

CHAPTER 2

# The diversity of sexual cycles

The first part of this chapter, printed in full in this book, characterizes the diversity of sexual cycles found among eukaryotes, paying special attention to the main features that are relevant to the evolution of sex-determination mechanisms; in particular, the relative importance of haploid and diploid phases, the timing and mode of sex determination, and the evolution of sexes and **dioecy**. A second part, available electronically at www.oup.co.uk/companion/beukeboom, covers the eukaryote phylogeny more systematically, describing how this diversity of sexual cycles and sex-determination systems is distributed throughout the domain (the eukaryote 'tree of sex'). Figure 2.1 summarizes a few terms important in describing sexual cycles and explains some symbols used in other figures in this chapter.

## 2.1 Sexual cycles among eukaryotes

### 2.1.1 Haploid versus diploid phases

The relative roles of haploid and diploid phases are extremely variable across eukaryote sexual cycles. At one extreme of the continuum lie haplontic cycles (TYPE 1; Fig. 2.2), where mitoses only occur at the haploid stage. The diploid zygote may act as a resting stage, but is followed by meiosis without further rounds of mitoses (zygotic meiosis). Haploid mitoses translate into asexual (clonal) reproduction in unicellular lineages (e.g. *Chlamydomonas*), and into somatic growth and cellular differentiation in multicellular organisms (e.g. *Volvox*). Multicellular haploid stages are often referred to as gametophytes and can take a diversity of forms (e.g. filamentous, branched, spherical, or parenchymatous). Gametes are usually formed in specific organs called **gametangia**. Purely haplontic cycles are scattered throughout eukaryote phylogeny, being widespread in Bikonta, for example among stramenopiles (Chrysophyta and Xanthophyta), alveolates (all Apicomplexa and most dinoflagellates), and Viridiplantae (Chlorophyta and Charophyta). Among Unikonta, some Amoebozoa (e.g. Dictyostelia) and fungi (e.g. Chytridiomycota, Zygomycota, Glomeromycota) also present purely haplontic cycles.

At the other extreme of the continuum are diplontic cycles (TYPE 3; Fig. 2.4), where mitoses only occur at the diploid stage. Meiosis takes place immediately before gametogenesis and zygote formation (gametic meiosis). Here again, diploid mitoses translate into asexual reproduction in unicellular lineages (e.g. diatoms, ciliates, kinetoplastids, metamonads) or into somatic growth and cellular differentiation in multicellular organisms (e.g. Metazoa). Multicellular diploid stages are often referred to as sporophytes and can also take a diversity of forms from simple filaments (as in oomycetes) to complex organisms with tissue differentiation (as in Fucales and Metazoa). Sporophytes produce haploid **meiospores** in specific organs called **sporangia**, where meiosis takes place. Purely diplontic cycles evolved many times independently among the main eukaryotic lineages, including Excavata (Metamonada and kinetoplastids), stramenopiles (all Bacillariophyta, Oomycota, and Opalinidae, as well as some Phaeophyta), alveolates (all ciliates and some dinoflagellates), some Chlorophyta (Bryopsidales), and all Metazoa. If diplontic cycles are alternatively defined by the absence of a free gametophyte, then several additional lineages may qualify as diplontic (or sub-diplontic), including some green algae in Dasycladales (*Acetabularia*) and Trebouxiophyceae (*Prasiola*), as well as all spermatophytes.

*The Evolution of Sex Determination*. Leo W. Beukeboom and Nicolas Perrin.

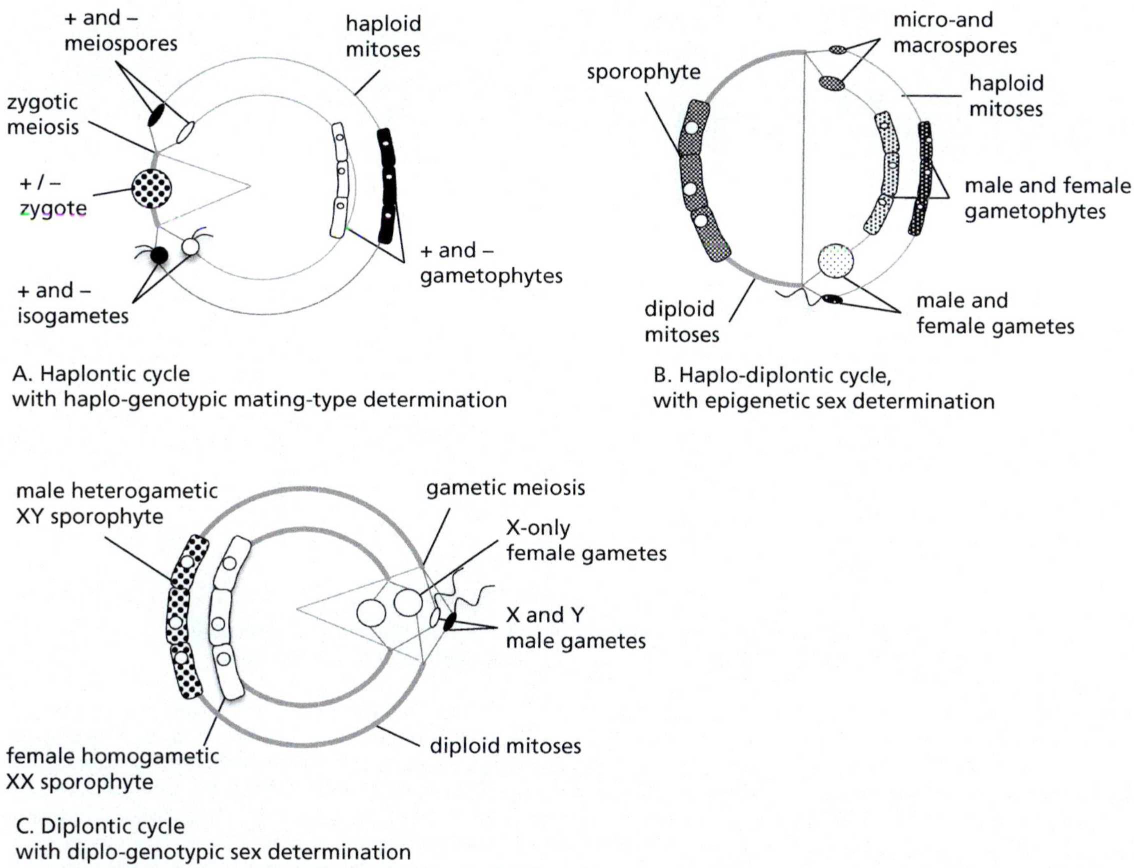


**Figure 2.1** Examples of life cycles explaining some terms and symbols. Life cycles are to be read clockwise. The thin-line parts of cycles represent haploid states, the thick-line parts diploid states. (A) **Haplontic** cycle with haplo-genotypic determination of mating types. The zygotic meiosis produces two + and two – meiospores, which develop through haploid mitoses into + and – gametophytes (**dioicy**). The complementary + and – gametes fuse to form a diploid zygote which is heterozygous at the MAT locus (harbouring a copy of both the + and – allele). Haploid states differing at the MAT locus are figured as unicolour (either black or white), while heterozygosity in the diploid state (zygote) is symbolized by a checkerboard appearance. (B) **Haplo-diplontic** cycle with epigenetic sex determination. The zygote develops via diploid mitoses into a sporophyte, which, in the case illustrated here, produces two types of meiospores: the smaller microspores develop into male gametophytes, and the larger macro- (or mega-) spores develop into female gametophytes (dioicy). The small and mobile male gamete fertilizes the large and immobile female gamete, restoring the diploid zygote. The sexual fate of gametophytes is thus not determined by genetic factors, but by the differential allocation of resources in the sporophyte (small versus large spores). The change in shading of the symbols along the gametophyte line symbolizes differentiation towards maleness or femaleness. (C) **Diplontic** cycle with diplo-genotypic sex determination. Sporophytes are separate (dioecy) and differ genetically, one being heterozygous and the other homozygous at the SEX locus (male heterogamety is illustrated here, with XY males and XX females). Accordingly, male meiosis produces both X and Y gametes, while female meiosis only produces X gametes. A sporophyte's gender is thus determined by the gamete received from its male parent. Homo- or hemizygous states are figured as unicolour (either black or white), while heterozygosity in the heterogametic sporophyte is symbolized by a checkerboard appearance.

In between lies a continuum of haplo-diplontic cycles (TYPE 2; Fig. 2.3), in which mitoses occur at both the haploid and the diploid stage. Both stages may remain unicellular (as in the budding yeast *Saccharomyces cerevisiae*), or may develop into a multicellular organism, either at one stage only (e.g. *Ustilago* produces unicellular yeasts when haploid, and hyphae when **dikaryotic**), or at both stages (e.g. many green, red, and brown algae). Cycles are called isomorphic when the sporophyte and the gametophyte are morphologically similar (as in *Ulva* or *Ectocarpus*) and heteromorphic when they differ

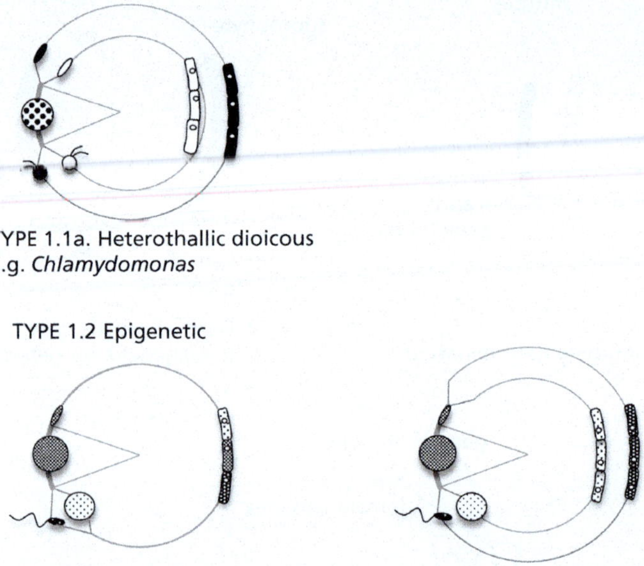


**Figure 2.2** Haplontic life cycles. Upper row: haplo-genotypic mating-type determination. TYPE 1.1a. In the cycle illustrated here, the gametes are isogamous and the haploid mitoses result in multicellular gametophytes. The system is dioicous (+ and – gametes are produced by different gametophytes with complementary mating types) and heterothallic (two gametophytes of complementary mating types stem from different meiospores). *Chlamydomonas* belongs to this TYPE 1.1a, but differs from the specific situation illustrated here by the absence of a multicellular gametophyte (haploid mitoses translate into asexual clonal reproduction). *Volvox* also belongs to this type but differs from the case illustrated here by having true sexes (anisogamy), which directly evolved from mating types. Lower row: epigenetic sex determination. TYPE 1.2a. In the cycle illustrated here, differentiated male and female gametes are produced by the same gametophyte (**monoicy**). Male and female organs (gametangia) are symbolized by different patterns. TYPE 1.2b. In this case, the male and female gametes stem from distinct male and female gametophytes, which however do not differ genetically, and might thus derive from the same meiospore (homothally).

morphologically. The dominant stage may be either the haploid gametophyte (as in mosses) or the diploid sporophyte (as in kelps). Asymmetry may be strong, and the haploid or diploid phase reduced to a few cells with no independent development (as for the gametophytes of angiosperms). This directly leads to transitions towards diplontic or haplontic cycles, respectively.

Additional stages are sometimes inserted within either the diploid phase (e.g. the **carposporophyte** of Florideophyceae; TYPE 2.1c, Fig 2.3) or the haploid phase (e.g. the *Codiolum* stage of Ulotrichales or the **schizont** of many benthic Foraminifera). Similarly, a dikaryotic (or di-haploid) stage is sometimes inserted between cell fusion (**plasmogamy**) and nuclear fusion (karyogamy), the latter being immediately followed by meiosis (TYPE 2.1e, Fig 2.3). Dikaryotic stages are mostly developed among higher fungi (the so-called Dikarya), in particular basidiomycetes, where dikaryotic mycelia may be extremely large and long-lived. Other groups (e.g. *Derbesia* in Bryopsidales, *Plasmodiophora brassicae* in Cercozoa) may insert a **coenocytic** (**heterokaryotic**) stage between plasmogamy and karyogamy. Dikaryotic and heterokaryotic cycles bear homologies with both haplontic and haplodiplontic cycles (strictly speaking, meiosis is zygotic, but occurs within a complex sporophyte built from two genomes).

### 2.1.1a Is diploidy better?

A widespread view holds that diploid states provide intrinsic advantages, which induce recurrent evolutionary trends towards diplontic cycles (with

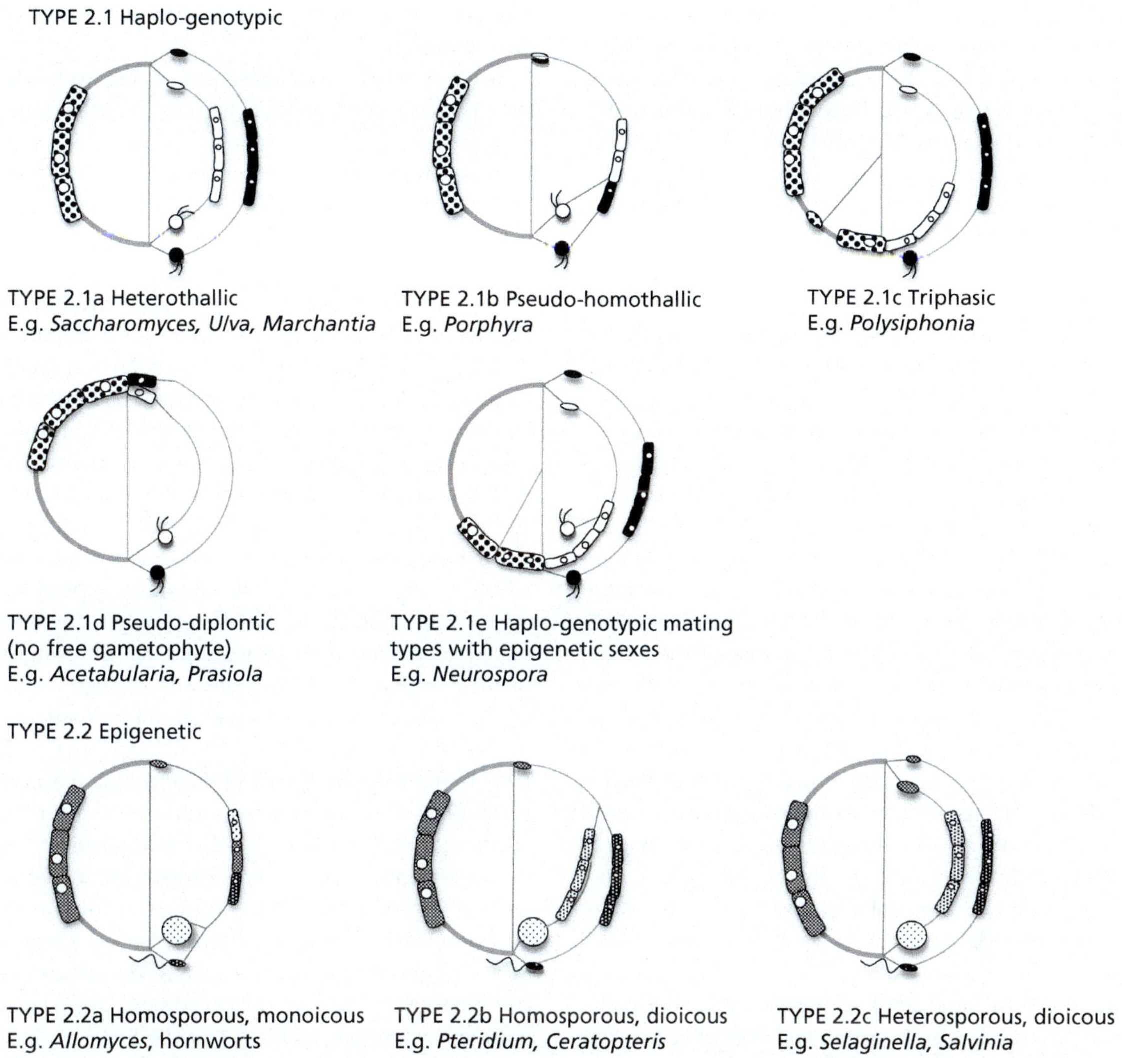


**Figure 2.3** Haplo-diplontic life cycles. Upper row: haplo-genotypic mating-type determination. TYPE 2.1a. In the cycle illustrated here, the gametes are isogamous and the haploid mitoses result in multicellular gametophytes. The system is dioicous (+ and − gametes are produced by different gametophytes with complementary mating types) and heterothallic (two gametophytes of complementary mating types stem from different meiospores). Second row: epigenetic sex determination. TYPE 2.2a. In the cycle illustrated here, differentiated male and female gametes are produced by the same gametophyte (monoicy), with male and female organs (gametangia) symbolized by different patterns. TYPE 2.2b. In this case, the male and female gametes stem from distinct male and female gametophytes (dioicy), which, however, do not differ genetically, and might thus derive from the same meiospore (homospory). TYPE 2.2c. The differentiated male and female gametophytes stem from distinct meiospores (micro- and macrospores respectively), which, however, do not differ genetically. Gametophytic sex is determined by a differential developmental programme (possibly triggered by differential resource allocation) initiated at the sporophyte level. Lower rows: diverse haplo-diplontic cycles with haplo-genotypic mating-type or sex determination. TYPE 2.1b: Isogametes of complementary mating types are produced by the same gametophyte, which is a genetic chimera resulting from the non-disjunction of meiospores (pseudohomothallism). TYPE 2.1c: Spermatia produced by male gametophytes fertilize the carpogonia of female gametophytes. A diploid carposporophyte develops on the female gametophyte and releases diploid mitospores that develop into diploid tetrasporophytes. TYPE 2.1d: Isogametes of complementary types are produced by haploid structures developing on the sporophyte. There is no free gametophyte (pseudo-diplontic cycle). TYPE 2.1e: The spermatia produced by one gametophyte fertilize ascogonia developing on a gametophyte of complementary mating type. In filamentous ascomycetes (e.g. *Neurospora*), karyogamy does not follow immediately cell fusion, resulting in a short-lived dikaryonic stage. Nuclear fusion is followed by meiosis.

a reduction of the haploid stage and concomitant expansion of the diploid stage). The patterns found among a variety of lineages might indeed be interpreted in this sense. In Phaeophyta (brown algae), the haploid phase is dominant or important in many Ectocarpales (which often have isomorphic gametophytes and sporophytes), but much reduced in Laminariales (e.g. with minute gametophytes and huge sporophytes in kelps), and absent in the diplontic Fucales and Ascoseirales. Among Chlorophyta, the basal Prasinophyceae are haplontic, while the more evolved Ulvophyceae comprise haplontic (Ulotrichales), haplo-diplontic (Ulvales, Bryopsidales, and Cladophorales), and diplontic cycles (e.g. *Codium*). Similarly, large amounts of morphological and molecular evidence support the 'antithetic' theory according to which land plants evolved from haplontic ancestors (namely, Charophyta). Gametophytes are still dominant in basal lineages (Hepatophyta, Bryophyta, and Anthocerophyta, in which the sporophyte grows at the expense of the female gametophyte), but small and short-lived (though still independent) in pteridophytes, and close to non-existent in spermatophytes (e.g. restricted to the pollen tube in the case of male gametophytes).

However, despite convincing evolutionary trends in some lineages, the notion that a diploid stage is intrinsically better is certainly biased by our own perception: we are diploids, and so are all macro-organisms that we meet on a daily basis (animals and plants). But a quick look at the eukaryotic tree shows that haplontic and haplo-diplontic cycles are widespread among most lineages, and coexist successfully with supposedly superior diplontic competitors.

What might be the respective evolutionary benefits of diploid and haploid stages? Diploids are expected to harbour more genetic diversity, with potentially twice as many beneficial mutations. If dominant, beneficial mutations might indeed increase the rate of adaptation in diploids (Orr & Otto 1994). If co-dominant, mutations may also confer heterozygote advantages to diploids (Goldstein 1992). If recessive, however, beneficial mutations are first masked in diploids, are thus more likely to be lost than in haploids. Not only is the fixation probability higher in haploids, but fixation also takes half the time. Haploid lines indeed adapt faster than diploid lines in *S. cerevisiae* populations exposed to a range of different environments (Gerstein et al. 2011).

Deleterious mutations also play a role in the balance between diploidy and haploidy, and certainly a larger one, being far more frequent than beneficial ones. If recessive, they will be fully expressed in haploids, but masked in diploids. This complementation might seem at first glance to benefit diploids, but for the very same reason, deleterious mutations will accumulate in the germ line: their equilibrium frequency is $\mu/hs$ in diploids, versus $\mu/s$ in haploids, where $\mu$ is the mutation rate, $s$ the **selection coefficient**, and $h$ the dominance coefficient. The mutation load is thus better purged from haplontic lineages, increasing their average equilibrium fitness over that of diploids (Crow & Kimura 1965; Kondrashov & Crow 1991; Perrot et al. 1991; Otto & Goldstein 1992). From this reasoning, recessive deleterious mutations affecting the germ line should be expected to favour haplontic cycles. There is a caveat, however: in a genetic mixture of individuals that differ in the relative importance of haploid and diploid stages, the benefits of haploidy (lower mutation load) will be shared with diploids, which will therefore benefit from both the lower load and the masking (complementation). Reciprocally, haploids within a diploid population will be strongly disadvantaged by the increased load. From this mechanism, we might expect an irreversible evolution towards diplontic cycles, despite their lower equilibrium fitness. Haploid cycles might only be maintained when sex and recombination are rare, and inbreeding or assortative mating is common (Mable & Otto 1998).

Different effects are expected when such deleterious mutations are somatic. If recessive, they are also masked in diploids, but not transmitted to the progeny. Diploidy might thus be advantageous to build large and long-lived phenotypes (and the more so in terrestrial or upper-tidal environments highly exposed to mutagenic UV radiation). There is an additional twist, however: if cell lineages compete during individual growth, somatic mutations will be eliminated by selection among cell lineages in haploid individuals, but be masked and persist in diploids. This would favour haploidy in relatively undifferentiated organisms (with a large opportunity for selection among cell lines), but diploidy if somatic tissues are highly differentiated, so that competition among cell lines is local and limited. These effects of somatic mutations would predict the predominance of a diploid phase in large, long-lived, and differentiated organisms and of a haploid phase in short-lived or less-differentiated ones. Although this prediction might seem to fit the patterns of dominant diploidy (or **di-haploidy**) found in many multicellular organisms (including animals, plants, macroalgae, and higher fungi), many short-lived unicellular lineages (e.g. kinetoplastids, ciliates, diatoms, or Opalinidae) also have diplontic life cycles, a fact which remains to be explained.

### 2.1.1b What maintains haplo-diplontic cycles?

In addition, the above genetic models cannot account for the maintenance of haplo-diplontic cycles: either haplontic or diplontic cycles should evolve, depending on whether the haploid or the diploid stage is fitter. Richerd et al. (1993) suggested an intrinsic benefit to haplo-diplonty, due to an alleviated cost of sex and meiosis. Assuming similar generation times for gametophytes and sporophytes (and no asexual reproduction), sex is half as frequent in haplo-diplontic cycles than in purely haplontic or diplontic ones. This might suffice to favour haplo-diplontic cycles for a significant range of haploid and diploid fitness values. This benefit is similar to an alternation of sexual and asexual generations.

Furthermore, if the encounter rate of gametes is low, inserting a diploid growth phase after zygote formation (rather than entering directly meiosis as in haplontic cycles) multiplies not only the number of spores produced, but also the number of meiotic products and the genetic diversity of gametophytes resulting from one single mating. This also opens opportunities for a selection-arena mechanism (Stearns 1987), assuming that sporophytes can differentially provision potential offspring depending on quality. Insertion of additional asexual reproduction events is frequent among parasitic lineages with low encounter rates (e.g. in Haemosporidia). They also occur among free-living forms: in the *Polysiphonia* type of cycle in Florideophyceae (in which fertilization rate is possibly limited by the absence of flagellated cells), the zygote is retained on (and provisioned by) the female gametophyte, where it develops into a diploid carposporophyte that releases numerous **mitospores** (TYPE 1.2c, Fig 2.3). Each spore may give birth to a diploid **tetrasporophyte**, which will then produce haploid meiospores. Carposporophytes constitute a key innovation that certainly contributed to the success of Florideophyceae by multiplying their reproductive capacity (Searles 1980; Fierst et al. 2005). Ulotrichales (Chlorophyta) also insert a haploid Codiolum stage after meiosis, producing many haploid mitospores, each with the potential to develop into a gametophyte. In Foraminifera, benthic forms often have trimorphic cycles: the diploid **agamont** produces several haploid schizonts meiotically that enter additional rounds of asexual reproduction before producing gamonts, which undergo gametogenesis (some planktonic forms, by contrast, drop the agamont stage, resulting in simplified haplontic cycles). Note that a change in ploidy level is certainly not required to enter an additional round of growth and reproduction, but just offers a handy opportunity to do so.

Another potentially important benefit of haplo-diplonty rests in the potential to express different genetic programmes depending on ploidy level (Coelho et al. 2007). Heteromorphic algae (e.g. *Porphyra*) tend to show a seasonal specialization of sporophytes and gametophytes, producing, for instance, a large morph in winter when mortality is low, and a small morph in summer when mortality is high (Bessho & Iwasa 2010). Morphological phases have been shown, in particular, to alternate with grazing pressure in macroalgae (Lubchenco & Cubit 1980; Slocum 1980). Similarly, the unicellular Coccolithophora alternate between a diploid planktonic stage and a haploid benthic stage, which are so different morphologically that they were first described as different genera. Heterolobosea also alternate between flagellate and amoeboid stages, which have been suggested to represent the haploid and diploid stages respectively (Droop 1962; Fulton 1993). Even when cycles are isomorphic, physiological differences have been found between haploid gametophytes and diploid sporophytes (e.g. Hannach & Santelices 1985; Clayton 1988; Destombe et al. 1993; Hughes & Otto 1999).

Differential patterns and opportunities for spore- and gamete dispersal also matter (Destombe et al.

1992). As noted by Bell (1997), the small gametophytes of macro-algae release gametes in low-velocity environments appropriate for mating and settlement, while large sporophytes release spores in high-velocity environments, favouring long-distance dispersal. The erect sporophyte of mosses and ferns also facilitate spore dissemination by the wind, while the crawling gametophytes are more exposed to the humid environment required for fertilization.

Parasites are also confronted by highly divergent environments, whenever cycles comprise more than one host or a free stage. Haplo-diplontic cycles might facilitate evolutionary responses to such disruptive pressures. In the dixenic Blastocladiales (Chytridiomycota), the haploid gametophyte parasitizes a copepod, while the diploid sporophyte parasitizes mosquito larvae. Similar cycles, with ploidy levels adapted to different hosts, are also found in Amblyosporidae (microsporidia) as well as in dimorphic rusts and smuts: *Ustilago* is saprobic as a haploid yeast, but parasitic as a dikaryotic mycelium. *Helicobasidium* is parasitic on other rusts when haploid, and on plant roots when diploid. The diploid stage of the *Plasmodium* life cycle (Apicomplexa) takes place in a mosquito (definitive host), and the haploid stage in the mammalian host.

The same argument actually holds for **haplobiontic** cycles. The sexual cycle in haplontic lineages is often initiated by worsening environmental conditions, the diploid zygote developing into a resting stage. As shown in *Chlamydomonas* (Box 3.5), the cell-wall building and encystment programme is directly triggered by the change in ploidy level. In *Trypanosoma*, gametic fusion initiates the developmental programme producing the metacyclic stage, adapted to the transfer to a vertebrate host.

Note again that disruptive environments do not directly select for an alternation of ploidy levels, but only for dimorphic cycles, which could be achieved by other means (as in the dimorphic larval and adult stages of insects or amphibians). However, a change in ploidy level provides a simple and handy molecular mechanism to switch on different sets of genes, adapted to different environments (Sections 1.2.1d and 3.2). This does not require that the haploid or diploid stages are intrinsically adapted to specific ecological circumstances.

### 2.1.2 (An)isogamy: from mating types to sexes

The sexual cycle in eukaryotes normally involves the fusion of two gametes, which are haploid cells specialized in mating. One important feature of life cycles pertains to the similarity or dissimilarity between the two interacting gametes. This trait is also continuous, often with a progressive differentiation from isomorphic gametes towards the highly specialized sperm and egg cells found, for example, in metazoans. Gametic (dis)similarity is a crucial feature since it defines sexes (as opposed to mating types): small gametes characterize maleness and large gametes, femaleness.

Isogamy refers to systems where gametes are morphologically similar, but not necessarily identical: they often possess different alleles at the mating-type (MAT) locus, express different genes, and may differ both physiologically and behaviourally. MAT alleles often encode complementary pheromone and receptor systems, allowing the identification and localization of potential partners, and triggering the gametic fusion and initiation of the diploid programme (Section 3.2.1). One benefit of isogamy is that gametes are little differentiated, and can resume **vegetative growth** into a haploid gametophyte if unsuccessful (e.g. in some Xanthophyceae and Phaeophyceae). Many lineages do not even produce specialized gametes: mating then involves either undifferentiated cells in unicellular organisms (as in dinoflagellates, coccolithophores, and cryptomonads) or somatic fusion in multicellular gametophytes (e.g. **conjugation** in Zygnematales, hyphal anastomosis in many fungi).

In case of **anisomotility**, the gamete displaying higher motility is sometimes identified as 'male'. Once released, the 'female' isogamete usually settles down and emits pheromones to attract its male partner (with few exceptions like filamentous ascomycetes, where the pheromones are emitted by the motile male gamete; Ni et al. 2011). Isogamy with anisomotility is widespread, being found, for example, among Chrysophyceae (Ochromonadales), Synurophyceae (Synurales), Xanthophyceae (Tribonemales), Phaeophyceae (Ectocarpales), Fragilariophyceae, and many other lineages.

A further evolutionary step away from complete isogamy is marked by morphological anisogamy (i.e. undisputable sexes). The female gamete is larger, but may remain motile, as found, for example, in *Codium fragile* or *Derbesia*. As a next step (referred to as oogamy), large and immobile female gametes (eggs) are fertilized by small and motile male gametes (sperm cells). Anisogamy and oogamy evolved many times independently (Baccetti 1985; Margulis & Sagan 1986), directly from mating types in some groups (e.g. Volvocales), or independently so in others, as in filamentous ascomycetes where mitospores devised for asexual reproduction and dispersal (**microconidia**) secondarily evolved the role of male gamete (spermatia; TYPE 2.1e, Fig. 2.3). Isogamy is generally the ancestral state, as documented in chlorophytes, where all the early-diverged (and **paraphyletic**) Prasinophyceae are isogamous (Lewis & McCourt 2004). Anisogamy often evolved in parallel with organismic complexity, as expected from theory (Box 2.1). In Volvocales, for example, the level of anisogamy (measured as the ratio of female to male gamete size) correlates with colony size, from the isogamous unicellular *Chlamydomonas* to the oogamous *Volvox*, culminating in *V. rouseletii* with up to 50,000 cells per colony (e.g. Knowlton 1974; Madsen & Waller 1983; Bell 1985; Randerson & Hurst 2001). In a few cases, however, a reverse trend occurred (e.g. oogamy represents the ancestral state in diatoms, nowadays found in all Centrales, and was secondarily lost in most Pennales).

A final step in the evolution of anisogamy consists in the retention of the female egg cell and provisioning of the zygote during development, either by the maternal gametophyte (as in many Archaeplastida: Rhodophyta, Charophyta, Bryophyta, and Coleochaetophyceae) or by the maternal sporophyte when sex differentiation is initiated at the diploid stage and the gametophyte is absent or reduced (as in angiosperms). This also evolves with internal fertilization (as opposed to broadcast sperm spawning). Among the keys to the success of angiosperms lies the mutualism they developed with insects to ensure pollen transport and internal fertilization, combined with zygote retention and provisioning by the maternal sporophyte. Internal fertilization and zygote provisioning also evolved high degrees of refinement in many animals, of which eutherian mammals are a prime example.

In some groups, pairing first occurs between so-called '**gamonts**', before gametogenesis is induced and true gametes fuse. This **gamontogamy** may occur in a wide variety of forms and a diversity of phylogenetic groups, involving either haploid or diploid

**Box 2.1 Evolution towards anisogamy**

The evolution towards anisogamy resulted from disruptive selection (Parker et al. 1972; Bell 1978; Charlesworth 1978a; Maynard-Smith 1978, 1982; Hoekstra 1982; Bulmer 1994; Bulmer & Parker 2002; Bulmer et al. 2002; Lehtonen & Kokko 2011). Given the opposing selective pressures to simultaneously maximize the number of gametes, their encounter rate, and the mass (and ensuing survival) of resulting zygotes, models show that under a wide set of circumstances, the fitness of both partners is maximized when one interacting gamete is small and mobile, while its large and sessile partner provides the resources required for zygote development. Intermediate gametes would do worse than small ones in terms of mobility and numbers, and worse than large ones in terms of provisioning. Furthermore, whenever the developing sporophyte (or di-haploid fruiting body in the case of filamentous ascomycetes) is directly provisioned by the female gametophyte, as is found in many algae, mosses, or ferns, the female gamete must remain attached to the maternal gametophyte (so that the male gamete has to actively search partners). Such developing constraints largely explain why sexes (at the gametic level) are two and only two, and why anisogamy independently evolved in many lineages.

Using game theory to formalize the effect of gamete and zygote masses on fitness, Bulmer & Parker (2002) showed, in particular, that both anisogamy and zygote size are expected to increase with organism size and complexity. Lehtonen & Kokko (2011) confirmed this expectation, and also showed that organism size and complexity, though necessary, are not sufficient: the evolution of anisogamy also requires either some gametic competition (the system is expected to return to isogamy in absence of local competition) or gamete limitation (e.g. due to low production rate, high gamete mortality, or low encounter rate).

mating partners. In the 'plastogamy' of gamontogamous Foraminifera, two haploid gamonts of complementary mating types attach to each other, and release gametes within the shared space. In the '**syzygy**' of Apicomplexa, the haploid male and female gamonts unite pair-wise and encapsulate before producing multiple gametes (which may be either iso- or anisogamous) which subsequently form multiple **oocytes**. In the case of diploid partners, this pairing is followed by meiosis and gametogenesis: in ciliates, conjugation involves two diploid partners of complementary mating types. After meiosis, two migratory (male) haploid nuclei are reciprocally exchanged. In pennate diatoms, gamontogamy also occurs between diploid partners; following meiosis, two male gametes are exchanged, either reciprocally as in ciliates (trans-behavioural anisomotility) or unidirectionally (cis-behavioural anisomotility).

### 2.1.3 Modes and timing of sex determination

Sex determination and differentiation result from complex molecular mechanisms, potentially affected by a diversity of interacting factors. Sex-determination systems are conveniently characterized by their position along a continuum between genetic and epigenetic factors. At one end of the continuum, determination is purely genetic (or genotypic; GSD) if the differentiation of sexes and mating types is primarily underlain by genetic differences (e.g. MAT in *Chlamydomonas* and the sex chromosomes of mammals). At the other end, sex determination is purely epigenetic if this differentiation is only or essentially due to the differential expression of a single genotype. Note that the term 'epigenetic' has been given different definitions. We use it in its broadest sense (Holliday 1990), which includes any individual or cellular differentiation triggered by non-genetic cues, be they intrinsic (e.g. positional) or extrinsic (e.g. environmental or social). Accordingly, sex determination is epigenetic in many turtles, because genetically identical offspring may develop into either males or females depending on incubation temperature. Similarly, the differentiation of male and female organs in hermaphroditic animals is epigenetic, because genetically identical cell lineages develop into either sperm or eggs, according to positional information. In between are systems where both genetic and epigenetic effects interact to determine sex (e.g. in several fish species). Examples of the whole continuum, from purely genetic to purely epigenetic, are found among all three main types of cycles (haplontic, diplontic, haplo-diplontic) and across most lineages, with frequent transitions, as further developed in Chapter 4.

A second important categorization relates to the stage in the cycle at which sex determination occurs; namely, at either the haploid or the diploid phase. In haplontic lineages, sex is always determined at the haploid (gametophytic) stage (TYPE 1; Fig. 2.2). Haploid sex determination is also common in haplo-diplontic lineages (TYPE 2; Fig 2.3). The system is said **monoicous** (not to be confused with monoecious; Zander 1985; Haig & Wilczek 2006) if a single gametophyte produces gametes of both sexes (as in *Porphyra* TYPE 2.1b, or hornworts TYPE 2.2a), and **dioicous** (not dioecious) if sperm and ova are produced by separate male and female gametophytes (as in many mosses, TYPE 2.1a). It is worth noting that dioicy may result from either genetic (as in many mosses TYPE 2.1a) or epigenetic sex determination (as in many ferns; TYPE 2.2b and c). Conversely (although it would seem a priori that monoicy might only result from epigenetic sex determination), examples exist of haplo-genotypic sex determination (H-GSD) in monoicous gametophytes (TYPE 2.1b). In *Porphyra purpurea* (Bangiophyceae), for instance, all four meiospores resulting from a single meiosis disperse as a single unit (**tetraspore**) and contribute to the same gametophyte, which is thus a genetic chimera, with different parts developing either male or female gametangia depending on which of the four meiospores they stem from. Haploid sex determination may also occur in pseudo-diplontic life cycles: in *Acetabularia* (Dasycladales) and *Prasiola* (Trebouxiophyceae), which have no free gametophyte, isogametes of complementary mating type are produced by haploid structures developing directly on the diploid sporophyte (TYPE 2.1d).

In diplontic lineages, however, sex is normally determined at the diploid (sporophytic) stage, as also occurs in some haplo- or sub-diplontic lineages (as in some ferns and all Spermatophyta). The system is said **monoecious** if a single sporophyte produces gametes of both sexes (e.g. *Fucus spiralis*; TYPE 3.2a; Fig 2.4), and dioecious (or gonochoric) if sperm

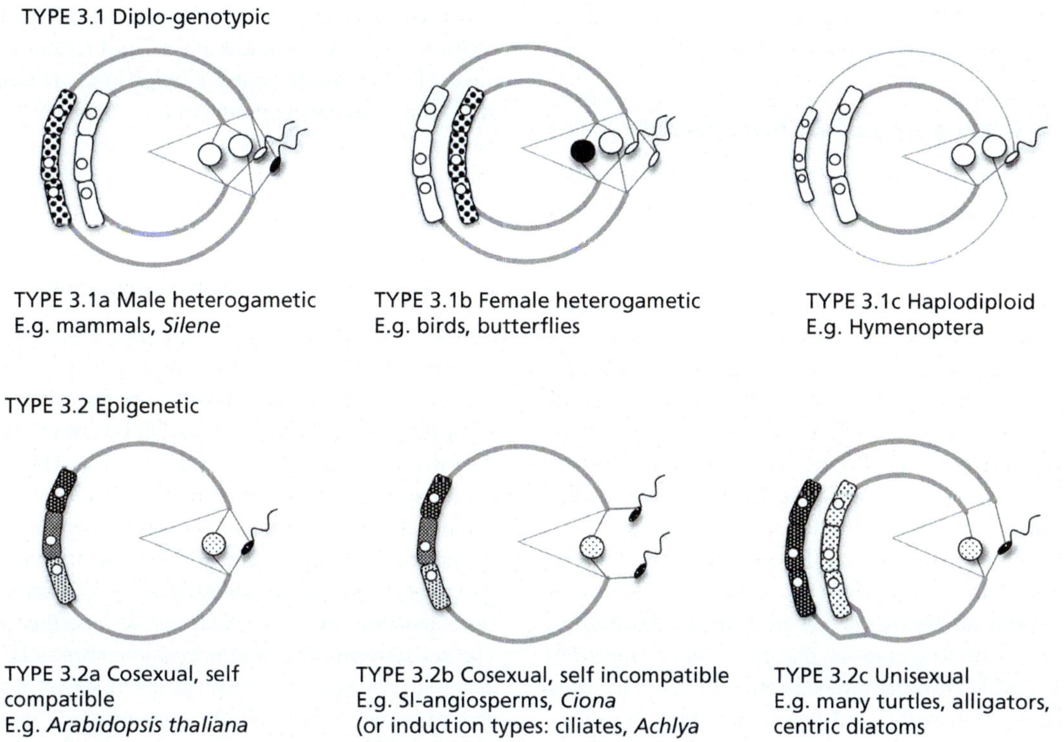


**Figure 2.4** (Sub-)diplontic cycles. Upper row: Diplo-genotypic sex determination. TYPE 3.1a Male heterogamety: the male sporophyte is heterozygous (XY) at the SEX locus (so that meiosis produces both X and Y gametes), whereas the female sporophyte is homozygous (XX) (so that meiosis only produces X gametes). TYPE 3.1b Female heterogamety: the female sporophyte is heterozygous (ZW) at the SEX locus (so that meiosis produces both Z and W gametes), whereas the male sporophyte is homozygous ZZ (so that meiosis only produces Z gametes). TYPE 3.1c Haplo-diploidy: the egg develops as a haploid male if unfertilized, and as a diploid female if fertilized. Epigenetic sex differentiation. TYPE 3.2a. Self-compatible co-sexual: genetically identical male and female gametes, produced by the same sporophyte, are self-compatible. TYPE 3.2b. Self-incompatible co-sexual: genetically identical male and female gametes, produced by the same sporophyte, are self-incompatible due to a secondary mating type (e.g. some angiosperms, some ascidians). Alternatively, gametogenesis and mating requires contact between compatible diploid individuals (e.g. *Achlya, Trypanosoma*). TYPE 3.2c. Unisexual (dioecious/gonochoric) with environmental or social determination of sex in the sporophyte.

and ova are found on separate male and female sporophytes or individuals (e.g. *Fucus vesiculosus;* mammals; TYPE 3.1). A further distinction is often made in angiosperms between hermaphroditic and monoecious species, depending on whether male and female organs are situated on the same flower or on different flowers from the same individual. The same distinction can be made, for example, in some diplontic Fucales, where **antheridia** and **oogonia** may occur either on the same or on different conceptacles of the same sporophyte. However, the term 'hermaphroditic' is also widely used to qualify monoicous or monoecious organisms outside angiosperms (as in ferns or animals respectively). Hereafter we will use it as a general term, in synonymy with 'co-sexual' or '**bisexual**'.

Dioecy (or gonochory) may result from either genetic or epigenetic sex determination (as in humans and alligators, respectively; TYPE. 3.1a versus 3.2c, Fig 2.4). Dioecy in Centrales (diatoms) also results from epigenetic processes (the same clone gives birth to both male or female cell lineages). Co-sexuality (monoecy and hermaphroditism) normally results from epigenetic sex determination, and is commonly found in multicellular organisms (as in most angiosperms) but may also occur in unicellular lineages: in ciliates, for instance, the male (migrant) and female (stationary) haploid **micronuclei** are genetically identical, resulting from a mitotic duplication of the same meiotic product.

### 2.1.4 Haplo-genotypic sex determination

GSD occurring at the haploid stage is termed haplo-genotypic (H-GSD). It may be associated with isogamy (i.e. mating types) as in *Chlamydomonas* and *Saccharomyces*, or anisogamy / oogamy (i.e. sexes) as in some green (Volvocales, Prasiolales, Oedogoniales) and brown algae (Dictyotales, Laminariales). It normally implies dioicy, with separate male and female gametophytes (but see the pseudohomothallism of *Porphyra*, TYPE 2.1b, Fig 2.3). H-GSD implies some genomic differentiation between the two interacting gametes (which is not the case for **diplo-genotypic sex determination** (D-GSD)). The sex-determining region may be limited to one single locus (as in *Chlamydomonas*), extend to longer non-recombining segments (as in **bipolar** basidiomycetes), or up to whole sex chromosomes (referred to as **U** and **V chromosomes**; Section 5.2.1) as in some mosses, liverworts, and fungi (e.g. *Microbotryum*). Under isogamy, H-GSD may involve several independent mating-type loci, and each can be multi-allelic, which may lead to hundreds of different mating types, as found in some Basidiomycota (as opposed to sexes, of which there are always two).

**Heterothallism** (as opposed to homothallism) refers to a mating incompatibility between genetically identical gametes originating from the same meiospore (i.e. from the same haploid clone), and directly results from H-GSD. Introduced by Blakeslee (1904) to describe mating systems in Mucorales, this term is now largely used throughout algae and fungi (including cases with anisogamy). More generally, H-GSD and heterothallism are widely distributed across most of eukaryote groups, including lineages with no free gametophytic stage (e.g. Dasycladales or Trebouxiophyceae). Among stramenopiles, haploid mating types are found in Synurophyceae (Synurales), Chrysophyceae (Ochromonadales), Xanthophyceae (Botrydiales), and Phaeophyceae (Ectocarpales, Scytosiphonales, Sphacelariales, Dictyotales). In alveolates, mating types exist in many dinoflagellates and possibly some Apicomplexa (gregarines, piroplasms, adeleorines). They also occur in Rhizaria (Foraminifera and possibly some Cercozoa, such as *P. brassicae*) and Hacrobia (Coccolithophora). Most Rhodophyta and Chlorophyta have H-GSD, as do many Charophyta. H-GSD is also found in land plants (many liverworts and mosses), Amoebozoa (Dictyostelia and Myxogastria) and most fungi (including Chytridiomycota, Ascomycota, Basidiomycota, and Zygomycota).

### 2.1.5 Diplo-genotypic sex determination

Genetic sex determination occurring at the diploid stage is termed **diplo-genotypic** (D-GSD). It implies a genomic differentiation between the interacting sporophytes (diploid individuals), but not between the interacting gametes, as under H-GSD. The diversity of sporophytes is often limited to two genders, usually males and females (i.e. sporophytes strictly specialized into the male and female function respectively), but not necessarily so (e.g. androdioecious systems like *Caenorhabditis elegans*; Box 5.15). The sex-determining region may encompass largely differentiated sex chromosomes (as in *Silene*), a short chromosomal region (as in *Carica papaya*; Liu et al. 2004; Box 3.11), or be limited to a single locus (as in the medaka fish *Oryzias latipes*), or even a single nucleotide polymorphism as in the case of the pufferfish (*Takifugu rubripes*) (Kamiya et al. 2012). Systems are called **male heterogametic** when males are heterozygous at the sex-determining locus (XY as in mammals; TYPE 3.1a, Fig 2.4), and **female heterogametic** when females are (ZW as in birds; TYPE 3.1b). Alternative ways of D-GSD are haplodiploidy (TYPE 3.1c) including **arrhenotoky** and **paternal genome loss** (Box 7.2), found in many arthropods, as well as XO or ZO systems. D-GSD may involve multiple alleles at the sex-determining locus (e.g. the swordfish *Xiphophorus maculatus*, where males are XY or YY, and females XX, WX or WY; Section 4.2) or a balance between diverse genic effects from multiple loci (as in the zebrafish *Danio rerio*; Liew et al. 2012).

D-GSD can actually be found in any cycle with a significant diploid phase, be it diplontic as in Fucales (e.g. *Fucus*, *Ascophyllum*, *Cystoseira*) and Metazoa (e.g. mammals and birds), or sub-diplontic as in Spermatophyta (many gymnosperms and a few angiosperms). With a few exceptions, it only evolved in cycles with differentiated male and female gametes (i.e. sexes). Possible exceptions include some pennate diatoms (e.g. *Seminavis robusta* or *Tabularia tabulata*), where mating actually occurs among gamonts, and isogamy evolved secondarily from oogamous ancestors.

## 2.1.6 Epigenetic sex differentiation

According to our definition, sex determination or differentiation is epigenetic when complementary mating phenotypes (be they sexes or mating types) can be produced by the same genotype. Epigenetic control may occur when sex differentiation is expressed at the haploid stage (e.g. in Charophyta, Monoblepharidales, Haemosporidia, or dinoflagellates) or diploid stage (e.g. in most angiosperms, many metazoans, diatoms and ciliates, or some Fucales). Homothallism (as opposed to heterothallism) refers to a mating compatibility between genetically identical gametes stemming from the same meiospore (same-clone mating), and results directly from epigenetic sex- or mating-type determination. Homothallism is widespread, for instance, among filamentous ascomycetes, where it often evolved from heterothallism by genic capture (homothallic strains often harbour copies of two mating-type alleles; Section 3.2.3b).

It is worth noting that ascomycetes also have sexes (represented by **microconidia** and **ascogonia**; TYPE 2.1e, Fig 2.3). Differential gametogenesis is likely to facilitate gene **imprinting**, and thereby the differential expression of alternative mating-type alleles in male and female gametes. As a matter of fact, epigenetic sex determination mostly evolved among anisogamous/oogamous lineages, with differentiated male and female gametogenesis. Among Chytridiomycota, for instance, Chitridiales tend to be both isogamous and heterothallic (H-GSD; e.g. *Dictyomorpha dioica*), while epigenetic sex determination with homothallism evolved in the oogamous Monoblepharidales (e.g. *Monoblepharis polymorpha*). H-GSD is also predominant among isogamous Chlorophyta, but epigenetic sex determination (monoicy) evolved in anisogamous or oogamous Caulerpales and Oedogoniales (Rawitscher-Kunkel & Machlis 1962). In Volvocales, all isogamous lineages (e.g. *Chlamydomonas*) have H-GSD, but several oogamous lineages evolved epigenetic sex determination (e.g. the dioicous *Volvox aureus*, or the monoicous *V. barberi*). In Charophyta, the isogamous Zygnematophyceae have H-GSD, while epigenetic sex determination evolved in Coleochaetophyceae and a few Charophyceae (all oogamous). In the diplontic diatoms, similarly, the Centrales are both oogamous and homothallic, while the derived Pennales evolved towards both isogamy and genetic sex determination.

Epigenetic sex differentiation allows a much wider diversity of genders than does GSD (basically, any possible stage along the continuum between complete maleness and complete femaleness). Simultaneous and sequential hermaphroditism, including both monoicy and monoecy (where male and female gametes are produced by the same gametophyte or sporophyte, respectively), are normally indicative of epigenetic processes, but exceptions may occur (e.g. the case of pseudohomothallism in *Porphyra*; TYPE 2.1b; *Acetabularia*, *Prasiola* TYPE 2.1d). Reciprocally, dioicy and dioecy may also result from epigenetic sex determination. *Plasmodium* (Haemosporidia), for instance, are dioicous (a given haploid **gametocyte** produces either male or female gametes) with epigenetic sex determination (a single clone can produce gametocytes of both sexes). Similarly, Centrales (diatoms) are dioecious (diploid gametangia produce either male or female gametes), with epigenetic sex determination (the same clone may produce both male and female lineages).

### 2.1.6a Simultaneous hermaphroditism

The proximate mechanisms of sex differentiation differ depending on whether hermaphroditism is simultaneous or sequential. In the former case, the process relies essentially on intrinsic cues, such as positional information in the developing embryo. The example of floral development in hermaphroditic angiosperms is presented in Section 3.3.1. Evolutionary causes also differ: whether hermaphroditism is simultaneous or sequential affects reproductive potential and mating opportunities. As already pointed out by Ghiselin (1969), simultaneous hermaphroditism is likely to be favoured by any feature that reduces mating opportunities (such as low population density, mobility, or mate search efficiency; Eppley & Jesson 2008). It is thus common in plants, but much scarcer in animals, where it is often associated with sessile habits (such as in

Porifera, Bryozoa, Entoprocta, and Ascidiaceae). In crustaceans, simultaneous hermaphroditism is normally found in groups that are either sessile or live in marginal and low-density habitats with limited mating opportunities. Note that the same conditions may select for clonal reproduction, as found in many unicellular lineages (where sex and reproduction may be totally independent) as well as in many plants and sessile animals.

In this context, simultaneous hermaphroditism facilitates the sexual cycle by ensuring that any conspecific met is a potential partner. It also opens opportunities for selfing, which are sometimes exploited (e.g. as a backup when no outcrossing opportunity arises), but might also be actively prevented by self-incompatibility (SI) systems (see Section 2.1.7). As an alternative to simultaneous hermaphroditism, some organisms with GSD have evolved permanent associations between male and female individuals. Young males of deep-sea angler fishes (Ceratiidae), for instance, quickly attach to the first female encountered, fuse with her tissues and progressively atrophy, finishing as nothing more than a pair of gonads. Similar evolutionary processes towards phoretic dwarf males are found in many other groups, including Rotifera, Crustacea (Copepoda), molluscs (ectoparasitic snails and commensal clams), Chlorophyta (e.g. *Oedogonium*), or Bryophyta (e.g. *Micromitrium*). Note also that the same result can be obtained via **social sex determination**, which ensures that potential partners develop into complementary sexes (Section 2.1.6d).

#### 2.1.6b Sequential hermaphroditism

Hermaphroditism is said to be sequential when an individual's gender (or mating type) changes during its life. This includes **protandry** (where individuals first develop as male, then female), **protogyny** (first female, then male), or multiple switches. The cues triggering such changes may be intrinsic (e.g. size or age) or extrinsic (e.g. growing conditions or social environment). Accordingly, switches may be mediated by a variety of molecules, including **aromatase** (or aromatase inhibitors) in several species of fishes (Kroon et al. 2005), or pheromones in the case of social determination (as in some oomycetes; Section 3.4.2b).

The ultimate causes favouring sequential hermaphroditism were also first formalized by Ghiselin (1969), who noted that the relative benefits of being male or female may depend on their own condition or social context. If, for instance, fecundity increases with size more in females than in males (as often occurs), then individuals should reproduce as male early in life, and as female later in life (the size-advantage model; see also Warner 1975, 1978). The resulting patterns of protandry are commonly found, for example, in fishes or snails. Reciprocally, in harem-breeding species, larger males may gain disproportionate benefits from controlling larger groups of females, which is expected to select for protogyny. Comparative analyses of sex change in wrasses (Labridae) indeed support a role for the size-advantage model (Kazancioglu & Alonzo 2010). Sequential hermaphroditism also occurs in plants (e.g. the protandric *Arisaema triphyllum*), for similar reasons: as pollen is cheaper to produce than seeds, young plants can more easily afford to reproduce as males than as females. Some plants may also switch sex recurrently from year to year (as do some fishes). In the epiphytic orchid *Catasetum viridiflavum*, which sex is expressed depends both on extrinsic (photoperiod, light intensity, nutrition, and temperature) and intrinsic causes (size and age; Zimmerman 1991). Usually, stressful conditions and small size trigger male development.

Social triggering of sex switches occurs when the benefits of playing male or female depend on the available partners. In the slipper limpet *Crepidula fornicata*, dispersing larvae that settle on a congener first develop as males, and later switch to a female role when additional larvae join the reproductive unit and develop as males. The switching decision depends on both intrinsic (body size) and social cues (mating partners). When two individuals of *Achlya bisexualis* (oomycete) meet, the one producing more of the pheromone A induces its partner to develop as male and to produce **antheridia**, while itself develops as female and produces oogonia (Section 3.4.2b).

A special mention must be made here to the process of mating-type switching, found in some lineages with either haploid (primary) or diploid (secondary) mating-type determination (e.g. yeasts and ciliates, respectively). In budding yeast, for instance, a given cell displays either the **a** or the α mating type, depending on which allele occupies the MAT locus. However, each cell also possesses a silenced copy of the alternative allele. Its activation via the specific gene *Ho* (for homothallism) allows generation of haploid cells of complementary mating types (Section 3.2.3a). Although not triggered by social cues, mating-type switching allows genetically homogeneous colonies founded by a single haploid cell to express compatible phenotypes. Many laboratory lines have lost the gene *Ho* and are thus unable to switch mating type; this underlines how thin are the borders between epigenetic and GSD.

Although sex change is widespread among animals, it is not common, which may be surprising given the many benefits it may potentially convey. Besides constraints imposed by the sex-determination system (it is normally prevented under GSD; but see mating-type switching as a counter-example), this rarity might largely be due to the associate physiological costs of sex change, which are bound to be large in species where male and female phenotypes are strongly dimorphic and subject to highly divergent selective pressures (Warner 1978; Policansky 1982).

#### 2.1.6c Environmental sex determination

Epigenetic processes may also trigger a definitive commitment to one or the other sex (thereby resulting in dioecy/dioicy). Several environmental factors have been shown to affect sex determination (e.g. pH, nutrient availability, and photoperiod), but temperature is certainly the proximate cue that is most-frequently co-opted. Temperature sex determination was first discovered in lizards (Charnier 1966) and turtles (Pieau 1971, 1972), and has since been documented in many other groups of animals, mostly reptiles (e.g. alligators, Lang & Andrews 1994; tuataras, Mitchell et al. 2006), and fishes (Ospina-Álvarez & Piferrer 2008). Temperature also affects gender- or mating-type determination in other groups (e.g. Xanthophyceae, Phaeophyceae, ciliates). As temperature strongly impacts the overall physiology and metabolism, it may directly affect the expression of genes or the activity of enzymes involved in the sex-determining pathway. However, temperature (like other environmental effects; Korpelainen 1994, 1998) is also likely to operate indirectly, by affecting the growing conditions during the sensitive period of development (e.g. Mittwoch 1996). The immediate cue for sex differentiation might thus be the size or stage reached at a critical time in development (Kraak & De Looze 1993; Korpelainen 1998; Kraak & Pen 2002), itself affected by temperature or any other environmental factor.

The ultimate causes favouring ESD are very much like those for sequential hermaphroditism (Charnov & Bull 1977). Specifically, ESD should evolve when environmental conditions differentially affect male and female fitness (and individuals have little control over which environment they experience). Specifically, if good growing conditions improve female fitness more than male fitness, then embryos should develop into females when experiencing a favourable environment, and into males otherwise. This prediction has received empirical support from several well-documented examples: in southern populations of the Atlantic silverside (*Menidia menidia*), offspring produced early in the season (at low temperature) develop into females, while those produced later in the season (in warmer conditions) develop into males (Conover & Heins 1987a). Females have thus more time to grow before the onset of reproduction.

In many fish species, females normally benefit more than males from a large body size (fecundity being proportional to body mass). Accordingly, low feeding rates or poor growth conditions tend to produce males (e.g. Roncarati et al. 1997; Davey & Jellyman 2005; Lawrence et al. 2008; Vandeputte et al. 2007; Chapter 4). Depending on mating system, however, large size may favour males more than females. In the monogamous Midas cichlid (*Cichlasoma citrinellum*), large males are better in defending the breeding territory against competitors (and are preferred by females). Accordingly, sex is determined by the relative size as a juvenile: individuals that are relatively larger than average members of their group tend to develop as males (Francis & Barlow 1993). Similarly, large males gain a disproportionate mating success in the amphipod *Gammarus duebeni*, due to mate-guarding behaviour. Early-produced individuals, benefiting from a longer growth period, develop into males, and later-produced ones into females. The main cue in this case is photoperiod (male development being triggered by the long-day periods of early summer, and female development by the short-day periods of late summer; Naylor et al. 1988).

Warner & Shine (2008) provided an experimental test of the Charnov–Bull model by investigating the differential effect of incubation temperature on male and female fitness in the short-lived agamid lizard *Amphibolurus muricatus*, a species with a **FMF** (female at low and high, male at intermediate temperature) pattern of temperature sex determination. By de-coupling sex- and temperature effects through hormonal manipulation of embryos, they found that male **lifetime reproductive success** was maximized when incubated at intermediate temperatures (27°C), while female lifetime reproductive success was maximized when incubated at either low (23°C) or high (33°C) temperatures, in full accordance with expectations.

ESD, however, also bears costs, by introducing potential biases in sex ratios (contrasting with GSD, which normally generates even sex ratios). Such costs might be large when life span is short and generations do not overlap, but are certainly alleviated in long-lived species, where large generation overlap dampens the between-cohort fluctuations in sex ratios. It might not be surprising, therefore, that clear benefits to ESD have only been identified in short-lived species, where they have to meet these predictably large costs. In long-lived species (turtles, alligators, tuataras), ESD might be close to neutral (Bull & Bulmer 1989).

#### 2.1.6d Social sex determination

Epigenetic sex determination might also be triggered by social cues (inter-individual interactions). As in the case for sequential hermaphroditism, such effects might evolve due to limited mating opportunities, or because the expected fitness returns from either male or female commitment depend on individual condition *relative to* potential partners. In the case of the Midas cichlid (Section 2.1.6c) mentioned, the sexual fate of a juvenile depends on its relative (not absolute) size in the group (Francis & Barlow 1993). Social effects are actually important determinants of sex in many fishes (Godwin et al. 2003), particularly among reef-dwelling species. In the presence of a conspecific adult of a given sex, juvenile *Gobiodon erythrospilus* rapidly mature into the opposite sex (Hobbs et al. 2004). Upon removal of the largest male, the largest female in the harem of the coral reef fish *Anthias squamipinnis* quickly changes into a male (Fishelson 1970). The reverse occurs in the protandrous anemone fish *Amphiprion*, where females are the largest and dominant members of social groups (Fricke 1979, Hattori 1991). Similar patterns have been described in invertebrates: the planktonic larvae of the green spoon worm *Bonellia viridis* (Echiura) that come into contact with a female adult develop as males, spending the rest of their lives as a sperm-producing parasite in her genital sac (Berec et al. 2005). Dispersing larvae of the polychaete *Osedax roseus* similarly develop into dwarf paedomorphic males if they encounter a female, but into females otherwise. Social sex determination also occurs in plants. In many **homosporous** ferns, gametophytic gender is determined epigenetically: isolated spores develop into hermaphrodites, and produce a pheromone (**antheridiogen**) that induces other spores settling in the neighbourhood to develop into males (Döpp 1950; Näf 1979; Näf et al. 1975; Banks et al. 1993). Culture experiments with *Woodwardia radicans* have shown that antheridiogen inhibits the growth of gametophytes, and that slow growth favours maleness, whereas fast growth favours femaleness, irrespective of the presence or absence of antheridiogen (Quintanilla et al. 2005). This supports the suggestion that the antheridiogen effect is mediated by its effects on growth (Korpelainen 1998), ultimately triggered by sex-differences in the fitness consequences of growth and body size (Haig &Westoby 1988). In mosses with epigenetic sex differentiation, the gametophytes also tend to develop into males as density increases (Kimmerer 1991).

#### 2.1.6e Maternal control

Epigenetic sex determination opens opportunities for a maternal control over offspring sex (Section 6.3.1). Temperature dependence might be exploited, for example, by fine-tuning oviposition sites. Females of *Chrysemys picta*, for example, may adjust offspring sex ratio by assessing the vegetation cover on oviposition sites (Janzen 1994). Female diamondback terrapins (*Malaclemys terrapin*) place larger eggs under warmer incubation conditions, so that large embryos develop into females (large size benefits females more than males, by shortening their maturation time; Roosenburg 1996). Similarly, growth- or condition-dependent sex determination may be exploited via differential egg provisioning. Increased yolk allocation feminizes offspring in the scincid lizard *Bassiana duperreyi*, in interaction with genetic and temperature effects (Radder et al. 2009). **Heterosporous** spermatophytes actually capitalize on this effect: the genetically identical **microspores** and **macrospores** produced by a sporophyte develop into male and female gametophytes respectively, mediated by this differential maternal provisioning (Korpelainen 1998). Maternal steroids deposited in egg yolk may also influence sex determination in reptiles (Bowden et al. 2000; Radder 2007).

If environmental conditions (resource quality or local mate competition) differently affect the fitness of male or female offspring, mothers should control brood sex ratio depending on their assessment (e.g. Charnov et al. 1981; Werren & Taylor 1984). In many parasitoid wasps, where offspring sex is under maternal control via haplodiploid sex determination (fertilized eggs develop as females, unfertilized eggs as males; TYPE 3.1c, Fig 2.4), the proportion of females in the progeny increases with host size, because resource level increases fecundity in females more than in males (e.g. Jones 1982; Kobayashi et al. 2003). Similarly, sex ratio is tuned to the level of local mate competition in parasitoid and fig wasps (Werren 1983; Herre et al. 1997; Section 6.3.1).

If offspring condition depends on maternal condition, mothers should influence offspring sex in accordance with their own state (Trivers & Willard 1973). Thus, whenever fitness increases more with body condition in males than in females, mothers in good condition are expected to invest in the production of sons, while those in poor condition should invest in daughters. Data from the polygynous red deer (*Cervus elaphus*) support the Trivers–Willard model: females produce more sons when in good condition, and more daughters when in poor condition (Clutton-Brock et al. 1984). In Barbary macaques (*Macaca sylvanus*), where social dominance is maternally transmitted, dominant females give birth to more sons, and subordinate females to more daughters (Paul et al. 1993). In European kestrels (*Falco tinnunculus*), early-breeding females produce more sons than late breeders, with more males among their first-laid eggs (whereas late breeders have more daughters in their first-laid eggs; Dijkstra et al. 1990). Early-born sons are able to obtain a territory the following spring whereas late-borns are not. For females no such birth date advantage was observed. In other raptors, the trend may be reversed depending on the size and age of first breeding relation (Daan et al. 1996). In the Seychelles warbler (*Acrocephalus sechellensis*), daughters often remain in their natal territory to help at the nest. This increases parental fitness, but only in high-quality territories (otherwise the effect is negative). Accordingly, pairs with relatively good territories produce more daughters than pairs with low-quality territories (Komdeur 1996). The control over sex apparently occurs before ovulation (Komdeur et al. 2002), but the exact mechanism is still unknown. The females of many hymenopterans also adaptively control offspring sex, so that their mode of sex determination also qualifies as epigenetic. In social species, queens can easily bias allocation towards daughters (but **secondary sex ratios** can also be manipulated by workers, via differential care allocation). Research on parasitoid wasps has been particularly instrumental for testing predictions of adaptive sex allocation theories (West 2009; Section 6.3).

#### 2.1.6f Parasite manipulation

Epigenetic sex differentiation paves the way to parasite manipulation (Sections 3.3.3 and 6.3.2). Many intracellular parasites have evolved means of feminizing their host, as a way to enforce vertical transmission (e.g. Terry et al. 1998; Cordaux et al. 2011). Bacterial endosymbionts (*Wolbachia*, *Cardinium*, *Rickettsia*, *Arsenophonus*, *Spiroplasma*), which occur ubiquitously in insects and some other ecdysozoans (isopods, spiders, nematodes) may increase their own transmission through male killing, host feminization, or parthenogenesis induction (Moran et al. 2008; Werren et al. 2008; Duron et al. 2008; Saridaki & Bourtzis 2010). Microsporidia (*Microsporidium*, *Nosema*) can also feminize their host (crustaceans) by inhibiting **androgenic gland** differentiation (Jahnke et al. 2013). Conversely, parasites that are transmitted horizontally by males may masculinize their host; such is the case of the pollinator-borne anther smut fungus (*Microbotryum violaceum*), which infects female *Silene* plants and masculinizes them to allow subsequent transmission (Ruddat et al. 1991). The ensuing costs to the hosts and biases in population sex ratios may induce genetic responses to escape or counteract manipulation, including switches in sex-determination systems (Cordaux et al. 2011; Werren 2011; Section 7.2.3c).

### 2.1.7 Self-incompatibility systems and induction types

Simultaneous hermaphrodites incur a risk of selfing, which may cause fitness losses via inbreeding depression in diplontic or haplo-diplontic lineages (but not in haplontic ones, because most genes are expressed in hemizygous state, which directly exposes deleterious mutations to selection). One way of preventing selfing is via sequential hermaphroditism or dioecy. An alternative is offered by SI systems; namely the evolution of physiological processes or morphological structures that prevent selfing. Such systems are best known among angiosperms, where they are often controlled by a single di- or multi-allelic locus (S) encoding one protein expressed in the pistil (the female determinant), and another one in the pollen (**gametophytic SI**) or the anther (**sporophytic SI**). Molecular mechanisms are detailed in Section 3.4.1a. Additional SI mechanisms rely on flower heteromorphy (Section 3.4.1a; Fig. 3.11): **distyly** is normally determined by a single S locus with two alleles (one recessive *s* and one dominant *S*) generating two self-incompatible genotypes (*ss* and *Ss*), while **tristyly** involves two di-allelic loci and a larger number of genotypes. Among simultaneous hermaphrodites in Metazoa, several ascidians evolved SI systems similar to angiosperms (Harada et al. 2008; Section 3.4.1b).

The secondary mating types that induce 'hormonal heterothallism' in oomycetes bear analogy with SI systems: 'heterothallic' strains of *Phytophthora* produce both male and female gametes, but pheromones from complementary strains are required to reciprocally induce sexual development (Section 3.4.2b). These 'induction types' (Ko 1988) apparently rely on the alternative expression or repression of two linked pairs of genes (forming a secondary MAT locus) encoding complementary pheromones and receptors. The mating types of ciliates are to be interpreted in a similar way: each conjugant produces both a male (migratory) and a female (stationary) haploid **micronucleus**, but exchanges only occur between partners of complementary mating types (Section 3.4.2b). Analogous processes seem to occur in other diplontic lineages such as Bacillariophyceae (trans-behavioural anisomotility) or kinetoplastids (e.g. *Trypanosoma*), where mating initiation requires the presence of different clones (Gibson 2001). It is worth noting that SI systems and induction types can also be determined epigenetically (as exemplified by the cytoplasmic and caryonidal inheritance of mating types in ciliates; Box 3.24), including via social effects (e.g. *Achlya*, oomycetes), and may also display mating-type switches, as documented, for example, in oomycetes (*Phytophthora*), ciliates, and possibly diatoms (*Seminavis*).

In line with the inbreeding-avoidance interpretation, SI systems and induction types only evolved in diplontic or sub-diplontic lineages. The point could be made that mating types also coexist with sexes in heterothallic filamentous ascomycetes, which spend most of their life as haploids (meiosis is zygotic). However, contrasting with the above examples (where SI systems or induction types were superimposed on pre-existing sexes; TYPE 3.2b, Fig. 2.4), sexes in ascomycetes were secondarily superimposed on primary mating types (TYPE 2.1e, Fig. 2.3). Microconidia first evolved as dispersal units (asexual reproduction), and secondarily took a function as male gametes. In this context, microconidia and ascogonia have evolved as part of a mating strategy allowing both long-distance fertilization (through microconidia) and nurturing of the di-haplontic fruiting body (ascogonia). In turn, this evolution may have facilitated the apparition of homothallism in many filamentous ascomycetes, mediated by the genic capture of complementary mating types and their differential expression in microconidia and ascogonia.

### 2.1.8 Evolutionary paths among sex-determination modes

The D-GSD systems found across a series of diplontic cycles did not evolve directly from H-GSD systems. Such transitions required intermediate stages with epigenetic sex determination, during which the timing of sex differentiation progressively shifted from the haploid to the diploid phase. Such a trend is nicely illustrated across the green lineage (Fig 2.5), in parallel with the increasing importance of the sporophytic stage. Basal Chlorophyta are haplontic with H-GSD (Fig 2.5a). In Ulvophyceae, sex determination is haplo-genotypic in most haplontic (Ulotrichales) and haplo-diplontic cycles (e.g. Cladophorales or Ulvales; Fig 2.5b), but becomes epigenetic in Bryopsidales, where some diplontic lineages also evolved D-GSD (e.g. *Codium*). The haplontic Charophyta (from which Embryophyta emerged) also have gametophytic sex determination, with evolution from haplo-genotypic to epigenetic in many Characeae. Bryophyta, with a dominant gametophytic stage, maintain gametophytic sex determination, also with a shift from haplo-genotypic in the primitively dioicous lineages (Fig 2.5b) to epigenetic in the derived monoicous lineages. Sex determination remains epigenetic in Pteridophyta (with a dominant sporophytic stage) but displays a transition from gametophytic (in homosporous lineages; Fig 2.5c) to sporophytic (in heterosporous lineages): the sporophytes of Selaginellales, Isoetales, and Salviniales produce differentiated macrospores and microspores, that give rise to female and male gametophytes respectively (Fig 2.5d). All spermatophytes, finally, have sporophytic sex determination, which remains epigenetic in the majority of angiosperms and Coniferophyta (both microspores and macrospores being produced by the same sporophyte; Fig 2.5e). From this ancestral situation, D-GSD (Fig 2.5g) evolved in many gymnosperms (including Cycadophyta, Gingkophyta, Gnetophyta, and some Coniferophyta) and a few angiosperms (e.g. *Silene*). Other angiosperms that maintained

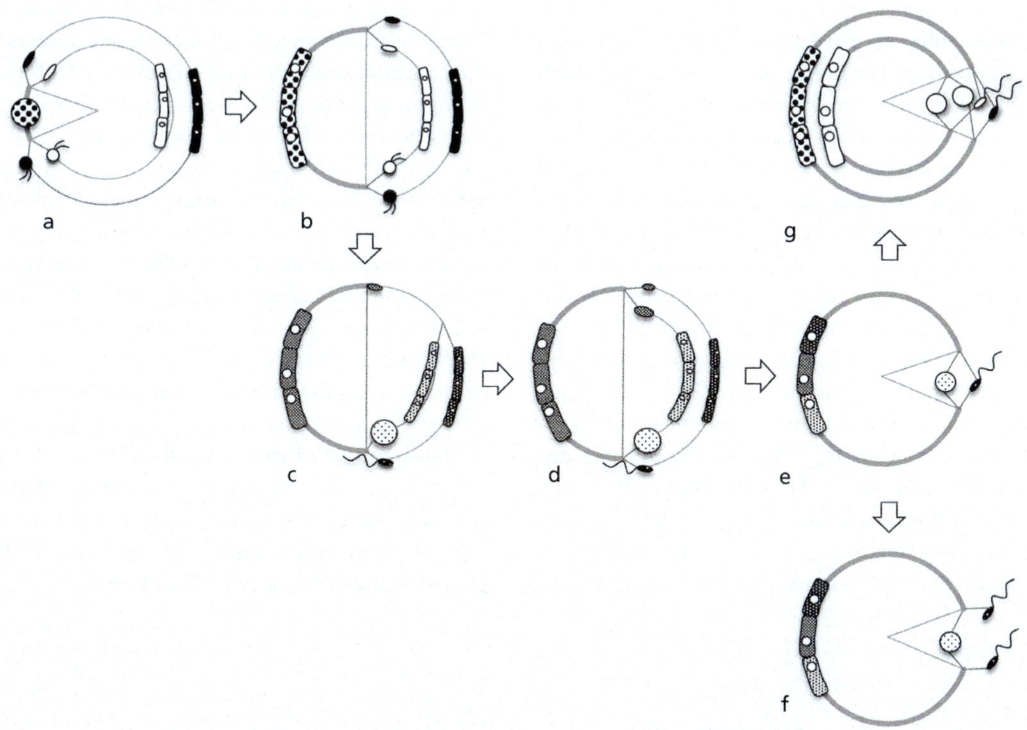


**Figure 2.5** Possible evolutionary paths among sexual cycles, examples from the green and the brown lineages. (a) Haplontic cycle with haplo-genotypic mating-type determination (*Chlamydomonas*, *Ulothrix*). (b) Haplo-diplontic cycle with haplo-genotypic mating-type determination (*Ulva*, *Cladophora*, *Ectocarpus*). (c) Haplo-diplontic cycle with epigenetic sex determination, dioicy and homospory (*Equisetum*, *Pteridium*, *Laminaria*). (d) Haplo-diplontic cycle with epigenetic sex determination, dioicy and heterospory (*Selaginella*, *Salvinia*). (e) Diplontic cycle, epigenetic sex differentiation, monoecy, self-compatibility (*Codium fragile*; *Arabidopsis thaliana*). (f) Diplontic cycle, epigenetic sex determination, monoecy, self-incompatibility (*Solanum*, *Plantago*). (g) Diplontic cycle, diplo-genotypic sex determination, dioecy, male heterogamety (*Codium coactum*, *Silene latifolia*, *Rumex acetosa*, *Cycas revoluta*, *Fucus vesiculosus*).

hermaphroditism alternatively developed SI systems (secondary mating types; Fig 2.5f), expressed either at the sporophytic or gametophytic level. A similar trend can be recognized in Phaeophyceae, where sex determination evolved from haplo-genotypic (found in many haplo-diplontic cycles such as in Ectocarpales, Sphacelariales, and Dictyotales) to epigenetic in groups where the gametophyte is either reduced (as in some Laminariales, Desmarestiales, and Sporochnales) or absent (e.g. Fucales and Ascoseirales). D-GSD then evolved from epigenetic systems in some of these diplontic lineages (e.g. *F. vesiculosus*, *A. nodosum*).

## 2.2 The eukaryote tree of sex

Eukaryotes, one of the three domains of life (Woese et al. 1990), diverged from Archaea and Bacteria some 1.6 to 2.1 (possibly up to 2.7) billion years ago. Our understanding of eukaryotic relationships was revolutionized by the use of molecular data, which have revealed their immense diversity, but also the complexity of their classification (e.g. Sogin et al. 1986). The current view of eukaryotic phylogeny is that of a small number of large 'supergroups', although with unequal support (some remain highly controversial; Keeling et al. 2005). Deep nodes, in particular, are poorly resolved. One classification opposes the bikonts (thought to originate from a two-flagellate ancestor) to the unikonts (descending from a uniflagellate ancestor). The latter comprise the Amoebozoa and opisthokonts (fungi, choanoflagellates, and animals), whereas the bikonts encompass the Hacrobia, Archaeplastida (Plantae), and the so-called SAR supergroup (stramenopiles, alveolates, and Rhizaria). The excavates might also form an early-diverged and paraphyletic group of bikonts. As deep phylogenies are not all settled yet, many uncertainties remain, and this broad picture is expected to change.

Eukaryotes present a series of unifying features, including the presence of mitochondria, resulting from a very early endosymbiosis with an alpha-proteobacterium. Similar symbioses have subsequently occurred in some lineages: the chloroplasts of archaeplastids (including green and red algae) result from a primary endosymbiosis with a cyanobacterium, while those of all other photosynthetic eukaryotes (e.g. brown algae, diatoms, dinoflagellates, coccolithophores, and euglenes) result from secondary symbioses with archaeplastids.

One of these unifying features of eukaryotes is meiotic sex (Fig 1.1). Although the sexual cycle has been lost from a few lineages and remains undescribed in many others, genomic analyses have revealed functional meiotic genes in members of all supergroups. This reveals that meiotic sex was clearly operational in the common ancestor of all eukaryotes. Unfortunately, the modalities of the sexual cycle (including how sex is determined) remain poorly understood in the vast majority of lineages. Outside animals and plants, information has mostly been gathered from a few model organisms, or from those of economic importance and that are easily amenable to laboratory analyses. As a result, important gaps in our knowledge of sexual reproduction and sex determination remain. We have tried to gather as much detailed information as possible on the diversity and phylogenetic distribution of sex-determination systems of existent phyla; this is provided electronically in tables and text at www.oup.co.uk/companion/beukeboom. On Plate 1 (central colour plate section), we present an overview of data for the main eukaryote supergroups. Plate 2 provides an overview of GSD in insects.

CHAPTER 3

# Molecular mechanisms of sex determination

Genomes must fulfil a variety of functions during an organism's life cycle. Accordingly, they encode a diversity of programmes that have to be either silenced or expressed, depending on situation. This also applies to sexes and mating types: genomes are fundamentally bipotent, having the potential to develop into either sex depending on specific cues. This statement is entirely true of systems with environmental sex determination (ESD): a turtle embryo may develop as male or female, depending on incubation temperature. It is almost true of systems with genotypic sex determination (GSD): an XY human or *Drosophila* carries copies of all the genes required to develop into a female; XX individuals can also develop into phenotypic males (which will be non-fertile however, lacking a few Y-linked genes required for proper testis development). One crucial issue, given this fundamental bipotentiality, is to make sure that an individual's commitment is unambiguous; sex differentiation must ensure proper development towards one or the other sex, and not towards intermediate phenotypes of reduced fertility. These mechanisms must also maintain sexual identity throughout life (the initial cue is often transient), or, alternatively, switch this identity in an efficient and adaptive way in sequential hermaphrodites.

In this chapter, we first describe the overall structure of sex-determination cascades and the function of the main upstream and downstream actors (Section 3.1). We then delineate specific pathways of haploid mating-type determination (Section 3.2) and diploid sex determination (Section 3.3). Finally, we outline the molecular mechanisms of self-incompatibility (SI) and induction types in diplontic or sub-diplontic lineages (Section 3.4).

## 3.1 Gene regulatory networks

The molecular mechanisms of sex determination encompass not only the initial trigger that directs an organism's developmental programme towards either the male or the female fate, but also the ensuing tiers of genes that organize primary sex differentiation. Once an initial decision is taken, it must be unambiguously conveyed to all effectors to ensure proper phenotypic development. Sex differentiation is a complex process, requiring sophisticated gene regulatory networks, which involve series of feedback loops and dynamic antagonisms, often with functional redundancy (Section 3.1.1). Key roles in this machinery are taken by *cis*- and *trans*-**acting** regulatory elements (**promoters** and **transcription factors**, which control gene expression in a cell-autonomous way), RNA-binding proteins (which control the post-transcriptional processes resulting in functional proteins), as well as hormones and pheromones (which convey information and trigger expression of relevant genes in other cells from the same or from different organisms).

### 3.1.1 Sex-determination cascades

Sex determination takes place through a network of interacting genes that are hierarchically ordered in a cascade. At the bottom of cascades lie major effectors, which integrate cues on sexual identity (running down the cascade) with positional and

*The Evolution of Sex Determination*. Leo W. Beukeboom and Nicolas Perrin.

developmental information, to orchestrate male or female development (i.e. switch on the correct differentiation programme at the right place and time). **DM-domain** genes fulfil this function in animals. In *Drosophila melanogaster*, for instance, the transcription factor DSX interacts with Hox genes and ecdysone-dependent mechanisms to control downstream effectors (e.g. *Bab1* and *Bab2*; (for *Bric à brac*) to produce male-specific patterns of pigmentation; Section 3.3.2a). Major effectors at the bottom of cascades are the most conserved components: DM-domain genes control sex differentiation in all animals, **MADS-box** genes in all angiosperms.

At the top of the cascade are the initial triggers, the proper sex determinants, which may be genetic (e.g. X:A ratio in *Drosophila*, *Sry* in mammals), environmental (e.g. temperature in turtles), social (e.g. density in some fishes or nematodes), or any combination of those. These triggers typically are the most variable part of sex-determination cascades, and may differ drastically even among closely related species. This primary information is processed via threshold effects (Chapter 4) into binary signals that relay instructions in an On/Off mode, similar to the 1/0 states in computation based on only two possible outcomes. It is accomplished by intermediate factors (the 'central gear') that convey the sex-identity decision to the main effectors in an unambiguous way, sometimes through a long cascade (e.g. *Caenorhabditis elegans*; Box 3.15). *Sox9* in vertebrates and *Transformer* (*Tra*) in insects take part in this function. These intermediate factors also constitute a memory device; the initial trigger is usually only expressed transiently, during a limited window of embryonic development, where it tips the balance towards the male or female fate. The downstream network then establishes an autoregulatory loop that keeps memory of the initial decision and ensures lifelong continuity. Such self-reinforcing feedback loops are established by *Sox9* in mammals (Section 3.3.2b), *Sxl* in *Drosophila*, or *Tra* in *Musca* and other dipterans (Section 3.3.2c). Last but not least, a major function of genes in the cascade is that of inhibiting the alternative pathway, in order to prevent the development of intermediate phenotypes of reduced fitness. Sex is a bistable equilibrium, and sex determination a battle between male and female specific programmes, occurring through a complex network of genes that repress each other (Kim & Capel 2006). In the fern *Ceratopteris richardii*, *Fem1* and *Tra* (with no homology to similarly named genes in insects and nematodes), which promote the differentiation of male or female traits respectively, antagonize each other; the initial cue (female-produced pheromone antheridiogen $A_{CE}$) tilts the balance in favour of the male fate by inhibiting *Tra* (Strain et al. 2001; Tanurdzic & Banks 2004). In vertebrates, sex antagonisms centre on the competition between *R-spondin*/*Wnt4* (which promotes female specific granulosa and theca cell fates) and *Sox9*/*Fgf9* signalling (which promotes male-specific Sertoli cell fate; Section 3.3.2b). Whatever equilibrium settles, it must be actively maintained lifelong; otherwise switches may occur, as happen in sequential hermaphrodites. To quote Sekido & Lovell-Badge (2008), 'sex determination is a story of opposing forces and crucial alliances but, although the winning team takes it all, its rule can be surprisingly tenuous'.

## 3.1.2 The main actors

### 3.1.2a Transcription factors

Transcription factors are proteins that control gene expression by binding to specific DNA sequences in the promoter region, inducing conformational changes of the DNA that enhances or reduces transcription. They may bind as monomers or oligomers (e.g. heterodimers) and form specific complexes with other proteins (co-factors), which activate or repress the recruitment of RNA polymerase. The promoter regions of genes often carry several binding sites for distinct transcription factors, which work in co-operation to allow specific and efficient expression. The DNA-binding domains of transcription factors are highly conserved, which allows the definition of a few major families. The other domains of the protein, responsible for recruiting co-factors to build specific transcription complexes, are much more versatile, allowing rapid evolution and adaptation to a diversity of roles in the control of cell fate and organism development.

One important such family is constituted of **homeobox** genes. Members of this family all have in common a ~180 bp homeobox, encoding a ~60 αα protein domain (homeodomain; HD) that binds DNA in a sequence-specific manner. HDs consist of two antiparallel N terminal α-helices (helices I and II), connected by a turn to a roughly perpendicular C-terminal α-helix (recognition helix III), which interacts directly with DNA. It is mainly four residues (at positions 47, 50, 51 and 54) that ensure recognition by contact with the bases in DNA's major groove. The N-terminal arm can also serve for HD dimerization. Homeobox motifs appeared early in the evolution of eukaryotes; they are commonly found in plants, animals, and fungi and are highly conserved over hundreds of millions of years of evolutionary history. The Hox genes, first described in *Drosophila*, constitute a subfamily of homeobox genes that play a key role in all metazoans in the establishment of body axes during embryogenesis. HD transcription factors take a key role in organizing mating processes in many fungi.

The MADS-box genes, which encode another important family of homeotic transcription factors, evolved in a common ancestor of all extant eukaryotes from the duplication of a gene encoding **topoisomerase** IIA subunit A (Gramzow et al. 2010). MADS-box genes control aspects of development or cell differentiation in plants, animals, and fungi (MCM1, for instance, is central to the transcriptional control of cell-type-specific genes and pheromone responses in the yeast *Saccharomyces cerevisiae*). Though their numbers have remained quite small in most eukaryote lineages, they are much diversified in land plants. There is just one MADS-box gene in chlorophytes, but more than 20 in mosses and around 100 in angiosperms. The functional diversification of MIKC genes is closely linked to the origin of developmental and morphological novelties in the sporophyte of seed plants, most spectacularly the floral organs and fruits of angiosperms, where they take homeotic functions similar to the *Hox* genes in animals (Box 3.6).

HMG (**high mobility group**) proteins constitute another important family of transcription factors. The overall structure of HMG-box domains is remarkably conserved in animals, plants, and fungi. This family includes SOX proteins, which play important roles in a variety of developmental processes during organogenesis in metazoans. The several SOX proteins recognize and bind similar DNA motives (A/TAACAA/T), and specificity in transcription is determined by the identity of the binding partners in particular tissues. Several members (*Sry*, *Sox9*, *Sox8*) are directly involved in the sex-determination pathway of mammals (Box 3.13). The α-box transcription factors often involved in the determination of mating types in ascomycetes (Section 3.2.2b) are related to the HMG family (Martin et al. 2010).

In the DM domain gene family of transcription factors, the conserved N-terminal binding site consists of two intertwined **zinc fingers**. DM factors bind in the minor groove as either homo- or heterodimers, without sharp DNA bending. The regulatory domains are C-terminal. Members of this family were first identified from flies and nematodes, where respectively *Doublesex* (*Dsx*) and *Male abnormal 3* (*Mab3*) play key roles in sex differentiation. The *Dsx and Mab3 related transcription factor 1* (*Dmrt1*) *was* later identified as a key player in the sex-determination pathway of vertebrates. Most animal species display multiple DM domain genes, usually with specific roles in gonadal development and sex differentiation (Section 3.3.2a). *Dmrt1* plays a major sex-determination role in birds, and **paralogues** of this gene have also taken a master sex-determination role in a few lineages of fishes and amphibians. No DM-domain genes have been identified outside metazoans.

A final family that can be mentioned here are the **bZIP-domain** transcription factors, widespread among all eukaryotes. They may take a diversity of roles in organism development. Some are involved in the sexual cycle (e.g. steroid hormone synthesis by endocrine cells), they play a role in transcriptional activation of the early *Sex-lethal* promoter in insects, and may also take a master sex-determination role (e.g. the *Mid* gene of Volvocales; Section 3.2.2e).

#### 3.1.2b Post-transcriptional regulation

Before resulting in functional proteins, transcription products must be processed in a number of ways. Post-transcriptional regulation involves RNA-binding proteins from different families, which often possess a RNA recognition motif that binds a specific sequence of the transcripts, and other domains fulfilling specific functions such as splicing or translation. Some of these families have both DNA- and RNA binding properties; this is the case of **zinc finger proteins**, which are among the most prevalent nucleic acid-binding motifs in eukaryotes (e.g. the transcription factor WT1, which upregulates *Sry* in mammals). Alternative splicing allows a single gene to produce multiple proteins, for example, by including or excluding particular **exons**, or varying exon size from the processed mRNA. Sex differentiation in insects involves the sex-specific splicing of the main effector *Dsx* by upstream RNA-binding proteins such as TRA.

#### 3.1.2c Hormones and pheromones

Hormones and pheromones are molecules released by some cells to affect gene expression in other cells, respectively from the same or from a different individual. Target cells only respond when they express a specific receptor to that hormone or pheromone. Binding of either of the two molecules results in the activation of signal transduction mechanisms that ultimately upregulate or downregulate specific genes. Hormones and pheromones play an essential role in the determination and/or differentiation of sexes and mating types. In plant, auxins, **cytokinins**, ethylene, and gibberellins (GAs) are instrumental in meristem differentiation and floral development (Section 3.3.1; reviewed in Chandler 2011; Plackett et al. 2011). Steroid hormones are particularly involved in the **sexual differentiation** of vertebrates (Section 3.3.2b). Being liposoluble (they are derived from cholesterol), they can pass through cell membranes. Once inside the cell, they bind with specific receptors found only in the cytoplasm of target cells. The receptor-bound steroid hormones then travel into the nucleus and bind to specific sequences on the DNA, where they regulate transcription. Androgens (e.g. testosterone), produced in the gonad of both sexes, are converted to oestrogen in the female gonad by the enzyme aromatase (encoded by the gene *Cyp19*). Changes in the patterns of sex determination often occur via the differential regulation of steroid hormones (e.g. by aromatase) and/or the localization of specific receptors, rather than changes in their structure (which is highly conserved due to their many **pleiotropic** effects).

The **anti-Müllerian hormone** (AMH, also called Müllerian inhibiting substance) is a member of the transforming growth factor beta (TGF-β) family. Non-steroid hormones exert their effects through membrane-bound receptors. Ligand binding drives receptor-specific cytoplasmic substrates (the Smad molecules) into the nucleus, where they act as transcription factors. AMH plays a key role in vertebrates by repressing the development and function of distinct reproductive organs (in mammals, it prevents the development of Müllerian ducts into a uterus). Changes in sex-determination patterns may thus also occur via the differential expression of AMH or AMH receptors.

Pheromones, which include a large variety of different molecules (not always peptidic), are crucial players in cell–cell recognition and mating-type differentiation in unicellular eukaryotes. They are also widely used for social communication during the sexual cycle in multicellular organisms, for example, to identify and localize mating partners (as in insects), or, more basically, to control sex differentiation in a potential partner (as in ferns and oomycetes; Section 3.4.2b).

### 3.1.3 Molecular mechanisms of transitions

The mechanisms of sex determination are surprisingly labile (Chapter 7). As already noticed by Wilkins (1995), however, this lability affects upstream much more than downstream components of cascades. In insects, *Dsx* keeps the function of main effector at the bottom of cascades in every species investigated so far; the first variations already become evident in its upstream regulator *Tra*, while the initial triggers at the top of the cascade display an astonishing diversity (Section 3.3.2c). This led Wilkins (1995) to propose that sex-determination cascades evolve from bottom upwards, because the recruitment of new genes into a sex-determination function is somehow easier at the beginning of the pathway. Downstream genes are more intensively integrated in gene networks and, thus, more resistant to change due to their pleiotropic effects. Divergence among biochemical pathways is similarly more likely in upstream intermediates than at the bottom, where enzymes generate the final product. Genomic studies are now revealing that changes in the cascade may a priori affect any of the above actors, including transcription factors, RNA-binding proteins, hormones, and receptors. Transitions may fall in several categories (which are neither exclusive nor exhaustive):

1. A gene duplication, followed or not by **neofunctionalization** or **subfunctionalization**. The additional copy of an existing gene is often enough to tilt the initial balance towards one or the other fate. The master sex-determination genes in medaka fish (*Oryzias latipes*) and clawed frogs result from a duplication of *Dmrt1* (Section 3.3.2a). In bees, *Csd* originated from a duplication of *Feminizer* (Box 3.21). In the Patagonian pejerrey (*Odontesthes hatcheri*), males have a duplicated copy of the AMH gene (Section 3.3.2b), while in cucumber, feminization results from the duplication of a gene coding for an enzyme that controls ethylene production (Box 3.8).
2. A change in the hierarchical position of an existing gene. In birds, the *Amh* gene is upstream of *Sox9*, and the ancestral basal gene *Dmrt1* moved higher up in the hierarchy to become the master sex-determination gene (Box 3.14). Such switches might be facilitated if these genes were first involved in a common self-reinforcing feedback loop (e.g. *Sox9* and *Dmrt1* in mammals; Box 3.12).
3. A mutation altering the molecular function of a gene product (e.g. its DNA-binding properties), affecting thereby the expression of target genes. Such mutations occurred in the evolution of *Sry* from *Sox3* in mammals (Box 3.13). Other examples include maternal sex determination in blowflies resulting from a loss of function of *Tra* (Box 3.20), and sex in pufferfish (*Takifugu rubripes*) determined by a single nucleotide polymorphism in the AMH receptor gene (Section 3.3.2b). Such mutations may cause the irruption of brand new genes in sex-determination cascades (e.g. the new splicing inhibitor PSI, which replaced TRA in *Bombyx*; Box 3.23).
4. A change in the transcriptional or translational regulation of a gene, affecting the level, timing, and/or localization of its expression. Masculinization in many metazoans results from the up-regulation of DM-domain transcription factors (Section 3.3.2a), while feminization in melon results from a transposon inserted in the promoter region of a transcription factor (Box 3.9). **Methylation** of the promoter region of aromatase induces maleness in sea bass (Section 3.3.2b). This last category encompasses many transitions to and from epigenetic determinants, including environmental (as in *Daphnia*, Box 3.16) and maternal effects (as in several flies; Box 3.20), or parental imprinting (as in *Nasonia*, Box 3.21).

## 3.2 Haploid mating-type determination

### 3.2.1 What genes are on mating-type loci?

Under haplo-genotypic sex determination (H-GSD), cell identity and mating processes are organized by one or more mating-type (MAT) loci, each carrying one or several genes. Autosomal genes (i.e. located in the recombining part of the genome, outside of MAT loci) are of course also involved in the downstream cascade. The main categories of genes found on MAT loci, or directly affected by MAT genes, fulfil the following functions.

First is the master sex- or mating-type determination role, normally taken by transcription factors that can belong to a diversity of families (e.g. b-ZIP, HD, HMG, and/or α-box; Section 3.1.2). Although many of these families are widespread throughout the eukaryote phylogeny, different ones have been installed at the top of the cascade depending on lineages, testifying to the intrinsic dynamics of sex-determination mechanisms. These factors are normally silenced when environmental conditions are favourable to vegetative growth, and activated by stressful conditions (e.g. nitrogen depletion). Once expressed, they take a key role during the mating process, both at the haploid and at the diploid stage. In the haploid cell, they may initiate gametogenesis by triggering the expression of mating-type specific pheromones and receptors that define cell identity, and contribute to co-ordinating cell fusion. After fusion, they may combine as heterodimers with complementary factors contributed by the mating partner to switch off the haploid programme, and switch on the diploid programme (e.g. activating meiosis genes in haplontic cycles). These are the only genes that are always found in MAT loci (Fig. 3.1), and thus fulfil the most basic function: initiate the correct developmental programme at the right time (Section 1.2.1).

The second set of genes involved in mating processes encodes pheromones and receptors. Cell signalling and recognition is crucial to localize potential

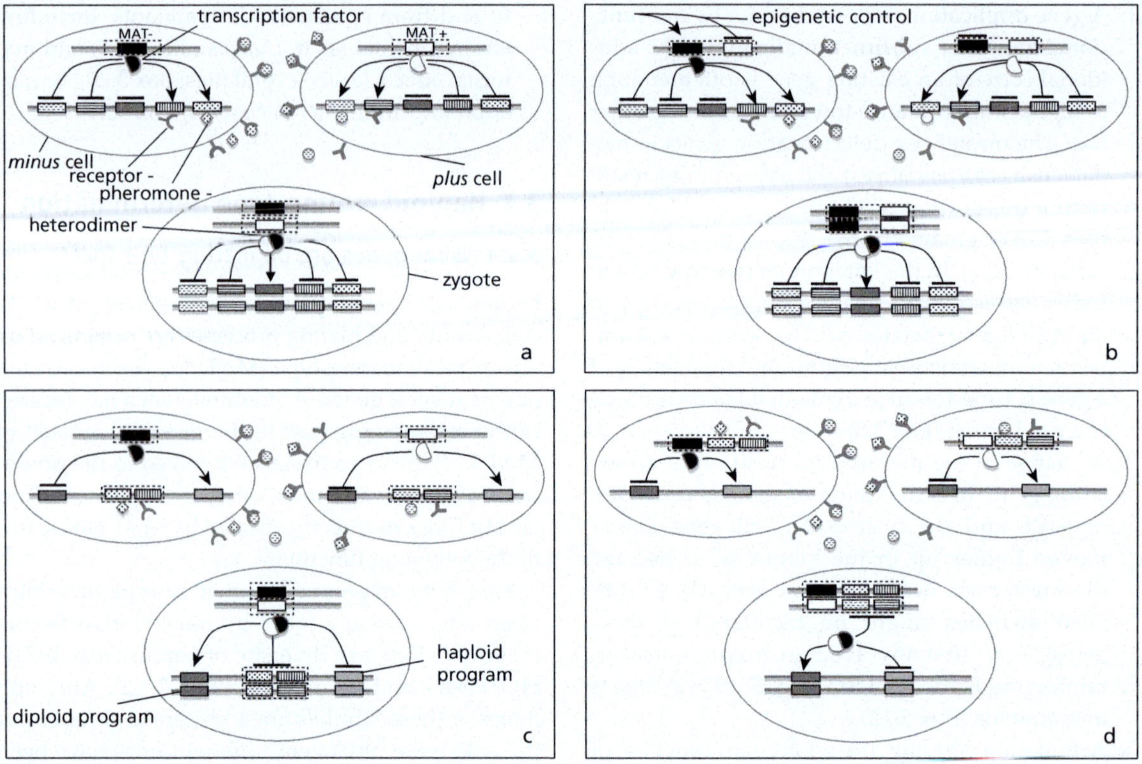


**Figure 3.1** Structure and functioning of different MAT loci. This figure does not display any specific system, but illustrates several main types of MAT loci structure and functioning. (a) Primarily bipolar system, as found in many heterothallic fungi (e.g. zygomycetes, ascomycetes) and algae (e.g. chlorophytes, phaeophytes). In the case illustrated here, the MAT locus (dashed box) only harbours a transcription factor (e.g. HD, HMG, or $\alpha$-box). The MAT– factor activates the *minus* mating function (pheromone – and receptor +) and represses the *plus* mating function (pheromone + and receptor –), while the MAT+ transcription factor does the reverse. Pheromone-receptor interactions trigger cell fusion. The *plus* and *minus* transcription factors then form heterodimers (e.g. HD1/HD2) that switch off the haploid programme (including the mating functions), and switches on genes involved in the diploid programme (dark grey box), such as meiosis genes which are constitutively repressed in the haploid phase. (b) Homothally is often achieved by genic capture, which brings two complementary MAT alleles into the same genome. Mating partners are genetically identical, but express different cell identities via epigenetic signals. In homothallic ascomycetes, for instance, a form of parental imprinting might occur via differential gametogenesis in male and female structures (microconidia and ascogonia respectively). (c) Tetrapolar system, as found in many basidiomycetes and Amoebozoa. One locus (MAT-1) encodes a transcription factor with two alleles, working as heterodimers in the zygote to switch on the diploid programme (dark grey box); the second and independent locus (MAT-2) encodes tightly linked pheromones and receptors. Both loci can be multi-allelic, and MAT-1 can be found in multiple copies; this functional redundancy increases the chance of compatibility between randomly mating partners. (d) Secondarily bipolar system, as found in some basidiomycetes (e.g. *Ustilago*, *Cryptococcus*). MAT-1 and MAT-2 have fused, forming one single large MAT locus, often encompassing other genes trapped in the non-recombining region. After cell fusion, heterodimers repress the mating function as well as autosomal genes involved in the haploid programme (light grey box), and switch on the dikaryotic programme (dark grey box).

partners, but also to correctly assess their identity and thereby make sure that the resulting zygote will inherit the complementary transcription factors required for proper development of the diploid stage. When autosomal, pheromone and receptor genes are triggered by transcription factors at the top of the cascade (Fig. 3.1a). They may also form an independent MAT locus, as in **tetrapolar** basidiomycetes and **syncytial** slime moulds (Fig. 3.1c), or localize on the same MAT locus as transcription factors, as occurs in bipolar basidiomycetes where the two MAT loci fused into a single one (Fig. 3.1d). Signalling from the pheromone-receptor interactions will then induce a cascade (e.g. the mitogen-activated protein kinase cascade in yeasts) resulting in cell fusion.

Third are genes that play a direct role in gametic fusion. These can be integrated to the MAT locus, such as *Fus1* in *Chlamydomonas*, or be autosomal and triggered by the pheromone-receptor interactions, as *Fus3p* (from the **MAPK** cascade) in *Saccharomyces*.

Fourth are genes that control uniparental transmission of organelles (mtDNA or clDNA). Uniparental inheritance of mitochondrial (mt) DNA from the *minus* parent, and chloroplast (cl) DNA from the *plus* parent in *Chlamydomonas* is controlled both by MAT genes (*Ezy1*) and autosomal HD transcription factors (*Gsp1*). In tetrapolar basidiomycetes, the two MAT loci interact to control mtDNA uniparental inheritance. In the secondarily bipolar *Cryptococcus neoformans*, a HD2 HD encoded by the MAT-a allele (*Sxi2a*) causes the selective elimination of MAT-a mtDNA within a few hours of fusion. In the acellular slime mould *Dictyostelium discoideum*, one MAT locus (MAT-B) is responsible for cell fusion, and an independent MAT locus (MAT-A) controls nuclear fusion and mitochondrial transmission (Section 3.2.2d). This, however, is not a universal function of mating types. In the slime mould *Didymium iridis*, mtDNA transmission, though uniparental, is random regarding mating types (Silliker et al. 2002; Scheer & Silliker 2006). In the yeast *S. cerevisiae*, both mating types transmit mitochondria, and potential conflicts are solved by mitochondrial fusion and mtDNA recombination (Takano et al. 2010).

Fifth, in lineages evolving towards anisogamy, MAT loci may carry genes involved in the morphological differentiation of gametes (e.g. in *Volvox carteri*), as otherwise expected from theory (genes that affect gamete size can only spread if they are linked to mating-type determinants; Charlesworth & Charlesworth 2010). Similarly, sexually antagonistic genes are expected to accrue to MAT loci or **U and V chromosomes** in systems displaying dimorphic male and female gametophytes (as in some mosses).

The final category includes genes that just happened to be trapped in the non-recombining region; these might be numerous, as in secondarily bipolar basidiomycetes where the MAT locus results from a fusion between two pre-existing MAT loci, or the U and V chromosomes of mosses and liverworts. Some of these genes might get secondarily involved in the sexual cycle.

### 3.2.2 Specific systems

Hereafter we detail the structure of MAT loci and the molecular mechanisms of mating for a few well-documented systems in model species of fungi, slime moulds, and green algae.

#### 3.2.2a Zygomycetes

Although Blakeslee first described mating types and heterothallism in Mucorales as early as 1904, only recently was the MAT locus of zygomycetes characterized in *Phycomyces blakesleeanus* (Idnurm et al. 2008). It comprises a single HMG transcription factor with highly differentiated *SexP* and *SexM* alleles, flanked by a triose phosphate transporter (TPT) and RNA helicase genes. *SexP* and *SexM* are only expressed during the mating phase, and likely involved in triggering both the cell signalling pathways and the diploid programme upon cell fusion (possibly as heterodimers; Lee et al. 2010b). Cell–cell recognition in zygomycetes is based on the conversion of β-carotene into trisporic acid, which triggers gametogenesis in both *plus* and *minus* cells. This conversion occurs via complementary paths in the two mating types. In both cells, β-carotene is first converted to 4-dihydrotrisporin. In the *minus* cell, this compound is then converted to trisporin and trisporol. Once released, these molecules are captured by *plus* cells and converted to trisporic acid. Reciprocally, *plus* cells convert the 4-dihydrotrisporin into 4-dihydromethyl trisporate, which is then released, and converted by *minus* cells into trisporic acid. The two cells possess the enzymes required for conversion along both pathways (Wetzel et al. 2009), but their differential regulation via the HMG transcription factors from the MAT locus ensures that gametogenesis only occurs in presence of the complementary mating type. It is worth noting that microsporidia harbour a MAT locus with the same genes in the same order (TPT, HMG, RNA helicase), suggesting that this highly diverged group descended from a zygomycete ancestor, and may have a genetically controlled sexual cycle (Lee et al. 2008, 2010a).

#### 3.2.2b Ascomycetes

Heterothallic ascomycetes also present one biallelic MAT locus, and are thus said bipolar (i.e. two complementary mating types are found among the spores resulting from one meiotic event). Although gene content and locus order are variable across lineages, its position in the genome is well preserved. Alternative alleles harbour transcription factors belonging to different classes, and are therefore often referred to '**idiomorphs**', to account for their apparent lack of homology. Gametogenesis and cell fusion have been well characterized for the mating types (**a** and α) of the budding yeast *S. cerevisiae* (Box 3.1). The MAT-α allele carries two genes: α1, encoding an α-box transcription factor, and α2, encoding a HD1 transcription factor. The alternative allele MAT-**a** (642 bp) carries a single gene (**a**1), encoding an HD2 transcription factor with no function in haploid cells. In other Saccharomycotina, this MAT-**a** allele harbours in addition an **a**2 gene (lost in budding yeasts and relatives), which seems to be the ancestral condition.

The fission yeast *Schizosaccharomyces pombe* (Taphrinomycotina) shows similar processes, but with different regulation pathways and different genes on the MAT locus, which is slightly larger than in *S. cerevisiae* (> 1 kb; Box 3.2).

Mating in filamentous ascomycetes (Pezizomycotina) is initiated by the development of an apical element on female ascogonia (the **trichogyne**). Directional growth, controlled by pheromones, contributes to the contact and fusion with the male element (microconidia, macroconidia, or simple hyphal cells). Fusion results in a multinuclear fruiting body, which develops into a dikaryotic stage (ascogenous hyphae), followed by meiosis and spore formation. In heterothallic species, male elements must fertilize ascogonia from a complementary mating type. The MAT loci only harbour HMG- and/or α-box transcription factors (Box 3.3).

### 3.2.2c Basidiomycetes

In basidiomycetes, the dikaryotic phase, resulting from nuclear exchange between two complementary

**Box 3.1 Mating-type determination in the budding yeast**

Cell signalling and recognition in *S. cerevisiae* are ensured by complementary mating pheromones and cell-surface receptors, encoded by autosomal genes: **a**-cells produce a α-receptor (STE2) and an **a**-pheromone (a 12-amino-acid oligopeptide with a C-terminal farnesyl group that makes it very hydrophobic), while α-cells produce an **a**-receptor (STE3) and a α-pheromone (an unmodified 13-amino-acid oligopeptide). The same two classes of oligopeptides are used as complementary pheromones throughout all ascomycetes. The **a**-pheromone and STE2 α-receptor are constitutively expressed in MAT-**a** haploid cells (activated by MCM1P, a general transcription factor from the MADS box family), while the α-pheromone and STE3 **a**-receptor are repressed. In MATα haploid cells, reciprocally, the α2 (HD1) transcription factor prevents MCM1P from activating the **a** mating function, while the α1 (alpha-box) binds co-operatively with MCM1P to induce expression of STE3 and α-pheromone. When STE3 surface receptors fix **a**-pheromones, the α-cell grows a projection (known as a 'shmoo') towards the cell source, and vice-versa. The two **a** and α-cells then come into contact, and fusion occurs, controlled by autosomal genes from the mitogen-activated protein kinase (MAPK) cascade (e.g. *fus3p*) that are upregulated by the pheromone-receptor interactions (Madhani & Fink 1998). Upon fusion, the **a**1 (HD2) and α2 (HD1) factors combine into **a**1/α2 heterodimers, which enter the diploid nucleus, switch off the haploid-specific functions (including the production of α1), and switch on diploid functions. This includes competence to enter meiosis upon starvation, because **a**1/α2 heterodimers also repress *Rme1*, a gene that inhibits initiation of meiosis (Galitski et al. 1999; Casselton 2002; Galgoczy et al. 2004; Lee et al. 2010b).

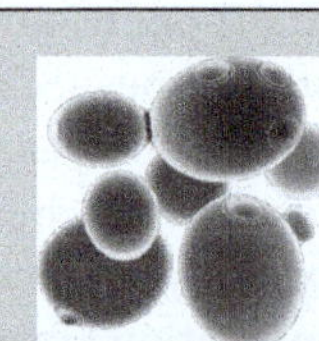

Experimental introduction of a MAT-**a** allele into a MATα haploid cell makes it behave like a diploid cell. It stops producing or responding to mating pheromones, and will attempt to undergo meiosis when starved (with fatal results). Similarly, deletion of either MAT-**a** or MAT-α in a diploid cell will cause it to behave like a haploid cell and seek mating. In the related *Candida albicans*, **a**1/α2 heterodimers are also required to repress mating competence: experimentally produced **a**/**a** or α/α diploids enter a mating programme, just as haploids do (Miller & Johnson 2002). Mating types in *S. cerevisiae* play no role in organelle transmission: both cells contribute their mitochondria, which fuse in the zygote, followed by mtDNA recombination (Takano et al. 2010).

**Box 3.2 Mating-type determination in the fission yeast**

The two genes on Mat1-P allele (*Mat1-Pc* and *Mat1-Pm*) of *S. pombe* are homologues of *S. cerevisiae* α1 and α2 (encoding α-box and HD1 transcription factors respectively), but those on the Mat1-M allele (*Mat1-Mm* and *Mat1-Mc*) differ from the *S. cerevisiae* MAT-**a**. Nutritional starvation induces expression of the autosomal *Ste11* (HMG transcription factor), which up-regulates *Mat1-Mc* (HMG) in M cells, and *Mat1-Pc* (α-box) in P cells. The latter acts co-operatively with MAP1 (the *S. pombe* homologue of MCM1P) to activate transcription of the P-cell pheromones and receptors (Nielsen et al. 1996). In M cells, similarly, MAT1-MC (HMG) acts co-operatively with STE11 to activate the M-cell signalling pathway. Pheromone signals promote the transcription of *Mat1-Pm* in P cells and *Mat1-Mm* in M cells (HD), which are both required in the zygote to switch on diploid functions. This includes triggering meiosis (via the *Mei3* gene), which rapidly follows cell fusion in this haplontic cycle.

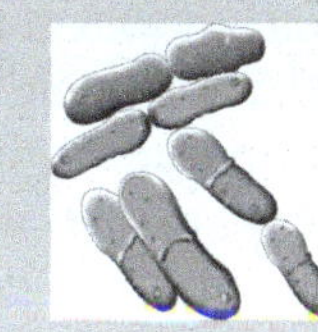

**Box 3.3 Mating-type determination in filamentous ascomycetes**

In heterothallic *Aspergillus* species (e.g. *A. fumigatus*, *A. oryzae*), one allele encodes a HMG, and the other an α-box. Both are required for proper completion of the sexual cycle, presumably acting as heterodimers in the zygote (as may also occur in homothallic species; Section 3.2.3b). Pheromones and receptors are autosomal. A similar MAT-locus structure occurs in the so-called dimorphic fungi and dermatophytes, all of which are heterothallic (Lee et al. 2010b).

In *Podospora anserina* and related species, the MAT+ allele (~3.8 kb) encodes a HMG transcription factor (FPR1, for Fertilization Plus Regulator, homologue of *S. pombe* MAT1-PC), which is required to regulate transcription of Mat+ pheromones and receptors. The MAT- allele (~4.7 kb) encodes an α-box transcription factor (FMR1, homologue of *S. cerevisiae* α1 and *S. pombe* MAT1-MC) and two other transcription factors (SMR1 and SMR2) required to complete the sexual cycle and only induced during mating. FMR1 and SMR2 form heterodimers that control the expression of proteins specific to MAT– nuclei (including pheromones and receptors). SMR2 (HMG, likely homologue of *S. pombe* STE11) maintains proper nuclear pairing during the development of ascogenous hyphae: nuclei from the two haploid gametophytes perfectly segregate in these dikaryons (i.e. each cell has one nucleus of each mating type), which obviously requires nuclear signalling and recognition (Coppin et al. 1997).

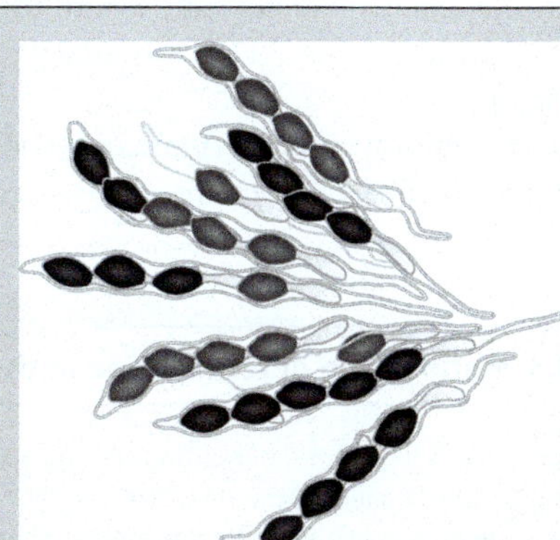

In *Neurospora crassa* (which also belongs to the Sordariacea and presents a very similar mating-type structure), experimental insertion of the two mating-type alleles *A* and *a* within the same nucleus enables nuclear fusion among identical nuclei, but further development of ascogenous hyphae is stopped (Coppin et al. 1997). Hence, complementary mating-type expression by the two nuclei is required to fulfil the dikaryon programme. In homothallic species (which possess copies of the two idiomorphs), this might be achieved via differential expression of mating-type alleles in male and female structures (see Section 3.2.3b).

gametophytes, is much longer-lived. Each cell of the mycelium harbours one nucleus of each partner; karyogamy and meiosis only occur in the final fruiting body (the **carpophore** of mushrooms). Mating-type determination is often tetrapolar (i.e. four mating types may occur among the spores resulting from a single meiosis), being encoded by two independent loci (Fig. 3.1c). One locus determines a pheromone/receptor system, while the other encodes transcription factors, which also act as heterodimers to regulate proper development of the dikaryon (Box 3.4). The two loci may be multi-allelic, and mating partners must differ at both loci to complete the mating process. These two loci also interact to control mtDNA inheritance.

In some lineages, this tetrapolar system has reduced to a secondarily bipolar system by fusion between the two loci (Fig. 3.1d), resulting in a largely expanded non-recombining region (Lengeler et al. 2002). In *Ustilago hordei* (a close relative of *U. maydis*), the MAT locus encompasses ~500 kb, with the *a* locus at one end, the *b* locus at the other end, and a series of genes trapped in between (Bakkeren & Kronstad 1994). This MAT locus expanded further to form dimorphic U and V chromosomes in the anther smut *Microbotryum violaceum*, where the non-recombining region spans ~ 1Mb (about 25% of the chromosome).

The homobasidiomycetes (or Agaricomycotina exemplified, for instance, by *Coprinus cinereus*) are also mainly tetrapolar, with two independent MAT loci: the *B* locus codes for the pheromone/receptor pathway, and the *A* locus for transcription factors. Interestingly, haploid mycelia may fuse and exchange nuclei independent of mating types. Pheromone stimulation (encoded by locus *B*) rather plays a role in organizing the growth of the dikaryon (nucleus signalling), ensuring that two nuclei of complementary mating types are present within each cell. This system organizes, in particular, clamp cell fusion, a process bearing similarity with the mating of two haploid gametes in *U. maydis* (Casselton & Olesnicky 1998). The HD1 and HD2 transcription factors (encoded by locus *A*) play a role both as homodimers in the gametophyte (where they express mating-type-specific genes and repress asexual sporulation; Tymon et al. 1992) and as heterodimers in the sporophyte, where they specifically repress the haploid programme and initiate the diploid transcription programme (development of the dikaryotic mycelium; Banham et al. 1995; Casselton & Olesnicky 1998; Spit et al. 1998). The potential for compatibility

**Box 3.4 Mating-type determination in the corn smut**

In *Ustilago maydis* (hemibasidiomycetes), mating specificity is ensured by the bi-allelic locus *a* (comprising two tightly linked genes that encode a mating-type-specific pheromone and the corresponding receptor), and mating-type-specific transcription factors are encoded by the multi-allelic locus *b* (~ 4 kb with >30 alleles known; Puhalla 1970). The *a*1 allele of locus *a* (~ 4.5 kb) includes *Mfa1* (for mating factor *a*1) and *Pra1* (for pheromone receptor *a*1), while the *a*2 allele encodes the corresponding genes *Mfa2* and *Pra2* (Banuett 1995). The *a*2 allele harbours two more genes (*Lga2* and *Rga2*) controlling gametic differentiation, and activated in response to pheromone stimulation (grow of conjugation types). In addition, *Lga2* is strongly upregulated after cell fusion (being a target of the bE/bW heterodimer complex, see below), where it mediates the loss of the *a1*-mtDNA during the dikaryotic (pathogenic) development (Fedler et al. 2009). Three alleles segregate in the related smut *Sporisorium reilianum*, each encoding two pheromones and one receptor, enabling each of the three *a* mating types to signal and respond to each of the two others, but not itself.

Mating-type-specific transcription factors are encoded by the multi-allelic locus *b* (~ 4 kb with >30 alleles known in *U. maydis*; Puhalla 1970). Each allele encodes two HD transcription factors, bW and bE, belonging to different functional classes (HD1 and HD2 respectively). After fusion of two compatible cells (decided by alleles at the *a* locus) the bW allele from one mating type binds to the compatible bE allele from its partner to form two active heterodimers (e.g. bE1/bW2 and bE2/bW1; Gillissen et al. 1992; Kämper et al. 1995). These heterodimers then enter the diploid nucleus to activate the zinc-finger transcription factor RBF1 (Heimel et al. 2010), which in turn controls the switch from yeast-like to filamentous growth, the maintenance of the dikaryotic state, and the subsequent pathogenic and sexual (meiotic) development (Heimel et al. 2010; Wahl et al. 2010). Importantly, heterodimers formed by HD1/HD2 transcription factors from the same MAT allele cannot trigger zygotic differentiation. In addition, the genes encoding the bW and bE factors are tightly linked (~260 bp of each other) and thus inherited together. This avoids generating haploid progeny that would produce active bW1/bE2 heterodimers, which would switch on the dikaryotic developmental programme in haploid cells.

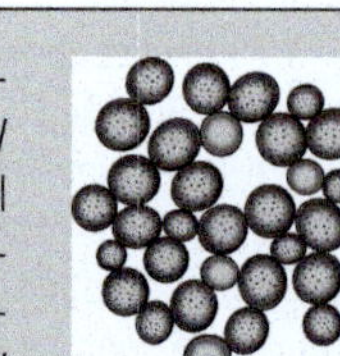

between randomly mating partners is multiplied in many basidiomycetes by several duplications of the MAT *A* locus, each copy encoding one HD1 and one HD2 factor. It is enough for the interacting gametophytes to be compatible at one copy only to produce active HD1/HD2 heterodimers (Casselton & Olesnicky 1998). This allows *C. cinereus*, for instance, to display as much as 160 mating types at its *A* locus (Kües & Casselton 1993), and up to 288 in *Schizophyllum commune* (Raper et al. 1958).

In the human pathogen *C. neoformans*, the *A* and *B* loci also fused into a single bi-allelic MAT locus, spanning >100 kb. Each allele (**a** and α) comprises a pheromone-receptor cluster linked to a HD transcription factor (HD1 in one allele, HD2 in the other). More than 20 additional genes are trapped in the non-recombining region, some of which are involved in meiosis and sporulation. Experimentally produced α/α zygotes cannot proceed further to complete the dikaryon stage. This requires the heterodimeric complex formed by the complementary transcription factors SXI1α (HD1) and SXI2**a** (HD2; Stanton et al. 2009). This transcription complex specifies the dikaryotic state by up-regulating genes required for the development of dikaryotic filaments, basidia, and spores, and repressing mating-specific genes (in particular the pheromone-encoding genes *Mf***a** and *Mf*α) to prevent additional fusion of dikaryons with haploids (Stanton et al. 2010). This heterodimeric complex also controls mtDNA inheritance (Yan et al. 2007). Although both types initially transmit their mitochondria, the MAT-**a** mtDNA is quickly eliminated a few hours after cell fusion. Disruption of *Sxi2a* results in biparental mtDNA inheritance. Similarly, progeny from artificially induced same-sex mating show biparental inheritance.

### 3.2.2d Amoebozoa

The single MAT locus of cellular slime moulds (Dictyostelia) is often multi-allelic, with, for example, three alleles in *D. discoideum*. The type-I and type-III alleles (thought to be ancestral) harbour one gene each encoding small hydrophilic proteins (respectively *Mat-A* and *Mat-S*). The type-III allele presents in addition a gene (*Mat-T*) coding for a larger protein with predicted signal peptides and anchor attachment sites. The MAT-A and MAT-S proteins are non-homologous (idiomorphs), and must interact to allow mating and subsequent formation of a diploid resistant zygote (macrocyst). The mating type II apparently emerged by capturing a copy of both type-I and type-III alleles. Its MAT-B protein (homologue of MAT-A) interacts with MAT-S when mating with type III, and its MAT-C protein (homologue of MAT-S) interacts with MAT-A when mating with type I. Importantly, *Mat-B* and *Mat-C* are sufficiently diverged from their original copies (*Mat-A* and *Mat-S*) that they cannot interact to foster zygotic development and macrocyst development in haploid type-II cells (Bloomfield et al. 2010; Bloomfield 2011).

Plasmodial slime moulds have haplo-diplontic cycles, where two haploid amoebal cells fuse to develop into a diploid **plasmodium**. *Physarum polycephalum* possesses three independent MAT loci. *Mat-B* (with 15 alleles known) controls the fusion between haploid amoebae (the probability to fuse is 2–4 orders of magnitude higher among hetero-allelic than among homo-allelic cells). *Mat-C* also marginally affects fusion, which occurs under a lower range of pH values if cells are homo-allelic at this locus (Kawano et al. 1987). After cell fusion, nuclear fusion and plasmodial development only occurs if mating cells are hetero-allelic at *Mat-A* (16 alleles known). If not, the two cells separate and continue to proliferate as amoebae (Youngman et al. 1981). Whether the complementary MAT-A products function as heterodimers to trigger the plasmodial developmental programme is unknown.

If (and only if) nuclear fusion occurs, all mitochondria from one parent will be selectively digested. Which parent transmits its mitochondria to the plasmodium is also decided by alleles at the *Mat-A* locus, which establish a linear hierarchy ('pecking order') among haplotypes. The exact mechanisms, however, are unknown, and apparently not perfect, resulting in paternal leakage or biparental transmission in some crosses (Moriyama & Kawano 2010). In the related slime mould *D. iridis*, however, mtDNA transmission, though uniparental, is random regarding mating types (Silliker et al. 2002; Scheer & Silliker 2006).

### 3.2.2e Chlorophyta

The model green alga *Chlamydomonas reinhardtii* presents a haplontic life cycle with H-GSD (Box 3.5).

**Box 3.5 Mating-type determination in *Chlamydomonas***

The two mating types ($Mt^+$ and $Mt^-$) of *C. reinhardtii* are determined by a bi-allelic MAT locus located on the left arm of chromosome VI. Recombination is suppressed over some 840 kb, comprising a ~190 kb central region (R) that shows rearrangements (four intra-chromosomal translocations, two **inversions**, and large deletions and duplications), flanked by a telomere-proximal region (T) on the left, and a centromere–proximal region (C) on the right (Ferris & Goodenough 1994). This MAT locus includes the master sex-determination gene *Mid* (for *minus dominant*), located on region R of allele $Mt^-$ and coding for a bZIP transcription factor (with a RWP-RK domain and a leucine-zipper in the carboxy-terminal region, involved in DNA binding and protein dimerization), as well as other genes involved in gametic cell fusion and uniparental inheritance of mitochondria and chloroplasts. Seven other genes are unrelated to the sexual cycle; these are housekeeping genes, expressed during several life-cycle stages, which happened to be trapped in the non-recombining segment. The sex-determination pathway also involves autosomal genes (Ferris et al. 2002).

Under normal trophic conditions, cells of the two mating types are phenotypically similar, expressing the same surface protein (agglutinin SAG1), coded by an autosomal gene. Sexual differentiation (gametogenesis) is initiated by worsening environmental conditions. Within 30 min of nitrogen depletion, the bZIP transcription factor MID is expressed in $Mt^-$ cells, and triggers two main developmental cascades. On one hand, it represses the autosomal *Sag1* and activates *Sad1*, located on the $Mt^-$ specific C domain. As a result, the two types of cells now express different surface proteins (agglutinins), allowing cell–cell recognition. When gametes are mixed, the *plus* (SAG1) and *minus* (SAD1) agglutinins, displayed on flagellar surfaces, mediate the initial adhesion reaction, during which the level of intracellular cAMP rises. The FUS1 glycoprotein, produced by the *Fus1* gene from the R domain of the $Mt^+$ allele, interacts with partner proteins from the $Mt^-$ mating type to trigger cell fusion and the formation of the binucleate quadriflagellated cell (Goodenough et al. 2007).

On the other hand, MID affects the expression of the autosomal *Gsp1* and *Gsm1* genes, which code for TALE HD-containing transcription factors (related to the class 1 KNOX and class 2 KNOX proteins used in land plants for switches in meristem development; Niklas & Kutschera 2010). *Gsp1* (*Gamete-specific plus1*), involved in gene regulation during the haploid stage (and expressed by cells of both mating types during vegetative growth) is then repressed in $Mt^-$ cells, while *Gsm1* (silenced in both types during the vegetative cycle) is activated. Thus, the $Mt^+$ and $Mt^-$ gametes contribute distinct transcription factors (GSP1 and GSM1 respectively) to the zygote. These two complementary molecules then form heterodimers that enter the diploid nucleus, repressing genes involved in the haploid programme, and activating other genes involved in the diploid programme (flagella resorption, zygote differentiation, cell wall building, and stress resistance).

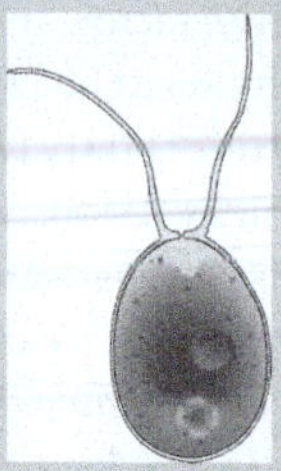

Both gametes initially contribute chloroplasts and mitochondria, but the $Mt^+$ mtDNA and $Mt^-$ clDNA are eliminated within 3–6 h of zygote formation, under control of *Ezy1* (*early zygote* 1), a gene located on the C domain of the MAT locus, and the autosomal transcription factor *Gsp1* (Nishimura et al. 2012). Thus, daughter cells only inherit mtDNA from the minus ($Mt^-$) parent, and clDNA from the plus ($Mt^+$) parent (Armbrust et al. 1993; Nakamura et al. 2003).

After a period of dormancy, zygotes returned to light in nitrogen-rich environment undergo meiosis, releasing four haploid meiotic products that resume vegetative growth. GSP1/GSM1 heterodimers are also involved in this meiotic/sporulation phase (Zhao et al. 2001). If heterodimerization fails, the zygote foregoes the meiotic pathway, and instead resumes vegetative growth as +/– vegetative diploids. If starved again, this zygote will differentiate as a *minus* gamete ($Mt^-$ being dominant over $Mt^+$) and subsequently mate with a haploid *plus* gamete, a deadly mistake given that triploid cells cannot complete meiosis (Ebersold 1967; Galloway & Goodenough 1985). Reciprocally, transformed haploid cells that constitutively express GSM1 and GSP1 do not differentiate into gametes, but directly enter the zygotic programme and form a cell wall (Goodenough et al. 2007). Similarly, in absence of a functional *Mid*, $Mt^-$ cells differentiate into pseudo-$Mt^+$ gametes and agglutinate with normal $Mt^-$ gametes, but cannot fuse (as they lack the *Fus1* gene specific to the $Mt^+$ allele, required for fusion).

Its MAT locus harbours 12 genes, five of which are involved in the sex-determination pathway, and only expressed in the gametic or zygotic phases of the life cycle.

A similar MAT locus appears widespread throughout Volvocales. The *Mid* gene arose early in this group as a regulator of mating type and kept this function in lineages that evolved multicellularity (e.g. *Gonium*; Hamaji et al. 2008) and anisogamy (e.g. *Pleodorina* and *Volvox*). In the oogamous *Pleodorina starrii*, with isomorphic male and female gametophytes, *Mid* only occurs in males, being expressed in male gametes, produced under nitrogen-poor conditions. This is clear evidence that sexes directly evolved from mating types in Volvocales, males deriving from the $Mt^{-}$ mating types (i.e. the MAT locus evolved into a SEX locus). In the multicellular dioicous *Volvox* species, this SEX locus expanded in size and evolved features of differentiated U and V chromosomes. In *V. carteri*, the non-recombining region is approximately five-times larger than in *Chlamydomonas* and includes many genes that are not part of the *Chlamydomonas* MAT locus (namely, over 70 protein-coding genes), as well as many more repetitive sequences. It also shows more divergence between the male (V) and female (U) copies, with multiple sex-specific and sex-regulated transcripts (Ferris et al. 2010). Sequences of the U and V alleles cluster by **gametologues** across the *Volvox* phylogeny, not by species. The *Mid* gene also occurs on the male allele (V), while the female allele (U) encodes a specific HMG transcription factor (a class also commonly used for sex determination in fungi and animals). This SEX locus also harbours genes involved in anisogamy, as expected from theory: if sexes evolved from pre-existing mating types, genes that affect gamete size can only spread if they are linked to mating types (Charlesworth & Charlesworth 2010). One strong candidate gene in this context is *Mat3*, mutants of which are known to produce tiny gametes in *Chlamydomonas*. In *V. carteri*, *Mat3* male and female gametologues are very divergent, with different structures (intronic sequences differ between sexes) and evidence of different selective constraints. Both alleles show many alternatively spliced mRNAs (Ferris et al. 2010). Gametogenesis in both sexes is induced by a diffusible sex-inducer protein released into the medium by males and triggering the switch to the sexual cleavage programme in the reproductive cells of male and female gametophytes (Gilles & Jaenicke 1984; Hallmann 2008).

### 3.2.3 Epigenetic mating-type determination

Homothallism refers to a mating competence between genetically identical cells, mitotically produced from the same meiotic haploid spore. This is to be distinguished from pseudohomothallism, such as found in some ascomycetes (e.g. *Podospora* or *Neurospora*) that manage to pack two haploid nuclei of complementary mating within a single dispersing meiospore (so that the di-haploid hyphae germinating from the same spore are self-fertile). Other forms of pseudohomothallism are found, for example, in *Porphyra* (Rhodophyta; Fig. 2.3): the four meiospores resulting from one meiotic event all contribute to the same gametophyte, which is thus a genetic chimera, producing both male and female gametes (a rare case of H-GSD with monoicy).

Genomic analyses reveal that truly homothallic species most often harbour copies of two complementary MAT loci in a single genome. Thus, most cases of homothallism (be it mating-type switching in yeasts, or self-compatibility in filamentous ascomycetes) are presumably underlain by the differential expression of MAT genes in complementary gametes (possibly mediated by parental imprinting). There seem to be few exceptions, such as the same-sex mating described in *C. neoformans* (Lin et al. 2005). In other instances, it might be that one of the alternative MAT loci is highly diverged and not yet identified. These few exceptions and the mechanisms underlying same-sex mating deserve further scrutiny.

#### 3.2.3a Mating-type switching

Several Saccharomycotina, including *S. cerevisiae*, have evolved the ability to switch mating types and undergo a self-fertile sexual cycle (Haber 2012). The current mating type of a haploid cell is controlled by the allele (MAT-**a** or MAT-α) expressed at the MAT locus (Section 3.2.2b; Box 3.1). However, all cells also possess (on the same chromosome) a silent copy of Mat-α on the left (called *Hml*, for *homology to MAT left*) and a silenced copy of MAT-**a** on the right (called *Hmr*, for *homology to MAT right*).

Mating-type switches are initiated by the gene *Ho* (for homothallism), only activated in the haploid phase (it is repressed by the **a**1/α2 heterodimer; Ezov et al. 2010). *Ho* encodes a highly specific endonuclease, which cleaves DNA at the MAT locus. Once cut, the DNA is degraded by exonucleases, and the resulting gap is filled by a copy of either *Hml* or *Hmr*. This **gene conversion** is strongly biased (up to 90%) towards the alternative mating type (i.e. α tends to become **a**, and vice versa; Wu & Haber 1996). Hence, a colony founded by a single haploid spore will soon restore both α– and **a**-mating types (many laboratory strains, which have lost the *Ho* gene, remain indefinitely either α or **a**). *Ho* was also lost in the related yeast *Kluyveromyces lactis*, and mating-type switches are mediated by a transposable element (Barsoum et al. 2010).

Similar mechanisms evolved independently in Taphrinomycotina. In the fission yeast *S. pombe*, where the two cell types P and M present different gene cassettes at their *Mat1* locus on chromosome II, the switch occurs through copying of the genetic information encoded by one of the two transcriptional silenced loci *Mat2-P* and *Mat3-M* into the expressed *Mat1* locus (e.g. Holmes et al. 2005). Pedigree analyses have shown that division of an 'unswitchable' cell leads to the formation of a 'switchable' and an unswitchable daughter cell (both of the parental mating type), while cell division of a switchable cell gives rise to an unswitchable daughter cell of the opposite mating type and a switchable daughter cell of the parental mating type. Mating-type switching has also been reported from *Agrocybe aegerita* (a basidiomycete), possibly also involving silenced mating-type cassettes (Labarère & Noël 1992), suggesting that this form of homothallism has evolved several times independently.

#### 3.2.3b Self-compatibility

Homothallism also evolved repeatedly from an ancestral heterothallic state in a diversity of filamentous ascomycetes (Fig. 3.1b). That homothallism is the derived state is supported not only by phylogenetic evidence (e.g. Coppin et al. 1997; Pöggeler 1999; Yun et al. 1999; Inderbitzin et al. 2005, 2006; Rydholm et al. 2007; Nygren et al. 2011), but also by modelling work showing that conditions for such a scenario are more easily met than for the reverse scenario (Nauta & Hoekstra 1992). Gene capture is the usual mechanism underlying this transition (Turgeon & Debuchy 2007). Experimental incorporation of the fused mating-type locus of homothallic taxa into heterothallic strains is sufficient to enable self-compatibility (Pöggeler et al. 1997; Yun et al. 1999).

In homothallic lineages of *Aspergillus* (such as the model species *A. nidulans*), gene capture apparently occurred by translocation of the chromosomal region harbouring the MAT locus. Accordingly, these lineages present two unlinked MAT loci, mapping to different chromosomes (e.g. Dyer et al. 2003). *Mat1*-1 harbours the HMG, and *Mat1*-2 the α-box. Both are required for proper mating: mutation in either *Mat1*-1 or *Mat1*-2 prevents the formation of fertile cleistothecia (Paoletti et al. 2007). Sexual recognition might be achieved by complex developmental regulation resulting in the differential expression of mating-type genes. *Mat1*-1, for instance, is suppressed in vegetative hyphae and upregulated in the reproductive tissues by both ***trans***- and ***cis*-regulatory elements** (Czaja et al. 2011). Similar patterns occur in related lineages, such as *Neosartorya* (Rydholm et al. 2007).

In Pezizomycotina, gene capture may have occurred through unequal gene exchange during recombination. Accordingly, the two idiomorphs often map to the same chromosome or the same locus (e.g. Yun et al. 1999). From a recent phylogenetic analysis, homothallism evolved more than six times independently in the genus *Neurospora* (Nygren et al. 2011). Pheromone receptor genes of the two mating types are conserved and functional in homothallic lineages (Nygren et al. 2012). Similarly, homothallic species of *Cochliobolus* (e.g. *C. heterostrophus*) carry both MAT idiomorphs, fused or closely linked on the same chromosome, whereas heterothallic species carry one or the other in each isolate (Yun et al. 1999). In homothallic species of *Stemphylium* the fused region contains an inverted copy of *Mat1*-1, joined to a forward-oriented *Mat1*-2 region (Inderbitzin et al. 2005). The four genes located within the MAT locus of the homothallic *Sordaria macrospora* are orthologues of the genes present in both *Mat1*-1 and *Mat1*-2 of the heterothallic model system *N. crassa*.

Interestingly, despite being genetically identical at their MAT locus, nuclei segregate according to parental origin in the binucleate ascogonial hyphae of *S. macrospora* (Heslot 1958; Coppin et al. 1997). Proper recognition of paternal and maternal nuclei likely implies a form of parental imprinting, which might be facilitated by the differential gametogenesis of microconidia and ascogonia. The evolution of sexes (differentiated gametes) might thus account for the propensity of ascomycetes to evolve homothallism. More generally, homothallism seems associated with anisogamy in other groups of fungi (e.g. the chytridiomycete *Allomyces macrogynus*, where both female and male gametangia are produced on the same haploid **thalli**) and many other kingdoms (see Section 2.1.6 for further examples).

## 3.3 Diploid sex determination

In addition to multicellularity and anisogamy, a few eukaryotic lineages (mostly among brown algae, green algae, land plants, and animals), evolved an expanded diploid sporophytic phase. Some reached a fully diplontic life cycle, with a haploid phase reduced to the male and female gametes. This had to be accompanied by a progressive shift of sex determination from the haploid to the diploid phase. In either case, sex determination can be genetic, but, as further developed in Section 2.1.8, diplo-genotypic sex determination (D-GSD) did not evolve directly from H-GSD; intermediate steps with epigenetic sex determination are required (as exemplified by the green lineage). Haplo- and diplo-genotypic sex-determination mechanisms fulfil different tasks and are subject to different constraints. Transcription factors on the MAT locus of haplontic or haplo-diplontic lineages organize gametogenesis and cell fusion, as well as the switch from a haploid to a diploid stage. To do this, they may capitalize on their homodimer/heterodimer bifunctionality (heterozygoty characterizes the diploid state). In diplontic lineages, sex-determination pathways also involve transcription factors, but their main function becomes that of triggering sporophyte development into either the male or the female gender. When sex is determined by a single locus, one gender normally has to be homogametic (ZZ or XX) and the other heterogametic (ZW or XY). We summarize the molecular pathways of sex determination in two such major lineages (angiosperms and animals) that independently evolved multicellularity, anisogamy, (sub-)diplontic cycles, and D-GSD (Sections 3.3.1; 3.3.2.).

### 3.3.1 Sex determination in angiosperms

Angiosperms differ from animals by their indeterminate growth and modular architecture, as well as the absence of germ-line differentiation: vegetative and reproductive organs develop from the same meristems. Floral meristems are nevertheless 'determined', in the sense that, once differentiated, cells will no longer divide. Flower development is initiated by a combination of photoperiod, vernalization, and autonomous pathways, the latter mediated by hormones, in particular GAs. These different pathways are integrated by SOC1, a MADS-box transcription factor, which initiates a change in meristem identity from vegetative to reproductive growth. The basic architecture of 'perfect' flowers consists of four concentric whorls; outside are the **perianth** verticils, formed by the sepals (whorl 1) and petals (whorl 2); inside are the reproductive verticils, formed by the stamens (whorl 3) and carpels (whorl 4). Four main classes of floral homeotic genes, referred to as A, B, C, and E classes, specify these four verticils (see Box 3.6).

Sex differentiation among unisexual flowers arises from the selective abortion of male or female organs (Irish & Nelson 1989). Abortion can occur at any stage of development, but tend to take place at similar stages in male and female flowers of a given species, pointing to the antagonistic roles of hormones in controlling alternative male and female developmental programmes (Diggle et al. 2011). A distinction is made between type II flowers, where abortion occurs at a very early stage (before initiation of stamens and carpels) and type I flowers, where abortion occurs at a later stage, by premature termination of the male or female programme.

Early abortion (type II) results from mutations that affect the regulatory modifiers of MADS box homeotic genes. In the dioecious *Thalictrum dioicum* and *Spinacia oleracea*, for instance, the homeotic B- and C-class floral identity genes show no allelic differences between genders, but are differentially expressed early in development. The B-class genes

**Box 3.6 The ABC model of floral development**

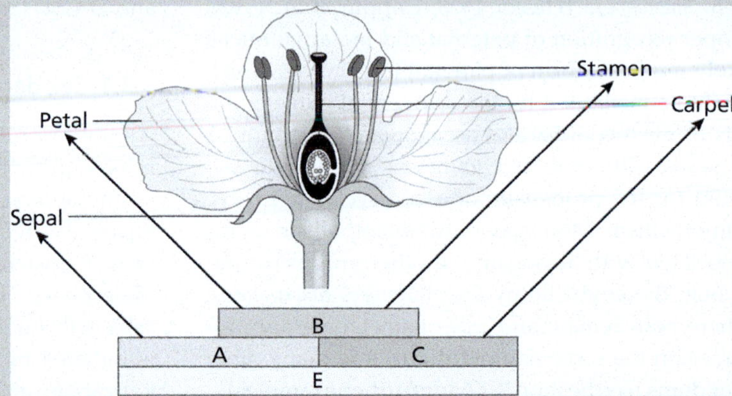


The ABC model was initially proposed based on the model plants *Arabidopsis* and *Antirrhinum* to formalize the interactions between homeotic MADS-box genes in determining organ identity within the floral meristems (Haughn & Somerville 1988). It has been since expanded into the ABCE model to account for the action of an additional class of MADS-box genes (Krizek & Fletcher 2005; Airoldi 2010; Liu & Mara 2010; Diggle et al. 2011). The A-class (*Ap1* and *Ap2* in *Arabidopsis*) and C-class genes (*Ag*) define the identity of the perianth and reproductive verticils, respectively. The B-class genes (*Ap3* and *Pi*) allow the differentiation of petals from sepals in the second verticil, and that of stamens from carpels in the third verticil. In addition, E-class genes (*Sep1*, *Sep2*, *Sep3*, and *Sep4*) are required to establish the floral context. All genes encode MADS-box transcription factors, except for *Ap2*, which encodes a transcription factor from a different family (ethylene-responsive element binding proteins). MADS-box proteins interact physically to form tetrameric complexes (quartets) that bind specific DNA sequences (CArG-boxes). A- and E-function proteins specify sepal development (whorl 1), A, B, and E petals (whorl 2), B, C, and E stamens (whorl 3), and C and E carpels (whorl 4; Airoldi 2010). E-class proteins seem required to stabilize tetramers; for example, AP3 and PI only show strong co-operative binding to DNA in presence of SEP3. The A and C functions (perianth and reproductive organs) are mutually antagonistic: if class-A genes are not functional, the C domain expands, resulting in a C, CB, CB, C pattern (i.e. carpels, stamens, stamens, carpels). Reciprocally, if C is non-functional, no sex organs will be formed, according to an A, AB, AB, A pattern (sepals, petals, petals, sepals). The determination of organ identity is followed by a large array of downstream responses involved in organ boundary delimitation and differentiation, mediated by hormone signalling.

*Ap3* and *Pi*, controlled by GAs, are expressed during stamen development in male flowers, but silenced in female flowers. RNAi knockdown in males results in normal development of female flowers.

Later abortion (type I) may result from a diversity of mutations or regulations that occur downstream of homeotic genes, in the developmental pathways leading to organ differentiation. In dioecious species, sex determination may take place via mutations affecting the tissue-specific synthesis or reception of hormones. In monoecious species, sex differentiation occurs in a uniform genetic background, via the differential expression of genes that regulate sexual development, controlled by the temporal and spatial dynamics of hormonal action. Such processes have been much studied with mutants that affect selective abortion in monoecious cultivated plants, such as maize (Box 3.7), cucumber (Box 3.8), or melon (Box 3.9).

Segregating mutations affecting the hormonal pathways are likely to provide the basis for gender determination in many dioecious species. In the annual mercury (*Mercurialis annua*), for example, male tissues require high levels of auxins and female tissues high levels of cytokinins (Durand & Durand 1991). However, sex determination in angiosperms is often less canalized than in animals (in line with the large phenotypic plasticity that underlies plant development in general) and quite complex, often being influenced by multiple nuclear, cytoplasmic, epigenetic, and environmental effects. Though mostly dioecious, *M. annua* sometimes produces monoecious or hermaphroditic plants. Genetic crosses between plants of different sex-determination types

## Box 3.7 Sex determination in maize

Monoecy in maize (*Zea mays*) results from the selective abortion of stamens in female inflorescences along the shoot (ears) and of pistils in the male terminal inflorescence (tassel). This pattern is underlain by steep gradients of feminizing (GA) and masculinizing (jasmonic acid) hormones along the shoot, responsible for the male-suppressing and female-promoting effects. Mutations at six unlinked loci are known to masculinize ears by preventing stamen abortion via controls of the GA pathways (resulting in andromonoecious plants). The *D1*, *D3* (for *dwarf*), and *An1* (for *anther ear*) loci encode enzymes involved in the synthesis of GA, while *D8* and *D9*, with cell-autonomous effects, encode transcription factors that weaken the response to GA. Direct GA application restores the **wild-type** phenotype (Irish 1996). *Tasselseed* mutations, in contrast, induce gynomonoecy by preventing carpel abortion in tassels. The gene *Ts1* encodes a lipoxygenase that affects jasmonic acid production (Acosta et al. 2009), *Ts2* a short-chain alcohol dehydrogenase that affects GA breakdown (expressed in the sub-epidermal layer of the **gynoecium** before abortion), and *Ts4* a regulatory **microRNA** (Banks 2008). Studies with double mutants typically show an increased phenotype (i.e. mutations have an additive component). There is also an environmental component to maize sex determination, mediated by hormone production: growing conditions, such as short days and cool nights, can also influence the sex of florets (Heslop-Harrison 1961).

## Box 3.8 Sex determination in cucumber

Cucumber plants (*Cucumis sativus*) are normally monoecious, with male flowers at the bottom and female flowers at the top of shoots. However, some individuals are **gynoecious** (female flowers only), others hermaphroditic, and still others andromonoecious (male flowers at the bottom and hermaphroditic flowers at the top of shoots). All of these patterns are under the control of two major loci, *F* and *M*. Sex differentiation along shoots is associated with a gradient of hormone concentration involving GA and auxin. These hormones affect expression of the *Acs1* gene (*F* locus), which encodes an enzyme that controls a rate-limiting step in the production of ethylene ($C_2H_4$). Ethylene has antagonistic effects on sex differentiation: on one hand it promotes the female function (carpel development is inhibited below a given $C_2H_4$ threshold); on the other hand, it represses the male function, via an *M* locus downstream of *F*, which inhibits stamen development above some threshold ethylene level. Thus, the shift from pure male to pure female flowers along the shoot depends on ethylene production by the *F* locus, itself controlled by the hormonal gradient (Fig. 4.3d).

Gynoecious plants have a duplicated copy of the *Acs1* gene at the *F* locus (*Acs2*), resulting in a higher production of ethylene. As a result, the female function is never turned off. Reciprocally, a recessive loss of function of the *M* gene (*mm*) prevents the male function from being turned off (Yamasaki et al. 2001). Thus, denoting by *F* the additional copy of *Acs2*, *M–ff* plants are monoecious (with a functional *M* but no *Acs2* copy), *M–F–* plants are gynoecious (female function always turned on by the high level of $C_2H_4$, and male function turned off by *M*), *mmF–* plants are hermaphroditic (the female function is always turned on, and the male function never turned off), and *mmff* plants are andromonoecious (female function turned off at low $C_2H_4$ concentration). Sex expression also shows phenotypic plasticity, being modulated by environmental factors (temperature and day length) that affect endogenous hormone levels.

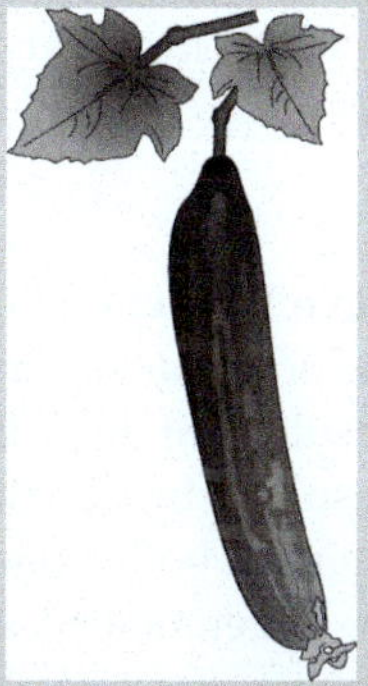

**Box 3.9 Sex determination in melon**

In monoecious melon plants (*Cucumis melo*), male flowers develop on the main stem and female flowers on axillary branches. Stamen abortion is controlled by *Cm-Acs7* (orthologue of cucumber *F* locus; Boualem et al. 2008), itself controlled by *Cm-Wip1*, which encodes a zinc-finger transcription factor. Cm-WIP1 has antagonistic effects on sex differentiation, by repressing both carpel development and ACS expression. Hence, flowers develop as male-only when *Cm-Wip1* is turned on, and as female-only when *Cm-Wip1* is turned off. Andromonoecy arises from a recessive mutation (*a*) of the active site of the Cm-ACS7 enzyme, which lowers $C_2H_4$ production and thereby prevents stamen abortion. Gynoecious mutants (*g*) have occurred from the insertion of a transposon in the promoter region of *Cm-Wip1*, preventing its expression (Fig. 3.2; Martin et al. 2009). Thus, *A–G–* individuals that carry a dominant allele of either gene are monoecious, *aaG–* individuals are andromonoecious, *A–gg* are gynoecious and *aagg* are hermaphrodite (Kenigsbuch & Cohen 1990). It is worth emphasizing that similar sex phenotypes in closely related species (*C. melo* and *C. sativus*) have arisen from different mechanisms.

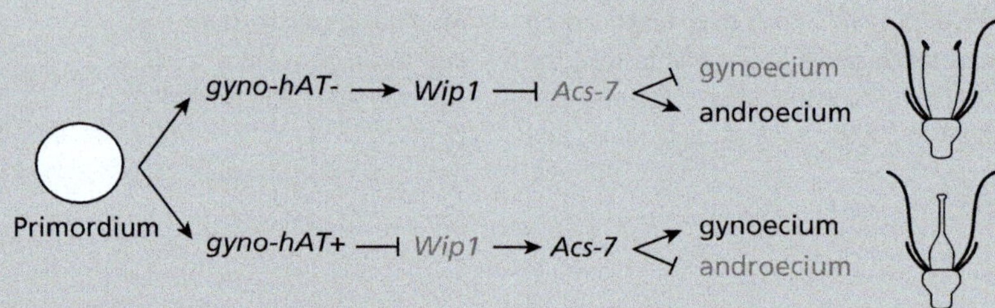


**Figure 3.2** Sex determination in melon. Male flowers normally result from the hormonal upregulation of *Wip1*, which represses *Acs7*, a gene required for gynoecium development (upper path). Gynoecious plants arise from the insertion of a transposon (*gyno-hAT*) in the promoter region of *Wip1* that inactivates this gene, resulting in female flowers only (lower path). Activation steps are indicated by arrows and inhibition steps by bars, active genes are in black and repressed genes in grey.

have frequently been used to infer the genetic basis of sex determination, but usually yield variable results that are difficult to interpret. This is well illustrated by the stinging nettle (*Urtica dioica*), which presents both monoecious and dioecious forms, with no differentiated sex chromosomes. Initial crosses between such forms showed a 'fortuitous' distribution of sex ratios among families, suggesting **multigenic sex determination** (Zuk 1970). A later study involving two closely related sub-species confirmed a strong family component, with some evidence of maternal (cytoplasmic) effects (Shannon & Holsinger 2007). Crosses between male and females plants also yielded different results depending on whether they stemmed from male and female parents or from a selfed monoecious parent. In addition, the two subspecies (*dioica* and *gracilis*) had distinct genetic mechanisms for monoecy. Self-pollination of monoecious plants from a Dutch population yielded female, male and monoecious plants in an approximate ratio of one female to three male/monoecious individuals (Glawe & de Jong 2009). This motivated the authors to propose a simple model where gender is determined by one major sex-determination locus with four alleles; males and monoecious plants have distinct genotypes but are both heterozygous at the sex-determination locus. Crosses showed that monoecy generally had Mendelian inheritance and was passed on to the next generation via both pollen and seeds, and that alleles from the dioecious system were often dominant. However, many exceptions also occurred, suggesting that dominance is incomplete and/or that more genes are involved in sex determination.

**Box 3.10 Evolution of dioecy in plants**

Dioecy may a priori evolve from hermaphroditism via either gynodioecy or androdioecy. A male-sterile (or female-sterile) mutation can only spread if it doubles the fitness return from seeds (or pollen), relative to that of hermaphrodites. Empirical evidence suggests that the mere reallocation of resources is often not enough to compensate for the loss of one of the two functions; Darwin (1877, in Charlesworth & Charlesworth 1978) already noted that the seed sets of female *Thymus serpyllum* is 1.79 times that of hermaphrodites, which is insufficient to account for male sterility in an outcrossing population. Under selfing, androdioecy is further disfavoured, because selfed ovules are not available for cross-fertilization. In contrast, inbreeding load may favour gynodioecy, because female-only plants avoid inbreeding depression (Charlesworth & Charlesworth 1978). Conditions for invasion are even less restrictive if the male-sterile mutation is cytoplasmic (e.g. Lewis 1941; Charlesworth & Charlesworth 1981; Gouyon et al., 1991; see below).

A modifier lowering female investment in hermaphrodites may subsequently spread in a gynodioecious population, because fitness returns from the male function increase with the frequency of female-only individuals. This may or may not lead to the fixation of dioecy (many plants show a continuum between gynodioecy and dioecy). The dynamics of this evolution is bound to depend on the initial **linkage disequilibrium** between the male-sterile and female-sterile mutations.

The pollination mode also matters: dioecy evolved more often in species pollinated by wind or by generalist insects, because fitness returns from the male function increase more with pollen yield. SI systems (which also evolved to prevent inbreeding, as an alternative to dioecy; Section 3.4) obviously suppress the selective pressure for male-sterile mutations. Accordingly, gynodioecy is more frequent in self-compatible groups (e.g. Labiatae). In contrast, SI systems might favour androdioecy: in *Phyllirea angustifolia*, a female-sterile mutation linked to the SI locus could spread because males are compatible with both SI alleles (which doubles the fitness return from pollen relative to hermaphrodites; Saumitou-Laprade et al. 2010).

Dioecy may also evolve from monoecy, as documented, for example, in *Mercurialis* and *Cotula* (Lloyd 1972). Under this scenario, inbreeding avoidance generates a disruptive selection in monoecious populations, such that some individuals progressively specialize in a male role, and others in a female role. From the sex-determination mechanism of cucumber (Box 3.8), for instance, dioecy would result from any mutation affecting the production of ethylene or response to it.

Evolution from hermaphroditism to dioecy requires a minimum of two mutations: one to abort stamens in female plants and another to abort carpels in male plants. These two mutations are unlikely to occur simultaneously, generating gynodioecy or androdioecy as intermediate steps in the evolution towards dioecy. Empirical work shows that the gynodioecy path is much more common (Fig. 5.2), and evolutionary models explain why (Box 3.10).

Selection should favour **linkage** between male-sterile and female sterile mutations (which might be mediated by chromosomal fusions; Charlesworth & Charlesworth 1980), in order to prevent the production of neuter individuals, bearing both mutations. Linkage is still incomplete in the nascent sex chromosomes of the subdioecious *Fragaria virginiana*, so that recombination sometimes generates hermaphroditic and neuter individuals (Spigler et al. 2008). Recombination arrest in the heterogametic sex will initiate sex-chromosome differentiation. If the initial male-sterile mutation is recessive (as expected from loss-of-function mutations), then a dominant female-sterile mutation will make the system evolve towards male **heterogamety** (XY males). This is the most frequent situation in dioecious angiosperms, as documented, for example, in *Silene* and papaya (Box 3.11). If the initial male-sterile mutation is dominant, then the system should evolve towards female heterogamety (ZW females) by recruiting a recessive female-sterile mutation.

Gynodioecy in plants often arose through mitochondrial mutations. Cytoplasmic male sterility (**CMS**) has been reported from over 150 plant species, including such crop species as beet and rice. More than a dozen different mitochondrial mutations causing CMS have been characterized (Chase & Gabay-Laughnan 2004, Hanson & Bentolila 2004). They may affect male function in a variety of ways. Some act very early by altering

**Box 3.11 Sex determination in papaya**

Papaya *(Carica papaya)* belongs to the Caricaceae, of which most members are dioecious. Papaya is somewhat unusual in being **trioecious**, with male, female, and hermaphroditic plants. Gender is determined by a pair of recently evolved sex chromosomes, with homogametic females (XX) and heterogametic males (XY). The non-recombining region (~7 Mbp) includes two sex-determination genes: a recessive stamen suppressor on the X copy and a dominant carpel suppressor on the Y copy. The stamen suppressor possibly is an upstream regulator of B-class genes (*Ap3* and *Pi*) because female flowers show no trace of stamens. The carpel suppressor, in contrast, is a downstream regulator: although aborted, the gynoecium is well developed in male flowers, and may even fully develop into fruits under optimal growing conditions. The non-recombining region shows two main inversions (possibly responsible for the recombination arrest) and accumulation of repeated elements. Besides a few functional genes, this region also harbours a gene required for embryonic development, with a recessive deleterious mutation on the Y copy making YY embryos abort within 25–50 days of pollination. The third gender (hermaphrodites) evolved only recently, being spread by human selection (their fruits are preferred over those produced by females). Hermaphrodites ($XY^h$) evolved from males by mutation of the Y chromosome; $Y^h$ differs from Y by the loss of function of genes that control carpel abortion and peduncle size (long in male flowers, but short in female and hermaphrodites). Because $Y^h$ carries the same lethal mutation as Y, self-fertilized hermaphrodites produce one-third of females and two-thirds of hermaphrodites ($Y^hY^h$ are lacking). Similarly, crosses between males and hermaphrodites give one-third each of females, males and hermaphrodites ($YY^h$ are lethal). Finally, crosses between females and hermaphrodites give 50% of each parental gender (Ming et al. 2007).

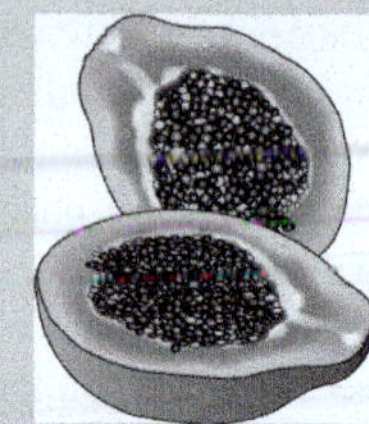

the expression of the ABCE floral identity genes (Box 3.6), so that stamens develop into carpels (if B-class genes are repressed) or petals (if C-class genes are repressed). Other may act much later, for example, by sterilizing pollen grains (reviewed in Chase 2007).

CMS mutations are favoured if resources are reallocated to the female function; it is sufficient that the total seed yield increases above that of a hermaphrodite. Sex investment will thus be biased towards the female function, which may induce a nuclear response to restore sex ratios. Fertility restoration is often polygenic. Genes restoring pollen fertility can be sporophytic (restoring all pollen grains) or gametophytic (restoring only grains that carry the restorer). Restorer genes have been identified from a number of species, such as *Petunia*, rice (*Oryza*) and sorghum (*Sorghum*); all encode a pentatricopeptide-repeat (PPP) protein family that affects the RNA-binding properties of mitochondrial transcripts. Crosses between lines that differ in their cytoplasmic sterility effects and nuclear restorers are often very informative about the underlying molecular mechanisms.

### 3.3.2 Sex determination in animals

Animals normally differ from plants by their unitary architecture, determinate growth, and canalized development, leaving less room for phenotypic plasticity. One trait that most animals also have in common is an early differentiation between **soma** and germ line, and the presence of sexually dimorphic gonads. The embryonic gonad is bipotential; in gonochoric species, it develops as either male or female depending on an initial signal, which may be genetic or environmental. The sex-specific programme must then be actively maintained lifelong (or switched in a relevant way in sequential hermaphrodites). In many animals (e.g. *Xenopus*, housefly, medaka fish), the germ line lacks cell-autonomous sexual identity, which is instead imposed by the somatic gonad. Others (e.g. *Drosophila* and mice) have partial cell-autonomous sex identity in the germ line (which renders sexes differentially sensitive to somatic cues), making germ line and somatic sex determination partly different.

In vertebrates, sex is normally first determined in the gonads, then hormones convey the gonadic sex decision to the brain and other organs (even though this gonad-dependent endocrine control is sometimes complemented by cell-autonomous components, as in marsupials and birds; Boxes 3.12, 3.14). In contrast, cell-autonomous sexual identity dominates in insects and nematodes (Section 3.3.2c). Whether endocrine control or cell autonomy prevails, transcription factors from the same family (DM-domain) take a central role in organizing sex differentiation (specification of the cell fates of the gonads).

### 3.3.2a DM-domain genes

DM-domain transcription factors contribute to sex differentiation across all the animal kingdom. The first member of this family to be discovered, *Doublesex* (*Dsx*) in *D. melanogaster*, was named from the phenotype of mutants, intermediate between males and females (Hildreth 1965; Burtis & Baker 1989); *Male abnormal* (*Mab3*) in *C. elegans* was later recognized as a homologue of *Dsx* (Shen & Hodgkin 1988; Raymond et al. 1998). Many more have been discovered since, constituting the DMRT family (for DSX/MAB3-related transcription factors). DM-domain genes are now known throughout all Metazoa, from cnidarians, platyhelminths, crustaceans, and insects, to tunicates, fish, amphibians, reptiles, birds, and mammals (Matson & Zarkower 2012). All lineages have multiple DMRT paralogues (7 in mice and humans, 11 in *C. elegans*). Although they may serve a variety of developmental functions, the most conserved one is that of organizing the sexual fate of cells, organs, or individuals. The association of DM family member expression with sexual development in cnidarians (Miller et al. 2003) suggests that this role evolved prior to the Cambrian, over 500 million years ago. It is maintained whether animals are hermaphroditic or gonochoric, whether the initial trigger is environmental or genetic, and whether sex determination is polygenic, male heterogametic, female heterogametic, or haplodiploid (Kopp 2012). However, the precise modalities vary among metazoans and have been altered multiple times with respect to position in the sex-determination cascade, molecular function, tissue specificity, and transcriptional regulation.

*Dmrt1* and its paralogues are most often expressed in the somatic gonads and/or the germ line. In mice, *Dmrt1* is expressed in the genital ridge of both sexes before sex-differentiation; expression then declines in ovaries but is maintained in males, both in Sertoli cells and in the germ line (where it activates spermatogonial differentiation). *Dmrt1* expression also correlates with testis development in teleosts, independent of their sex-determination system; in sequential hermaphrodites, it parallels the development of testes in protogynous species, and their regression in protandrous species. It also accompanies testis development in ESD reptiles (Kettlewell et al. 2000; Sinclair et al. 2002; Shoemaker et al. 2007). In insects, *Dsx* is not active in the germ line, but the male **isoform** ($DSX^M$) is expressed in somatic gonad precursors cells, where it is required for recruitment of additional precursors from the mesoderm; *Dsx* is also required for testis development in *Daphnia* and its loss gives testes an ovary-like morphology.

*Dmrt* genes play a central switch role, by turning on the male programme and turning off the female programme. *Dmrt1* fulfils both functions in vertebrates (Box 3.12): it inactivates female development (by repressing the *Foxl2* and *Rspo1/Wnt4/β-catenin* pathways) and activates male development, both by triggering the downstream male-specific effectors (together with *Sox9* and *Fgf9*, *Dmrt1* directly controls hundreds of target genes in Sertoli- and germline cells, responsible, for example, for sex hormone production and ensuing somatic differentiation), and by maintaining male identity over time, through a feed-back loop on *Sox9/Sox8* (loss of *Dmrt1* in embryonic male mice induces transdifferentiation of Sertoli cells into granulosa cells; Matson et al. 2011). *Dmrt1* thus selects and maintains sexual fate, by recruiting different co-factors to the enhancers or repressors of different DM-regulated genes.

The same antagonistic functions are fulfilled in insects by the alternative splicing of *Dsx*. The male ($DSX^M$) and female ($DSX^F$) isoforms share a common N-terminal DNA-binding (DM) domain, but have different C-terminal domains, with distinct effects on target gene expression: $DSX^M$ promotes male development and represses the female programme, while $DSX^F$ does the reverse. Yolk protein genes, for instance, are repressed by $DSX^M$, but transcription is stimulated by $DSX^F$. The enzyme desaturase-F (involved in pheromone synthesis) is

**Box 3.12 Sex determination in mammals**

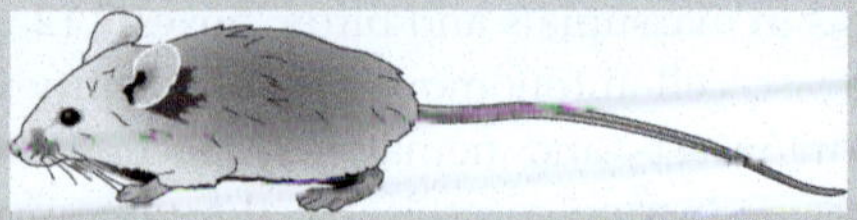

The first molecular step along the male pathway in therians is taken when a few upstream actors (including the transcription factor WT1) upregulate *Sry* in pre-Sertoli cells. In mice, SRY expression starts 10.5 days *post coitum* (dpc), reaches a peak at 11.5 dpc, and ends at 12.5 dpc. Correct timing and expression level are crucial: SRY has to reach a certain threshold within a precise time window for the full testis development to proceed (Jakob & Lovell-Badge 2011). SRY and SF1 together activate *Sox9* by binding to TESCO (testis-specific enhancer of *Sox9* core), a 1.4 kb sequence located 13 kb upstream of *Sox9* and highly conserved across mammals (including monotremes, which lack *Sry*). *Sox9* in turn upregulates *Dmrt1* via *Fgf9* (*fibroblast growth factor 9*). This initiates an auto-regulating *Sox9*/*Fgf9*/*Dmrt1* feed-forward loop (that also involves SF1), which both activates the male-specific programme (including *Amh* and the induction of androgen-producing **Leydig cells**) and inhibits female development (Kim et al. 2006; Matson et al. 2011; Xu et al. 2012; Fig. 3.3).

The female programme (which is the default in mammals) is controlled by the transcription factor *Foxl2* (*forkhead box L2*). In absence of the initial SRY-driven peak of *Sox9* expression, *Foxl2* will take over and inhibit the *Sox9*/*Fgf9*/*Dmrt1* feedback loop. Instead the auto-regulating *Rspo1*/*Wnt4*/*β-catenin* signalling pathway will start, leading to the development of follicle cells and ovaries, and the activation of aromatase. Sex determination is thus organized by two competing self-regulating pathways; normally the canonical *Wnt* signalling wins over *Sox9*/*Fgf9*, but *Sry* tips the balance in favour of the latter (Herpin & Schartl 2011b). Thus, the first decision that directs gonad development towards a male or a female fate is determined by the ability of *Sox9*/*Fgf9*/*Dmrt1* to establish an autoregulatory feed-forward loop. *Sry* provides the initial kick-off, but this primary decision has to be reinforced lifelong by *Dmrt1*, which is required to maintain testis cell identity and inhibit the alternative programme (its loss in embryonic male mice induces the transdifferentiation of Sertoli cells into granulosa cells; Matson et al. 2011).

Once gonadal sex is established, hormones will convey the sex-determination decision to the rest of the body. However, significant cell autonomous components may still exist: in marsupials, *Sry* controls gonadal sex, but the development of a pouch or scrotum depends on the number of X chromosomes: XO individuals are female with ovaries, but a scrotum instead of a pouch, while XXY individuals are male with testes, but with a pouch.

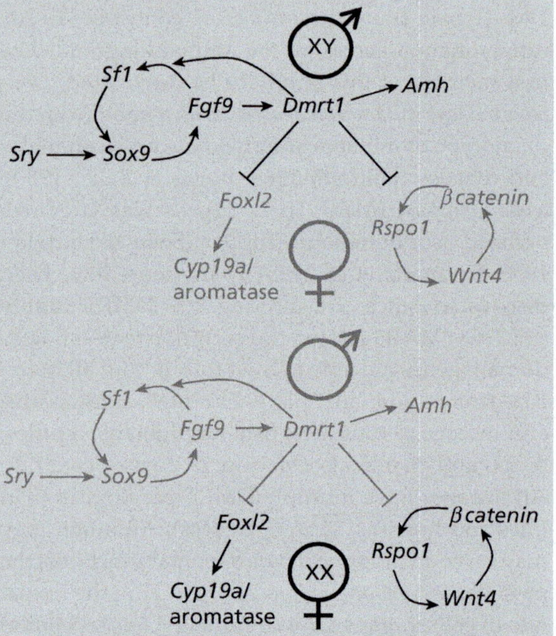


**Figure 3.3** The eutherian sex-determination pathway. Males (top) carry the Y chromosome with the dominant *Sry* gene, which activates autosomal *Sox9* in co-operation with *Sf1*. *Sox9* turns on *Fgf9*, which in turn upregulates *Dmrt1*; this will repress the *Rspo1*/*Wnt4*/*β-catenin* pathway, activate *Amh* and lead to testis development. Transient *Sry* expression is required to activate the *Sox9*/*Fgf9*/*Dmrt1* feed-forward loop (the memory device). In the absence of *Sry* in XX females (bottom), *Sox9* and *Fgf9* are not expressed at levels sufficient to inhibit *Foxl2*, which takes over, further inactivates the *Sox9*/*Fgf9*/*Dmrt1* loop, and activates instead the *Rspo1*/*Wnt4*/*β-catenin* pathway, leading to aromatase expression and ovary development.

**Box 3.13 *Sox* genes and the origin of *Sry***

How sex is determined in mammals, in particular humans, has been the subject of numerous studies and recent reviews; notwithstanding these intense efforts, many details remain unknown (Chadwick & Goode 2002; Wilhelm et al. 2007; Sekido & Lovell-Badge 2008). Surprisingly, it was as late as 1959 that the Y chromosome was shown to contain the testis-determining factor in humans. In 1990 the *Sry* gene (for *Sex-determining region Y*; Sinclair et al. 1990) was identified, based on single base pair mutations and deletions that result in male to female **sex reversal** (Berta et al. 1990), and translocations to the X (via **non-homologous recombination**) that cause female to male sex reversal (Li et al. 2001). *Sry* soon appeared both necessary and sufficient to induce male development, based on ectopic expression in XX mice (Koopman et al. 1991). Sekido & Lovell-Badge (2008) provide a good account of the history of mammalian sex-determination research.

*Sry* is a small intronless gene encoding a 204 αα-protein from the SOX (for *Sry-related HMG Box*) family of transcription factors, which bind to the minor groove of DNA with sequence (A/T)ACAA(T/A). Both binding and bending are necessary for proper transcriptional activity. Many genes from the *Sox* family have fundamental roles in the development of mammals and other vertebrates, typically acting as cell differentiation switches (reviewed in Bowles et al. 2000). In mammals, SOX1 is a neural factor and SOX2 is required for pluripotency. SOX3 is associated with sexual differentiation in fishes, amphibians, and reptiles. It is expressed in developing ovaries of *Xenopus*, and throughout gonadogenesis in both male and female chicken, as well as in fish during primordial germ-cell differentiation (Yao et al. 2007). *Sox9* has a pivotal role in testis determination in all vertebrates, being upregulated in Sertoli cell precursors by the primary sex-determination triggers (Bagheri Fam et al. 2010).

The SRY HMG-box specifically binds to a *Sox9* enhancer (TESCO), inducing a 60–85° bend. Outside of this conserved domain, the gene undergoes rapid sequence divergence (Whitfield et al. 1993) except for a conserved regulatory sequence in its promoter, which is a binding site for transcription factor CP2 (TFCP2). The functional significance of regions outside of the HMG-box is still unclear, but they are likely to provide conformational structure affecting DNA binding affinity, thereby explaining the species–specific differences in SRY timing and tissue affinity (Waters et al. 2007; Zhao & Koopman 2012).

*Sry* came into existence through a fusion between *Sox3* (which still exists on the oldest stratum of the mammalian X chromosome) and part of the first exon of *Dgcr8* (for *DiGeorge syndrome critical region gene 8*) that was inserted upstream of *Sox3* (Sato et al. 2010). *Tfcp2* originally functioned as a transcription factor for the α-globin gene. *Sry* is thus an example of a gene that emerged through a unique recombination event followed by the creation of a new promoter region, which led to the gain of a novel function by changing its target sites and binding affinity. Alteration of this gene on an ancestral autosome presumably initiated the evolution of the sex chromosomes of therian mammals (Wallis et al. 2008).

*Sry* is restricted to therians; orthologues exist on the Y chromosomes of both placentals and marsupials, which diverged some 148 Mya. It is absent from platypus and echidna (Wallis et al. 2007), which further details its origin to 148–166 Mya. Monotremes, however, present a TESCO site, implying that *Sox9* is regulated by another (still unknown) transcription factor. Given the crucial role of *Sry* in eutherian sex determination, it is remarkable that some lineages have lost it. This is the case for two species of mole voles (*Ellobius lutescens* and *E. tancrei*) and the Japanese spiny country rat (*Tokudaia osimensis*), which have lost the Y chromosome altogether (whereas closely related *Ellobius* and *Tokudaia* species have a normal Y chromosome and show *Sry* expression; Just et al. 1995, 2002; Soullier et al. 1998; Sutou et al. 2001). How the *Sox9* autoregulatory feedback loop is initiated in these species is still unknown. Reciprocally, some rodents (e.g. *Mus minutoides* and *Myopus schisticolor*) may develop as females despite a functional *Sry*. This results from a chromosomal rearrangement of the normal X that interferes with the sex-determination cascade downstream of *Sry*, preventing *Sox9* upregulation (Veyrunes et al. 2010; Section S2.2.9f).

only activated by $DSX^F$. Alternative splicing subdivides the regulatory activity of the ancestral *Dsx*, leading to a more flexible control of sex-specific development. This insect-specific innovation was likely superimposed on a more ancestral mechanism based on spatially regulated transcription, such as found in branchiopod crustaceans (Kopp 2012).

Its bifunctional regulatory activity confers a key role in organizing **sexual dimorphism** on *Dmrt* genes. In vertebrates (where sex differentiation is mostly gonad dependent), the sex-specific development of non-gonadal tissues is indirectly orchestrated by *Dmrt1* via endocrine signalling: Sertoli cells secrete sex hormones and control the development of other hormone-producing cell types in the gonad. However, cells may also display some autonomous sex identity. Many DM-domain genes are expressed in somatic tissues, including the brain, where they may organize sex-specific behaviour: male mice mutant for *Dmrt4*, for instance, tend to display same-sex copulatory behaviour.

In insects and nematodes (where sex determination is largely cell-autonomous), DM-domain genes integrate information about sex, position, and time to direct somatic tissues towards male or female fate. In *Drosophila* males, for instance, *Dsx* triggers the development of sex combs (male-specific sensory bristles on the first pair of legs that are used in mating), based on spatial cues provided by the Hox gene *Scr* (*sex combs reduced*), temporal cues from ecdysone-dependent mechanisms, and sex-identity information conveyed by cell-autonomous *Dsx* splicing. Similarly, $DSX^{M}$ and the Hox gene *abdominal B* (*Abd-B*) collaboratively control the strong pigmentation in the posterior abdomen of males *D. melanogaster* by regulating the expression of *Bab1* and *Bab2*. In the central nervous system, $DSX^{M}$ is expressed in discrete neuron clusters that control courtship song, ejaculation, and other sex-specific behaviours. Sexual dimorphism is achieved via differentiation of sex-specific neurons, and by controlling neuroblast proliferation and death. In *Daphnia magna*, the development of male-specific organs (larger eyes, thoracic hooks, longer antennae) is also controlled by local expression of *Dsx* (Kato et al. 2011). In *C. elegans*, similarly, the DM transcription factors *Dmd3* and *Mab3* integrate information on sexual identity (conveyed by *Tra1*), developmental stage, and position (provided by posterior Hox genes and localized *Wnt* signalling) to induce differentiation of the male tail. Thus, sexual dimorphism in insects and other ecdysozoans is largely underlain by the local and temporal patterns of regulation of DM-domain genes.

In a few lineages, *Dmrt1* or paralogues have moved up the regulatory hierarchy from a downstream position to a top sex-determination role. In birds, *Dmrt1* is located on the **Z chromosome** and absent from the W (even in the early diverged ratites with otherwise homomorphic sex chromosomes). It is expressed early in the bipotential gonad and at a higher level in males than in females, consistent with its double dose (Raymond et al. 1999). As in mammals, *Dmrt1* promotes testis development by activating *Sox9* and suppressing ovary development genes such as *Foxl2* and *Wnt4* (Smith et al. 2009; Chue & Smith 2011; Box 3.14). *Dmrt1* dosage is thus likely the initial trigger for gonadic sex determination in birds (though accompanied by an independent cell autonomous component that contributes to somatic sex; Box 3.14). In the medaka fish, a recent duplication of a 43 kb segment including *Dmrt1* has become the dominant male-determining locus (Nanda et al. 2002; Matsuda et al. 2002; Kondo et al. 2004, 2006). The newly derived paralogue *Dmrt1bY* (or *DmY*) originated by incorporating existing *cis*-regulating domains carried by transposable elements, which led to a change in its expression (Herpin & Schartl 2011a). The duplication is absent from closely related *Oryzias* species, indicating its recent evolutionary origin (Kondo et al. 2003). *Dmrt1* also belongs to one of the major sex-linked quantitative trait loci (QTLs) in the zebra fish *Danio rerio*, which has **polygenic sex determination** (Bradley et al. 2011), and is likely to play a similar role in many other fishes. In amphibians, *Dmrt1* is sex linked in a diversity of lineages, with a likely role in sex determination (Brelsford et al. 2013.). In the female heterogametic *Xenopus laevis*, sex is determined by a partial duplication of *Dmrt1* on the W (*DmW*; Yoshimoto et al. 2008). The duplication has conserved the DNA-binding N-terminal domain, but not the transcription-promoting C-terminal domain. As a result, DMW interferes with the normal DMRT1 pathway, preventing it from activating or repressing its transcriptional targets (Yoshimoto et al. 2010). In this case too, the sex-determination role of the new *Dmrt1* paralogue is a recent innovation; *DmW* is not found in closely related clawed frog lineages.

The overall picture that emerges is that DM-domain genes have a very ancient role in sex determination among animals, but that their precise mode of action may change quite rapidly. Consistent with our categories of evolutionary innovations

## Box 3.14 Sex determination in birds

The molecular details of avian sex determination are remarkably poorly known. A priori, sex in female heterogametic (ZW) lineages can be decided either by a dominant female determiner on the W, or by the dosage of Z chromosomes. Most evidence to date points to a dosage action, with *Dmrt1* as the prime candidate gene. It is Z-linked, not dosage compensated, and expressed at higher rate in males than in females prior to and during gonadal sex differentiation. *Dmrt1* is absent not only from the highly degenerated and gene-poor **W chromosome** of neognathous birds, but also from the W of ratites, which is largely homomorphic to the Z. *Dmrt1* expression starts as early as 3.5 dpc in embryonic male chicken (*Gallus g. domesticus*), while *Sox9* expression in pre-Sertoli cells and gonadal differentiation start at 6 dpc (Smith et al. 2009). The current most likely scenario is that *Dmrt1*, through *Sox9* activation, suppresses both the *Rspo1*/*Wnt4*/*β-catenin* and the *Foxl2*/aromatase pathways, resulting in the differentiation of testes (Fig. 3.4). The single dose of *Dmrt1* in females is insufficient to repress β-*catenin* and aromatase expression, resulting in ovary development. There is less evidence for the alternative scenario of a W-linked dominant feminizer, but there are reports of a functional interaction between the Z and W chromosomes (e.g. Kupper et al. 2012). Resolution of these issues awaits further overexpression and knockdown studies of sex-determination genes in different bird species.

Hormones are important in somatic sexual development: ZW embryos can develop as males with testes when aromatase is experimentally blocked early in development (references in Chue & Smith 2011). However, there is also evidence for a cell-autonomous action affecting colouration and plumage structure. This cell-autonomy can cause gynandromorphs where individuals present one side male and one side female, resulting from a failure of polar body exclusion during oocyte formation (Zhao et al. 2010).

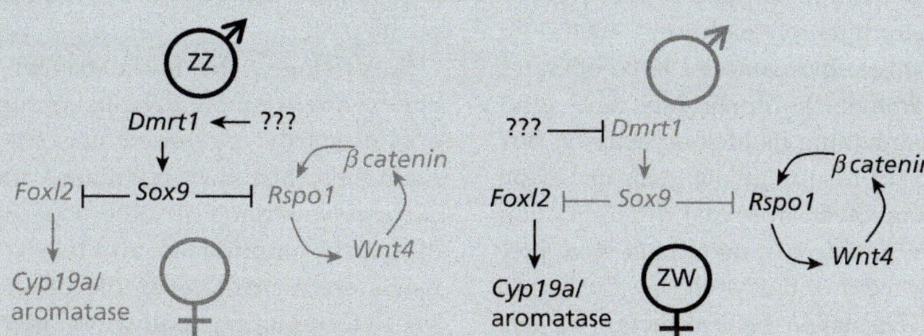


**Figure 3.4** Possible avian sex-determination pathway. Males (left) have two Z chromosomes and thus a double dose of *Dmrt1*. Another unknown Z-linked gene may lie upstream of *Dmrt1*, indicated with ???. *Dmrt1* suppresses, possibly through *Sox9*, both the *Rspo1*/*Wnt4*/*β-catenin* and the *Foxl2*/*Cyp19a* pathways, resulting in androgen production and testis development. Females (right) have a single Z chromosome and thus a single *Dmrt1* copy, possibly with an upstream W-linked inhibiting factor. Suppression of Sox9 results in active *Rspo1*/*Wnt4*/*β-catenin* and *Foxl2*/*Cyp19a* pathways, oestrogen production and ovary development.

(Section 3.1.3), the *Dmrt1* genes or paralogues have changed their position along the sex-determination hierarchy (evolving into the master-switch gene in some groups), have changed their molecular function in other groups (e.g. bifunctional regulator in vertebrates, transcriptional repressor in *C. elegans*), or have undergone changes in their regulation (e.g. alternative splicing in insects).

### 3.3.2b Sex determination in vertebrates

In all vertebrates, the undifferentiated gonad arises from a ridge of cells on the embryonic kidney (the genital ridge). Despite the impressive variety of initial triggers that determine the ultimate sexual fate of individuals, the histology of gonad differentiation is almost indistinguishable between mice, turtles, and fish; the programme of cell differentiation is almost unchanged and governed by much the same genetic pathways. When receiving a male sex-determination signal, the precursor cells from the genital ridge develop into Sertoli cells, which assemble into testis cords (precursors of adult spermatogenic tubules), enclosing the **primordial germ cells** that were formed at gastrulation in the epiblast and then migrated towards the genital ridges. Stimulation by Sertoli cells makes germ cells develop into spermatids, arrested in the G1 phase, and interstitial cells into androgen-producing Leydig cells. In females, precursor cells from the genital ridge develop instead into granulosa cells (the nurse cells in ovarian follicles that nurture germ cells), and interstitial cells into theca cells, also engaged in steroid pathways.

The same set of genes appears to be involved in sex determination across all vertebrates. Along the male line, *Sox9*, *Fgf9*, and *Dmrt1* inhibit the female pathways and drive Sertoli cell formation, tipping the originally bipolar gonad towards testis differentiation. The nuclear receptor SF1 (steroidogenic factor 1) regulates the transcription of many genes along the sex-determination pathways, including those involved in steroidogenesis; it co-operates with SRY in mammals to upregulate *Sox9* (and contributes to maintaining its lifelong activity; Box 3.12) and with *Sox9* to upregulate *Amh* in Sertoli cells, a hormone involved in sexual differentiation throughout all vertebrates. In mammals, *Amh* prevents the development of the Müllerian ducts into the uterus, and regulates the production of sex hormones. AMH receptors are also expressed on neurons in the brains of embryonic mice and likely contribute to the development of gender-specific behaviours.

Along the female line, the main players are *Foxl2* and the *Rspo1/Wnt4/β-catenin* pathway. In co-operation with *Sf1*, *Foxl2* upregulates *Cyp19a*, a gene encoding the aromatase that converts testosterone into oestrogens. *Foxl2* and *Rspo1* inhibit the *Dmrt1/Sox9* feed-forward loop, preventing testis development. Thus, *Foxl2* and *Dmrt1* lie at the heart of two opposed maintenance networks that uphold the initial sex-determination decision, as nicely illustrated in mammals (Box 3.12).

Sex determination in non-mammalian vertebrates involves the same genes, but often with variable places in the cascade (e.g. Smith et al. 1999; Smith & Sinclair 2004; Shoemaker et al. 2007). In living archosaurs (e.g. chicken, alligator), *Sox9* is controlled by *Amh* (Oreal et al. 1998; Western et al. 1999), which seems a derived state (in fishes, *Sox9* appears to act upstream of *Amh*, as in mammals; Rodriguez-Mari et al. 2005; Waters et al. 2007). In birds, *Dmrt1* has moved up the regulatory hierarchy from a downstream position to a top sex-determination role (Box 3.14).

The initial trigger may be genetic (as in birds and snakes), environmental (as in many turtles, lizards, and teleosts), or a mixture of both. Gonad differentiation in non-mammalian vertebrates depends on the sex-steroid ratio (androgen to oestrogen), and therefore on the activity of aromatase; changes in *Cyp19a* expression often accompany and likely mediate the effects of environmental temperature on sex ratios in ectothermic vertebrates: in reptiles with temperature sex determination (**TSD**), gonadal *Cyp19a* is upregulated at female-promoting temperatures, and downregulated at male-promoting temperatures (while *Dmrt1* shows the exact opposite pattern).

Surprisingly, the initial trigger (whether genetic or environmental) appears highly stable and evolutionarily conserved in some groups (as in mammals, birds, crocodilians, and snakes), but extremely variable in others (as in many groups of lizards, amphibians, and teleost fishes). Teleost fishes are a prime example of such lability. The sex-determination pathways involve the same genes as in other vertebrates, but their role may differ from species to species, with an impressive variety of expression patterns and regulatory controls. Sex determination is often polyfactorial, with a mix of genetic and environmental influences (Chapter 4). Many species show sequential hermaphroditism; testes and ovaries derive from the same precursor tissues (anlagen), from which they can flexibly initiate differentiation at various life stages, depending on hormone expression. Different reproductive strategies thus evolve via subtle alterations of the normal pathways that regulate the timing and course of hormone expression and gonad development.

Androgens are synthesized in males by interstitial testis cells analogous to Leydig cells. In females, testosterone is first produced by the ovarian thecal layer and subsequently converted to oestrogen via aromatase in granulosa layers (Devlin & Nagahama 2002). Upstream of hormonal synthesis, *Sox9* and *Dmrt1* are the key players along the male pathway. Both show differential expression in testis consistent with a role in sex determination and/or testis differentiation (Schartl 2004). *Sox9* sometimes occurs in multiple copies, with a clear sex-determination function, albeit with variable roles in the regulatory network (Guerrero-Estevez & Moreno-Mendoza 2010). *Dmrt1* has also been identified in many teleosts, and although the details of its regulatory control have not yet been elucidated, it clearly plays a role in sex determination and differentiation in many fishes (Herpin & Schartl 2011a); its expression parallels the development and regression of testes in sequential hermaphrodites such as *Acanthopagrus schlegeli* (He et al. 2003), it occurs on one of the major sex-linked QTLs in the zebra fish (Bradley et al. 2011) and has taken the master switch role in medaka fish (see below).

Along the female pathway, FOXL2 (together with SF1) upregulates gonadal aromatase (encoded by *Cyp19a*), the enzyme responsible for androgen conversion into oestrogens (Baroiller et al. 2009). Inhibition of aromatase by fadrozole classically induces female to male reversal (Kitano et al. 2000; Kwon et al. 2000). In tilapia (*Oreochromis niloticus*), which has mixed (G × E) sex determination, *Cyp19a* is upregulated in XX females during ovarian differentiation, but downregulated at high masculinizing temperatures. Similarly, *Cyp19a* expression is elevated in pejerrey and Japanese flounder (*Paralichthys olivaceus*) at feminizing temperatures, preceding histological differentiation. In the European sea bass (*Dicentrarchus labrax*), both genetic and temperature effects on sex determination are mediated by a differential methylation of the promoter region of *Cyp19a*, as nicely shown by Navarro-Martin et al. (2011): intrinsic levels of methylation are higher in genetic males than in genetic females, which prevents upregulation of aromatase by *Foxl2* at normal temperatures. Elevated temperatures during the sensitive period of development increase methylation levels in both sexes, which restrains aromatase expression in genetic females. Oestrogens are thereby kept at levels insufficient to

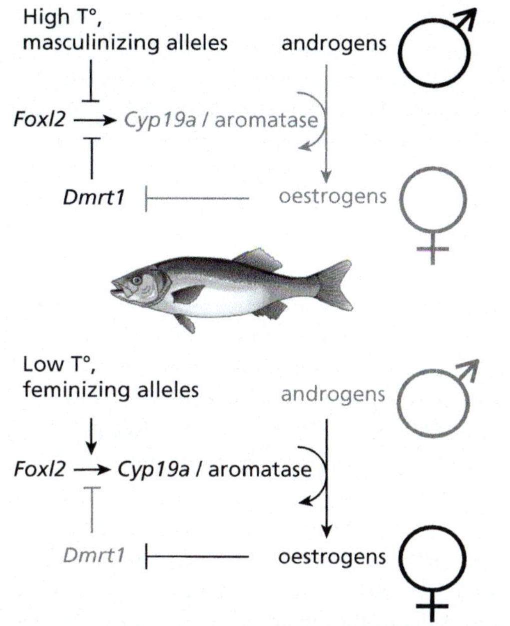


**Figure 3.5** Sex determination in sea bass. Up-regulation of *Cyp19a* by *Foxl2* depends on the differential methylation of its promoter by a mix of environmental and genetic effects. High temperature and/or masculinizing alleles (top) lead to high methylation levels that inhibit aromatase expression; the resulting androgens and *Dmrt1* expression drive male development. Conversely, low temperature and/or feminizing alleles (bottom) result in low methylation levels, so that aromatase expression converts enough androgens into oestrogens to inhibit *Dmrt1* expression and drive ovary development.

induce ovary development (Fig. 3.5). Such a mechanism might reveal quite general: in all TSD fish analysed so far, high temperatures have masculinizing effects, which are also invariably caused by inhibited *Cyp19a* expression and aromatase activity.

Reciprocally, developmental expression of *Dmrt1* perfectly correlates with rearing temperature in TSD species, being upregulated at male-determining temperatures (e.g. in the pejerrey; Fernandino et al. 2008). A negative feedback loop between *Dmrt1* and *Cyp19a* likely mediates antagonisms between the male and female pathways in fishes (and possibly other non-mammalian vertebrates): *Dmrt1* down-regulates *Cyp19a*; reciprocally, biologically active oestrogen down-regulates *Dmrt1* expression (Herpin & Schartl 2011b).

Mutations in a variety of genes along the male or female pathway may hijack the initial control over sex determination. The XY system of medaka fish arose some 10 Mya by duplication/translocation of an autosomal region containing a copy of *Dmrt1*, which thus occurs in three copies in males. The new gene (*DmY*) possibly contributes to specify maleness through a dosage effect (as in birds, Box 3.14), but it also acquired different patterns of expression. *DmY* is expressed specifically in the pre-Sertoli cells, where it regulates primordial germ-cell proliferation and differentiation during early gonad development, while *Dmrt1* is expressed after testicular differentiation (20–30 days after hatching) in **spermatogonium**-supporting cells, where it regulates spermatogonial differentiation (Nanda et al. 2002; Matsuda et al. 2002; Kobayashi et al. 2004). *DmY* also occurs in the related *Oryzias curvinotus*, but, surprisingly, was already replaced some 5 Mya by an alternative sex-determination mechanism in *O. luzonensis*, which belongs to the same **clade**. The new sex-determination gene arose by mutations in the promoter region of *Gsdf* (*gonadal soma derived growth factor*), which does not encode a transcription factor but a hormone from the TGF-β superfamily, originally identified as a factor controlling the proliferation of primordial germ cells and spermatogonia in rainbow trout (*Oncorhynchus mykiss*; Sawatari et al. 2007). Though identical to the X copy, the Y-encoded protein is expressed at higher level in pre-Sertoli cells, which is enough to tip the balance towards a male fate (Myosho et al. 2012). *Gsdf* also affects testis differentiation in medaka fish, but is autosomal and acts downstream of *DmY*.

In the Patagonian pejerrey, the master sex-determination gene results from a duplication of the gene encoding AMH, also from the TGF-β hormone family (Hattori et al. 2012). In the pufferfish, the only polymorphism associated with phenotypic sex is a single nucleotide polymorphism (C/G) in the *Amhr2* gene (*AMH receptor type II*), which changes one amino acid in the kinase domain. Females are homozygous His/His, while males are heterozygous His/Asp; sex is most likely determined by a combination of the two alleles (Kamiya et al. 2012). There is no sign of recombination arrest between the X and Y chromosomes (hence, male and female genomes only differ by one single base substitution). In the rainbow trout, another completely novel master switch gene has evolved that is not related to any known existing sex-determination genes. This Y-borne *SdY* encodes a protein with similarity to the C-terminal domain of the immune-related *interferon regulatory factor 9* (*Irf9*), expressed only in the testis and predominantly during testicular differentiation (Yano et al. 2012).

#### 3.3.2c Sex determination in Ecdysozoa

Ecdysozoans differ from vertebrates by a strong cell-autonomous component of sex differentiation. However, in all systems investigated so far (including nematodes, crustaceans, and insects), DM-domain genes also play a nodal role, at the crossroads between the male and female fate. In nematodes and crustaceans, masculinization results from upregulation of *Dmrt1* homologues (*Mab3* in *C. elegans*; *Dsx1* in *D. magna*). This holds true whether the initial trigger is genetic (as in *C. elegans*; Box 3.15) or environmental (as in *D. magna*; Box 3.16).

In insects, sex differentiation is organized by sex-specific splicing of *Dsx*, at the bottom of a cascade of regulatory genes (a mechanism that seems unique to this class; Verhulst et al. 2010b; Gempe & Beye 2011). *Dsx* has been reported from over a dozen species spanning the orders of Diptera, Lepidoptera, Coleoptera, and Hymenoptera (Salvemini et al. 2011) to be functionally conserved as a key effector in the pathway. *Dsx* splicing is controlled in most cases by the upstream *Tra* gene (Box 3.17), with some exceptions, however (Box 3.23)

Similar to *Sox9* in vertebrates, *Tra* acts as a central gear in sex-determination networks, accommodating a diversity of primary signals that range from dominant male or female determinants in houseflies (Box 3.20), X counting in *Drosophila* and *Sciara* (Boxes 3.18 and 3.19), to the diversity of mechanisms underlying haplodiploidy in hymenopterans (Box 3.21). Still other modes of functional haplodiploidy independently evolved in coccids (Box 3.22), and several of the above systems evolved towards maternal and/or environmental control (Boxes 3.19 to 3.22). As more species will be investigated, additional *Tra* regulatory mechanisms are likely to be discovered.

## Box 3.15 Sex determination in the roundworm

Androdioecy in *C. elegans* recently evolved from gonochorism through germline-specific changes in sex determination. The XX hermaphrodites are anatomically female but able to produce some sperm for self-fertilization. XO males result either from a mating between a male and a hermaphrodite or from rare, spontaneous non-disjunction of the X in the selfed offspring of hermaphrodites (Meyer 2000). The X:A ratio provides the initial trigger. Similar to *Drosophila* (Box 3.18), the number of X chromosomes is counted by polygenic X signal elements (**XSEs**); thus far two genes, *Sex1* and *Fox1*, have been identified in addition to two X-chromosomal regions (Akerib & Meyer 1994; Nicoll et al. 1997; Carmi et al. 1998). Individuals develop as males at X:A ratios 0.67 or below, and as hermaphrodites at X:A ratios 0.75 and above. This initial cue is transmitted to the DM-domain effector genes (*Mab3* and *Dmd3*) through the post-translational regulation of a complex cascade of genes that are alternatively activated (Fig. 3.6). In XX individuals, XSEs first repress *X0-lethal* (*Xol1*), allowing expression of the three downstream genes *Sdc1*, *Sdc2*, and *Sdc3* (for *sex determination and dosage compensation*). These genes in turn inhibit *Her1*, allowing expression of *Tra2* and *Tra3*, which then inhibit *Fem1*, *Fem2*, and *Fem3*. This finally allows *Tra1* to be released and translocate to the nucleus where is represses the DM-domain genes *Mab3* and *Dmd3*, activating the hermaphroditic programme. In the male pathway, XSE dosage is insufficient to repress *Xol1*; *Her1* can bind to the *Tra2* receptor; this releases the *Fems* which sequester TRA1, ultimately ending up in activation of the DM-domain genes required for the expression of male-specific developmental and behavioural genes, and suppression of female functions such as vitellogenin transcription (Yi et al. 2000).

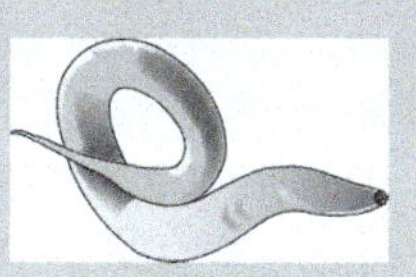

Switches to alternative sex-determination systems (including female heterogamety and TSD) have been obtained by mutations of several of the key genes in the cascade (Hodgkin 2002). The strain CB5362, for instance, produces ~100% hermaphrodites at temperatures below 15°C and ~100% males above 20°C. This results from a single base mutation in *Tra2*, which substitutes a leucine by a proline at αα-residue 127. This substitution occurs in an extracellular loop of the TRA2 protein (a cell membrane receptor) thought to interact with *Her1*, a repressor of TRA2 activity. The mutation presumably causes a conformational change in the protein that mimics HER1 binding at high temperatures, rendering the protein inactive, and leading to male development (Chandler et al. 2009). The strain also possesses a loss-of-function mutation in *Xol1*, enhancing the *Tra2* sex reversal and killing XO individuals (i.e. males derived by GSD rather than TSD).

*Xol1* is a very rapidly evolving gene that shows little sequence homology with its homologue in the closely related *C. briggsae*, despite a conserved function. In addition, the sex-determination pathways in *C. briggsae* reveal several differences; some genes, like *Fem1* and *Fem2*, are absent, suggesting that hermaphroditism has evolved independently and in a different way than in *C. elegans* (Haag & Doty 2005).

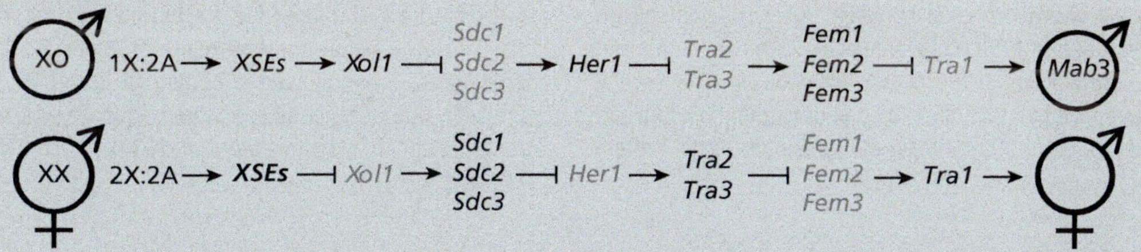


**Figure 3.6** The *Caenorhabditis* sex-determination pathway. Males carry a single X chromosome which is counted by several genes (*XSEs*) and their low dose activates *Xol1*. This sequentially leads to repression of *Sdc1–3*, activation of *Her1*, repression of *Tra2–3*, activation of *Fem1–3*, and repression of *Tra1*, resulting in male development. *Mab3* is a downstream gene required for male differentiation. Embryos with two X chromosomes have a double dose of *XSEs* which suppress *Xol1* and the opposite alternation of activation and suppression follows. Active *Tra1* is required for hermaphrodite development. During hermaphrodite development, *Tra2* is temporarily inhibited resulting in male germline development. Activation steps are indicated by arrows and inhibition steps by bars, active genes are in black and repressed genes in grey, the higher dose effects of XSEs in hermaphrodites is depicted in bold.

**Box 3.16 Sex determination in the water flea**

The branchiopod crustaceans form a sister group to insects. Water fleas (*Daphnia*) are characterized by **cyclical parthenogenesis** and ESD. Females clonally produce daughters by **ovovivipary** during the favourable growing season. Under worsening conditions, diploid males are also produced clonally; though genetically identical to females, they differ phenotypically by their smaller size, shape of first antennae, and the presence of a hook on the first thoracic appendage that allow clasping females. In the meanwhile, females meiotically produce haploid oocytes that require fertilization. The resulting zygote will survive desiccation and freezing, before giving birth to a new parthenogenetic female as conditions improve.

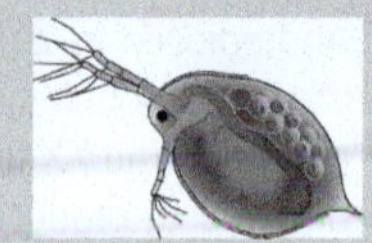

In *D. magna* the male phenotype is determined by up-regulation of *Dsx1* in specific tissues (gonads and sexually dimorphic organs) at day 3 of embryonic development (corresponding to birth). *Dsx1* expression appears directly controlled by the juvenile hormone produced by the brain in response to environmental cues, via juvenile hormone receptors in its promoter region (Kato et al. 2011).

**Box 3.17 Evolution of the *transformer* gene**

First identified in *D. melanogaster* as a major actor in the sex-determination pathway (Baker & Ridge 1980; Butler et al. 1986), the splicing regulator *Tra* has now been found in many insects, mostly dipterans and hymenopterans (Sanchez 2008; Verhulst et al. 2010b), but appears to be absent from the lepidopteran *Bombyx mori* (Box 3.23). It has also been found in the crustacean *D. magna*, but with no role in sex determination (Kato et al. 2010). It is not known outside arthropods (*transformer* genes in nematodes are not homologues). This places the origin of the *Tra-Dsx* regulatory axis somewhere between the common ancestor of the Hexapoda and the basis of the Holometabola (Fig. 2.7), with subsequent loss in at least the Lepidoptera.

TRA is a fast-evolving protein with an exceptionally high sequence divergence even between closely related species (O'Neil & Belote 1992; Kulathinal et al. 2003). A comparison of *Tra* orthologues in the *Drosophila* genus reveals several conserved domains involved in sex-determination function (Kulathinal et al. 2003), while the remainder of the protein evolves relatively unconstrained (McAllister & McVean 2000). Alignment of orthologues from all species examined to date also reveals a number of conserved domains with a shared Arg/Ser domain and a Pro-rich region, as well as a dipteran and hymenopteran specific domain (Verhulst et al. 2010b). In addition, a putative autoregulatory domain is present in all sequences except that of *Drosophila*, in which *Sex-lethal* has co-opted this function upstream of *Tra* in the pathway (Box 3.18).

In *Apis*, *feminizer* (the *Tra* orthologue in Hymenoptera) is controlled by hetero/**hemizygosity** at *Csd* (Box 3.21), a paralogue of *feminizer* (Hasselmann et al. 2008; Gempe et al. 2009). The *Csd* gene has also been identified in a number of ants and bumblebees (but not in *Nasonia*; Werren et al. 2010), which places the *Tra* duplication back to more than 120 Mya, before the divergence of Aculeata. The two genes have since followed a pattern of concerted evolution, likely due to constraints on their sequence similarity for proper functioning (Schmieder et al. 2012).

### 3.3.3 Molecular mechanisms of sex manipulation by parasites

Endosymbionts may actively disrupt sex determination at different levels in their host regulatory networks (Beukeboom 2012). Manipulation may occur at the top of cascades by affecting host chromosomal constitution, when this is used to trigger sex development; many parthenogenesis-inducing *Wolbachia*, *Cardinium*, and *Rickettsia*, for instance, feminize their haplodiploid hosts via gamete duplication (Huigens & Stouthamer 2003; Hagimori et al. 2006). In the wasp *Encarsia hispida*, removal of *Cardinium* results in reversal of diploid females into diploid males rather than haploid males (Giorgini et al. 2009), as commonly observed in antibiotic curing studies of haplodiploids (Huigens &

**Box 3.18 Sex determination in *Drosophila***

As already noted by Bridges (1925), sex in *Drosophila* is determined by the balance between X chromosomes and autosomes, not by the presence of the Y. Individuals with an X:A ratio 1.0 or higher develop into females, while those with an X:A ratio 0.5 or lower develop into males. Triploid individuals with 2X and 3A (X:A = 0.67) are sexual mosaics, with mixtures of male and female cells. A recent view, however, is that *Sxl* is regulated by ploidy-dependent differences in developmental rates (rather than autosomal contribution); which is, in essence, counting the number of X chromosomes (Erickson & Quintero 2007).

As in *C. elegans* (Box 3.15), the cascade is initiated by XSEs, such as *Scute*, *SisA*, *Runt*, and *Unpaired* (*Upd*), which act in a dose-sensitive fashion on a single target gene, *Sex-lethal* (Fig. 3.7). *Sxl* has two promoters, early and late. Only the former responds to the primary X:A signal, by releasing a first female-specific SXL product that activates a positive autoregulatory loop maintaining *Sxl* in a ON state (a device memory for sex determination; the primary X:A signal is not required anymore afterwards). In males, the dose of XSEs is below the threshold required to activate the early promoter, so that *Sxl* remains functionally off.

After the blastoderm stage, the late *Sxl* promoter is activated in both sexes, but products will be differently spliced in presence of the initial female specific transcript. The male-splice form ($SXL^M$) contains an additional exon 3 with stop codons, giving rise to a truncated, non-functional protein. In presence of early SXL products this exon is spliced out and the female transcripts give rise to a functional RNA-binding protein $SXL^F$ that will in turn control the splicing of the pre-mRNA from the downstream gene *Tra*. The female-specific transcript of *Tra* also lacks a male-specific exon with stop codons; a full-length TRA protein is synthesized, which in turn combines with the constitutively produced TRA-2 protein (Inoue et al. 1990; Amrein et al. 1988) to direct the female-specific splicing of *Dsx*, $DSX^F$, at the bottom of the cascade (Hoshijima et al. 1991; Ryner & Baker 1991). In absence of *Sxl* and *Tra*, *Dsx* by default is spliced in the male mode ($DSX^M$). The two alternative proteins $DSX^M$ and $DSX^F$ share the same N-terminal DM domain (and thus bind to the same DNA sequences) but differ in their C-terminal domain (exon 4 is excluded from the male splice form), which provides specific regulation properties (Section 3.3.2a). The TRA–TRA2 complex also controls the sex-specific splicing of *fruitless* (*Fru*), involved in development of the male central nervous system, where it triggers male courtship behaviour. We refer to Salz & Erickson (2010) and Cline et al. (2010) for recent reviews.

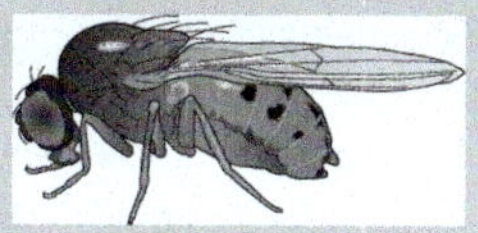

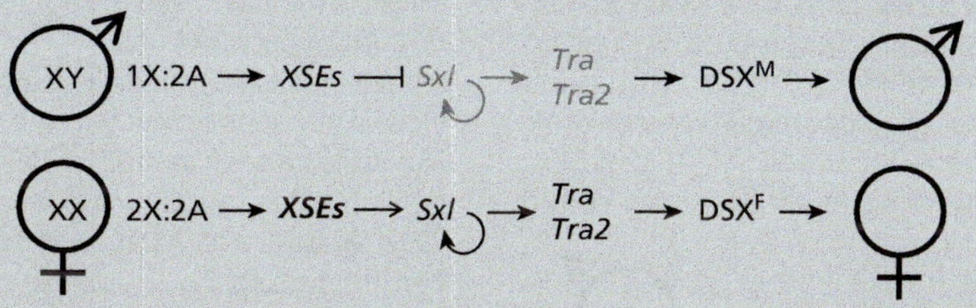


**Figure 3.7** The *Drosophila* sex-determination pathway. Males carry a single X chromosome which is counted by several genes (XSEs) but their dose is too low for activation of *Sxl*. This results in male-specific splicing of *Dsx* and male development. Embryos with two X chromosomes have a double dose of XSEs which activates *Sxl*. *Sxl* is autoregulated and induces in turn the female-specific splicing of *Tra* and of *Dsx*. *Dsx* transcriptionally regulates downstream sex differentiation genes. Activation steps are indicated by arrows and inhibition steps by bars; active genes are in black and repressed genes in grey; the higher dose effects of XSEs in females are depicted in bold.

Stouthamer 2003). Apparently, in this case, the bacteria somehow feminize diploid offspring. Weeks et al. (2001) found that curing of *Cardinium* in the asexual haploid mite *Brevipalpus phoenicis* changes haploid daughters into haploid sons, which also suggests that feminization is accomplished through altering the host's sex determination at a lower level in the cascade. In diplodiploid organisms, symbionts could conceivably manipulate the chromosomal constitution of embryos by altering segregation of the sex chromosomes in the heterogametic sex; as males in XY systems do not transmit the bacteria, this is only expected in female heterogametic systems, such as Lepidoptera. To our knowledge,

## Box 3.19 Sex determination in fungus gnats

Sciaridae and Cecidomyiidae are XX/XO, with sex also determined by the X:A ratio. The number of X chromosomes in the developing embryos is under maternal control: male sperm provides two X chromosomes, of which either one or both are eliminated, resulting in daughters (XX) or sons (XO) respectively (Sanchez 2008). The maternal factor controlling this process recognizes the parental origin of X chromosomes and then eliminates one or two; the term 'imprinting' was actually introduced by Crouse (1960) to describe this phenomenon in *Sciara*.

Sex ratios vary much among families within populations. Some species are monogenic, with females specializing into production of a single sex. In *Sciara coprophila*, androgenic females are XX and gynogenic females XX'. The mutated X' chromosome bears an inversion preventing recombination with its X homologue. Androgenic females always eliminate the two paternal Xs (resulting in XO males), while X'X females eliminate a single one. The X' and X otherwise show Mendelian segregation, so that X'X females produce a 1:1 ratio of androgenic and gynogenic females. Thus, the X' chromosome only occurs in females, while the X spends part of its time in males. In *S. ocellaris*, sex ratios vary with temperature: more females are produced at high temperature, due to a higher rate of single paternal X elimination.

## Box 3.20 Sex determination in other dipterans

The regulation of sex determination in other investigated dipterans is comparable to that of *Drosophila*, with two important exceptions: *Sex-lethal* is not part of the cascade, and a dominant male determiner, rather than X:A ratio, acts as the primary signal (Saccone et al. 2002; Dübendorfer et al. 2002; Sanchez 2008; Gempe & Beye 2011). In the housefly (*Musca domestica*), the downstream *Dsx* gene (*Md-dsx*) is turned into a female mode of expression (Md-DSX$^F$) by functional products of the *transformer* homologue *Md-tra*. Initial expression of *Md-tra* is triggered by maternal Md-TRA and Md-TRA2 products provided to the egg. Once activated, a self-sustaining feedback loop will maintain the *Md-tra* female-promoting functions (Hediger et al. 2010). In XY embryos, the *Md-tra* loop is interrupted by the action of a hitherto unidentified Y-linked masculinizing factor (*M*). This sets *Md-dsx* into the male mode of expression (Md-DSX$^M$) and turns on male development (Fig. 3.8). Similar ways of *Tra* and *Dsx* regulation seem to occur in other dipterans such as the medfly (*Ceratitis capitata*) and *Aedes* mosquitoes (Pane et al. 2002; Salvemini et al. 2009). In the blowfly (*Chrysomya rufifacies*), the masculinizing *M* factor has been replaced by an upstream regulator of *Tra* (*f*) (D. Bopp, pers. com.). Homozygous *ff* females are androgenic, producing only *ff* males; heterozygous *Ff* females are gynogenic, providing eggs with enough maternal TRA for the self-sustaining loop to be initiated. Following Mendelian segregation, half of their daughters are gynogenic *Ff*, and half are androgenic *ff* (Ullerich 1984).

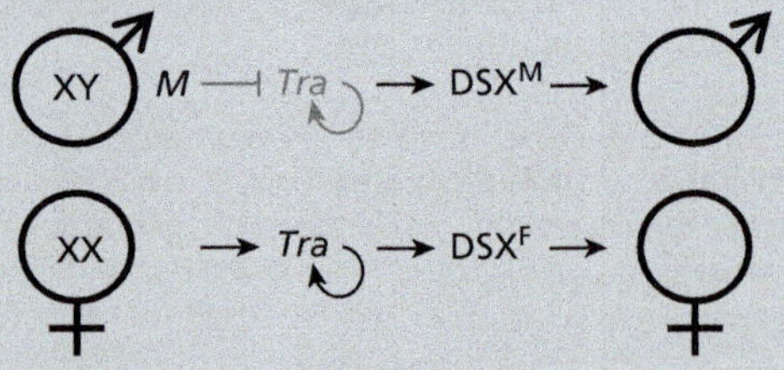


**Figure 3.8** The dipteran sex-determination pathway. Males carry a Y chromosome with a dominant male determiner *M* which blocks the *tra* autoregulation loop, resulting in male-specific splicing of TRA and DSX and male development. In embryos without a Y chromosome, female-specific splicing of TRA and of DSX causes female sexual differentiation. Activation steps are indicated by arrows and inhibition steps by bars; active genes are in black; repressed genes in grey.

The housefly is further characterized by a polymorphism in sex-determination mechanisms. So-called 'standard' populations are male heterogametic, with the dominant male determiner locus (*M*) on the Y, and *Md-tra* on chromosome IV. In some populations, the *M* factor is located on one of the five autosomes or on the X; individuals can even be homozygous for *M* or carry multiple *M* factors on different autosomes (reviewed in Feldmeyer et al. 2008). In such

*continued*

**Box 3.20** *Continued*

populations, a dominant constitutive mutation of *Md-tra, Md-tra^D* is also present, which triggers female development even in the presence of several *M* factors (Dübendorfer et al. 2002; Hediger et al. 2010). Females are thus effectively the heterogametic sex in this situation. Although of recent origin, autosomal *M* factors appear now well established in some populations (Franco et al. 1982; Kozielska et al. 2008; Hamm & Scott 2008). They show a clinal distribution, being more abundant at lower latitudes with higher temperatures (Kozielska et al. 2008; Feldmeyer et al. 2008), for reasons not yet understood.

*Md-tra^man* is a loss-of-function of *Md-tra*; homozygotes develop as males even in the absence of *M*. Presence of a wild-type *Md-tra* allele in this strain acts as dominant female determiner. *Arrhenogenic* (*Ag*) is a variant of an autosomal *M* factor, too weak for blocking *Md-tra* expression in somatic tissues (so carriers are females), but strong enough to prevent expression of maternal Md-TRA products in the germ line (so offspring are male). By this mutation, *M. domestica* has changed from a zygotic to a maternal effect type of sex determination, a system that is standard for some other dipterans (Dübendorfer et al. 2002).

**Box 3.21 Sex determination in hymenopterans**

All hymenopterans, including ants, bees, wasps, and sawflies, reproduce by haplodiploidy: haploid males develop parthenogenetically from unfertilized eggs, and diploid females from fertilized eggs. Almost all knowledge about the genetics of haplodiploid sex determination comes from the honey bee (*Apis mellifera*) and the jewel wasp (*Nasonia vitripennis*). Interestingly, although the downstream cascade is well conserved (a functional TRA^F is only produced in females, leading to female specific splicing of *Dsx*), the upstream regulatory pathways differ drastically.

In the honey bee, female development depends on heterozygosity at the *complementary sex determiner* gene (*Csd*); hemi- or homozygosity leads to male development (Beye et al. 2003; Hasselmann et al. 2008; Gempe et al. 2009). High levels of gene diversity are thus required in order to prevent the formation of homozygous diploid males, which are sterile. The active complementary sex determination (**CSD**) complex induces female-specific splicing of *Tra* (called *feminizer* in bees; Fig. 3.9); how the two allelic CSD variants interact to form an active complex is not yet solved. This so-called 'single locus CSD' (sl-CSD) has been reported for over 60 hymenopteran species, mostly based on the observation of diploid males, but the molecular details have not yet been elucidated outside honey bees (reviewed in Van Wilgenburg et al. 2006; Heimpel & De Boer 2008).

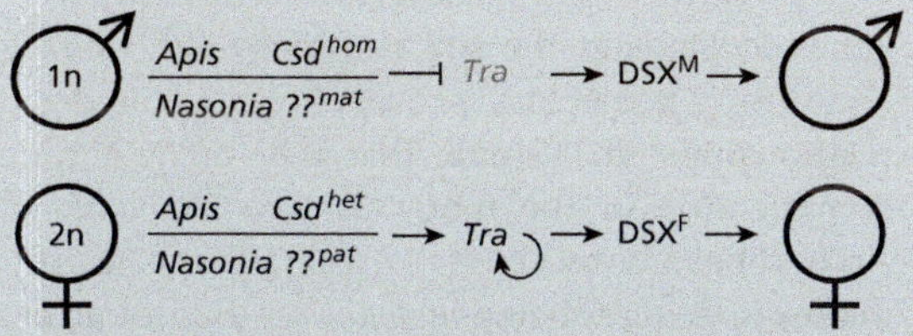


**Figure 3.9** The hymenopteran sex-determination pathway. Male embryos are haploid and carry a single allele of the *Csd* gene (*Apis*) or a maternal imprint on an unknown factor (*Nasonia*) that results in male-specific TRA and DSX and male development. Diploid female embryos carry two different *Csd* alleles (*Apis*) or an active paternal copy of an unknown factor in addition to the silenced maternal copy (*Nasonia*). This activates the autoregulation of *Tra* which induces the female-specific splice form of DSX and female sexual differentiation. Activation steps are indicated by arrows and inhibition steps by bars; active genes are in black; repressed genes in grey.

In the chalcidoid wasp *N. vitripennis*, female development is triggered (as in many other insects, including the honey bee) by a *Tra* self-sustaining feedback loop, initiated by the maternal TRA and TRA2 products provided to the egg (Beukeboom & Van de Zande 2010). However, the zygotic activation of the maternal *Tra* allele is prevented through a hitherto unknown mode of imprinting of a putative upstream regulator of *Nv-tra* (Verhulst et al. 2013). As a result, the TRA loop cannot be sustained in haploid eggs, because they only

*continued*

**Box 3.21** *Continued*

possess the maternal complement. In contrast, embryos from fertilized eggs also have the active paternal *Nv-tra* copy, and so can transcribe Nv-TRA in sufficient quantities to initiate and maintain female specific splicing (Verhulst et al. 2010a; Fig. 3.9). This mechanism has thus far only been documented for *Nasonia*; it remains to be seen how widespread it is among haplodiploids.

**Box 3.22 Sex determination in hemipterans**

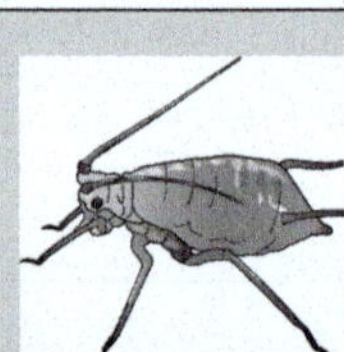

Hemiptera also display a remarkable lability of sex-determination systems, containing all the variation that is present at the level of the insect class. To mention just a few: primitive coccids (scale insects) are XX/XO, but in some derived lineages, the whole paternal set of chromosomes is either inactivated (Lecanoid coccids) or eliminated (Diapsidid coccids) in embryos destined to become males, via an unknown maternally controlled mechanism. Males are thus functional haploids, and only transmit their unrecombined maternal genome. Progeny sex ratios fluctuate widely among females, varying with environment and maternal age. Some species reproduce parthenogenetically, but still produce males at low rates.

Aphids show cyclical parthenogenesis similar to water fleas (Box 3.16). The switch to sexual reproduction is triggered by photoperiod and temperature, the signal being transmitted from the brain to ovarioles by the juvenile hormone. Males are produced by elimination of one X (they are otherwise genetically similar to females). This unusual mode of transmission of the X, coupled with cyclical parthenogenesis, results in similar effective population sizes and predicted levels of genetic diversity for X chromosomes and autosomes under **neutral evolution** (Jaquiéry et al. 2012).

**Box 3.23 Sex determination in the silkworm**

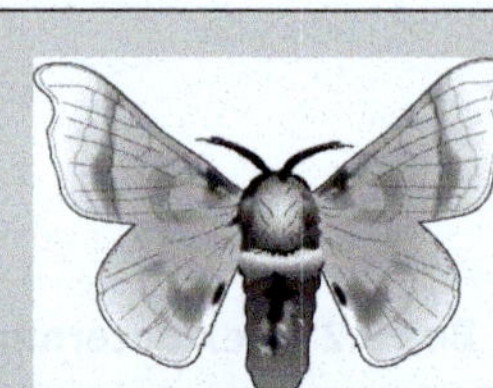

The silkworm (*Bombyx mori*) is a model organism for lepidopteran development and genetics. Like all Lepidoptera, it has a ZZ-ZW sex-determination system, with a dominant feminizing factor on the W chromosome. At the bottom of the sex-determination cascade, *Dsx* serves as the sex-specific regulator of development. In contrast to other insects, however, its sex-specific splicing is not regulated by *Tra*, but by the splicing inhibitors *Bm-Psi* and *Bm-IMP*, which prevent the inclusion of a female-specific exon in $DSX^{M}$. *Psi* (for *P-element somatic inhibitor*) is a known splicing inhibitor in *Drosophila*, but has thus far only been shown to have a sex-determination function in *Bombyx* (Suzuki 2010).

such 'meiotic-drive' symbionts have not yet been documented.

Alternatively, symbionts can directly interfere with expression of specific genes along the cascade. Most scope is expected at the primary signal level, because the cascade is more conserved and integrated towards the bottom. Symbionts could interfere with dominant female- or male-determining genes located on the sex chromosomes, or target X-counting genes to alter sex chromosome dosage relationships in systems that rely on X:A ratios. Feminization of the moth *Ostrinia scapulalis* (the Adzuki bean borer) by its *Wolbachia* parasite occurs via sex-specific splicing of *doublesex* (Sugimoto et al. 2010; Sugimoto & Ishikawa 2012); the female form is produced in ZZ individuals, which develop as females. Interestingly, ZW individuals become males upon antibiotic curing, indicating that the host has evolved a dependence on the endosymbionts for sex determination.

Molecular data on microbial manipulation of sex-determination pathways are still scant. Beukeboom (2012) has argued that the *transformer* gene may be a particular strong candidate for microbial manipulation of insects for several reasons: first, it plays a central role in sex-determination cascades and can process many different primary signals; secondly, silencing of *transformer* by RNAi in females results in sex reversal, and thirdly *transformer* is both maternally and zygotically expressed in many species, which makes the window of opportunity for manipulation large. Another speculative option would be that symbionts acquire host sex-determination genes by lateral transfer. As endosymbionts continue to be discovered in other arthropod groups, such as crustaceans, spiders, and mites, as well as in nematodes (Noel & Atibalentja 2006; Duron et al. 2008; Martin & Goodacre 2009; Perlman et al. 2010; Cordaux et al. 2012), studies of their actions on host reproduction and possible counteractions by their carriers, are expected to yield new mechanisms of sex-determination manipulation and be very rewarding for our understanding of sex-determination evolution.

## 3.4 Self-incompatibility systems and induction types

Secondary mating types evolved among simultaneous hermaphrodites in several diplontic or sub-diplontic lineages, as a means of preventing inbreeding (TYPE 3.2b, Fig. 2.4). Two main mechanisms are to be recognized, relying on either SI systems (which prevent fertilization between male and female gametes from the same individual) or induction types (where meiosis and gametogenesis are induced by recognition of non-self)

Both SI systems and induction types were secondarily superimposed on pre-existing sexes. They play no role in switching in the diploid developmental programme, or in uniparental transmission of organelles. These secondary MAT loci mostly encode cell-signalling molecules (e.g. pheromone-receptors) to ensure discrimination of self- from non-self. The genes coding for these complementary molecules must be tightly linked within the locus, to prevent their recombination. An elegant solution to this problem is offered by *Euplotes* (ciliate), where the pheromones and receptors produced by a mating type result from the alternative splicing of a single gene (Vallesi et al. 2005). In the ascidian *Ciona*, the gene coding for the female surface molecule gene is positioned within the first **intron** of the gene coding for the sperm surface molecule, but transcribed in the opposite direction (Harada et al. 2008). As for primary mating types (e.g. the conversion pathways of β carotene into trisporic acid in zygomycetes; Section 3.2.2a), complementarity may be mediated by differential enzymatic activity: some ciliates (*Paramecium*) and oomycetes (*Phytophthora*) use the pheromone released by their partner as a precursor to produce their own pheromone. As another similarity with primary mating types, SI genes and induction types often show high levels of polymorphism and sequence variability, resulting from the same ultimate causes (frequency-dependent selection; Section 1.2.2). Similarly, they may also evolve towards epigenetic control, as exemplified by the cytoplasmic or caryonidal mating-type inheritance in ciliates, or the self-induction ('homothallism') found in some oomycetes.

### 3.4.1 Self-incompatibility systems

In several angiosperms and animals (e.g. ascidians), gametic fusion is controlled by complementary signal-ligand molecules expressed at the cell surface. Other angiosperms and many hermaphroditic animals (e.g. many earthworms and snails) rely on morphological adaptations to limit or prevent selfing.

#### 3.4.1a Self-incompatibility systems in angiosperms

Several hermaphroditic angiosperms have evolved SI systems that allow recognition and rejection of self, so that fertilization requires pollen and eggs of different genotypes. Early inheritance studies have revealed SI systems to be genetically controlled by a single di- or multi-allelic S locus. This locus encodes a female determinant, expressed in the pistil, and a male determinant, expressed either in the anther or in the pollen. Interaction between male and female determinants from the same haplotype prevents pollen germination or arrests pollen tube elongation. SI-systems have evolved several times independently, with different molecular mechanisms of regulation (reviewed in McClure et al. 2011).

Under gametophytic self-incompatibility (GSI; e.g. found in Solanaceae, Rosaceae, and Plantaginaceae), the male determinant, a F-box protein encoded by the *Slf* gene (for *S locus F-box*), is only expressed in the haploid pollen grain. The female determinant is a S-RNase expressed in the stigma (Fig. 3.10), which specifically recognizes pollen

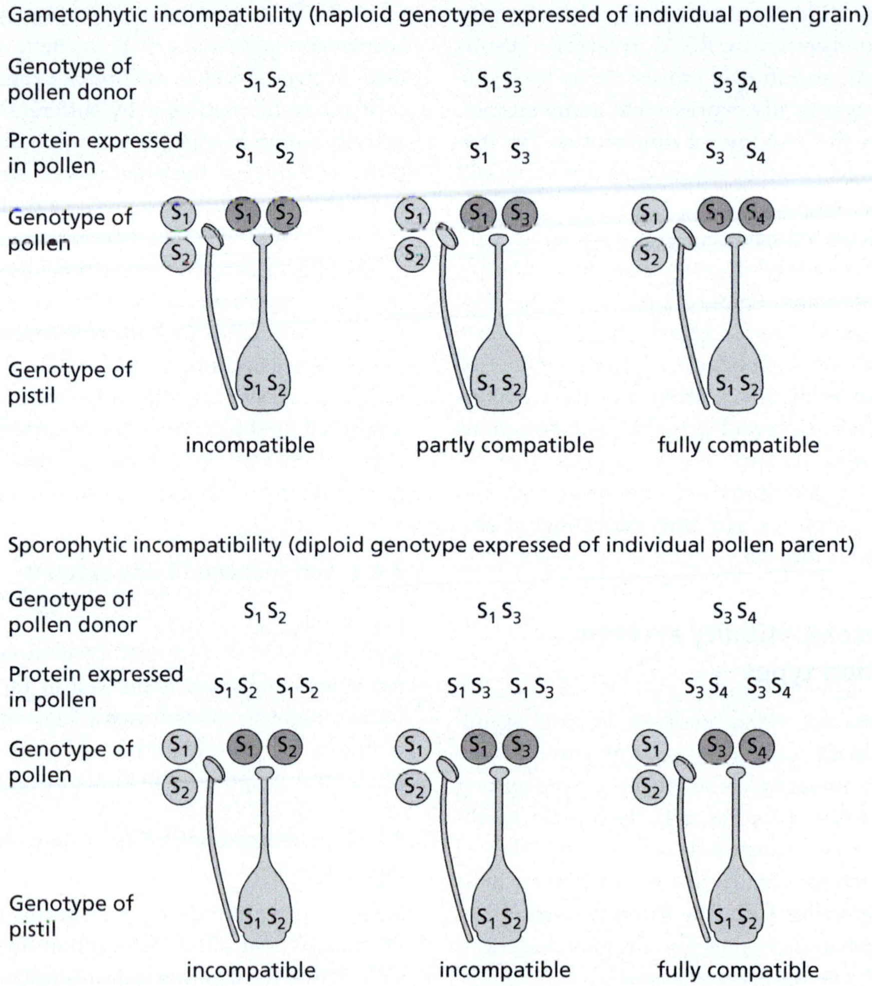


**Figure 3.10** Gametophytic (GSI) and sporophytic (SSI) incompatibility. Self-incompatibility is controlled by a multi-allelic S-locus that encodes the pollen and pistil S-specificity. Under GSI, incompatibility occurs if the haploid genotype of the pollen matches that of either allele of the pistil, and under SSI if either allele of the pollen parent matches the pistil genotype. Modified after Franklin-Tong & Franklin (2003).

grains expressing F-box proteins from the same S allele, and degrades their rRNA. If the pistil is from an $S_1S_2$ heterozygote, S-RNases from the $S_1$ and $S_2$ alleles inhibit growth of pollen expressing the $S_1$ and $S_2$ F-box, respectively. When the donor is $S_2S_3$, only the $S_2$ pollen grains are inhibited.

Frequency-dependent selection is responsible for the high number of S alleles and their maintenance over long evolutionary periods. This explains their high molecular divergence, reflected in the many amino-acid differences between **RNases** (Stone 2002). Although structural differences exist between S-RNases of different lineages, the genes appear to be derived from a common ancestor. Exceptions include Papaveraceae, where the pollen determinant (a transmembrane 20 kDa protein) interacts with the pistil determinant to trigger a $Ca^{2+}$ -dependent signalling network resulting in pollen inhibition and programmed cell death (Wheeler et al. 2009). In addition to variation through allelic differences, several modifier genes of the one-locus SI mechanisms have also been identified. They fall into three classes: those that form complexes with S-RNase and SLF products to affect their

S-specificity, genes required for **SI** but not pollination, and genes required for SI and other pollen-pistil interactions (McClure et al. 2011). Finally, two loci rather than one (S and Z) are responsible for SI in grasses (Poaceae), adding further to the complexity of SI-regulatory mechanisms (reviewed in Stone & Goring 2001). We refer to McClure et al. (2011) and De Franceschi et al. (2012) for recent reviews of the molecular details of GSI, and to Castric & Vekemans (2004) for an attractive treatise on the population genetics of allelic polymorphism and the consequences of different interactions among alleles.

Under **sporophytic self-incompatibility** (SSI; e.g. in Brassicaceae), the male determinant is expressed in the diploid anther (Fig. 3.10). The *Scr* gene (for *S-locus cysteine rich*) encodes ligands that are exposed at the surface of the pollen grain and interact with the pistil receptor SRK. Receptor-ligand interaction triggers a signal transduction cascade that leads to the inhibition of pollen growth. Due to sporophytic expression, a $S_2$ pollen grain from an $S_2S_3$ donor potentially expresses both $S_2$ and $S_3$ SCR ligands. It may however fertilize a $S_1S_2$ pistil if $S_3$ is dominant over $S_2$. A minimum of three alleles is required (as opposed to two under GSI), but frequency dependence often maintains higher levels of polymorphism (for molecular details and references see the review by Stone & Goring 2001; Nasrallah 2005; Rea & Nasrallah 2008).

Several angiosperms evolved **heteromorphic self-incompatibility**, which adds a physical barrier to self-fertilization. The style-stamen length polymorphism is controlled by a set of closely linked genes ('supergene') that are inherited together (Barrett 1992; Lewis & Jones 1992). In distyly (e.g. as found in *Primula* or *Linum*), the S locus determines both the SI phenotype and the floral phenotype of two floral morphs. One is heterozygous (*Ss*) for the dominant *S* haplotype and has short-styled flowers with long stamens. The other is homozygous (*ss*) for the recessive *s* haplotype and has long-styled flowers with short stamens. *S* and *s* haplotypes comprise several tightly linked genes, including pistil and pollen S genes as well as several genes that determine floral morphs (Barrett 1992; Ushijima et al. 2012). Tristyly (e.g. in some species of *Oxalis* and *Lythrum*) involves three style-stamen length polymorphisms. In one morph, the pistil is short and the stamens are long and intermediate; in the second morph, the pistil is intermediate and the stamens are short and long; in the third morph, the pistil is long and the stamens are short and intermediate. It is encoded by two di-allelic loci, S and M (Fig. 3.11), but their molecular characterization is still poorly known.

SI systems may directly interact with sex-determination systems. Androdioecy (a rare mating system in plants; Section S2.2.6) could evolve in *Phyllirea angustifolia* (Oleaceae) thanks to its di-allelic GSI system. The sex-determination (female sterile) mutation is linked to the S locus (either physically or through a modifier somewhere in the genome). Males could spread because they are compatible with both SI alleles, and therefore can fertilize any hermaphrodite (whereas hermaphrodites are only compatible with half of the population; Vassiliadis et al. 2000; Saumitou-Laprade et al. 2010).

#### 3.4.1b Self-incompatibility systems in ascidians

Ascidians also possess a well-developed SI system (reviewed in Harada & Sawada 2008) that attracted the attention of Morgan early on (1942). In the solitary ascidian *Ciona intestinalis*, SI is governed by two highly polymorphic loci (named *Themis A* and *B* after the Greek goddess who prohibits incest) that act co-operatively. Each locus contains a tightly linked pair of genes that code for a sperm surface protein (S-THEMIS, a polycystin 1-related transmembrane receptor) and an egg vitelline coat protein (V-THEMIS, a fibrinogen-like ligand). The *v-Themis* gene is actually positioned within the first intron of the *s-Themis* gene, but transcribed in the opposite direction (Harada et al. 2008). Both A and B S-THEMIS proteins must be bound by the corresponding female ligand to prevent fertilization. Only S-THEMIS-B is structurally complete with 11 trans-membrane domains; the S-THEMIS-A protein is truncated, suggesting that two distinct signalling cascades are activated and must co-operate to block fertilization (Harada et al. 2008).

### 3.4.2 Induction types

In contrast to SI systems, induction types do not necessarily prevent selfing. They only allow meiosis and gametogenesis to be induced by the recognition

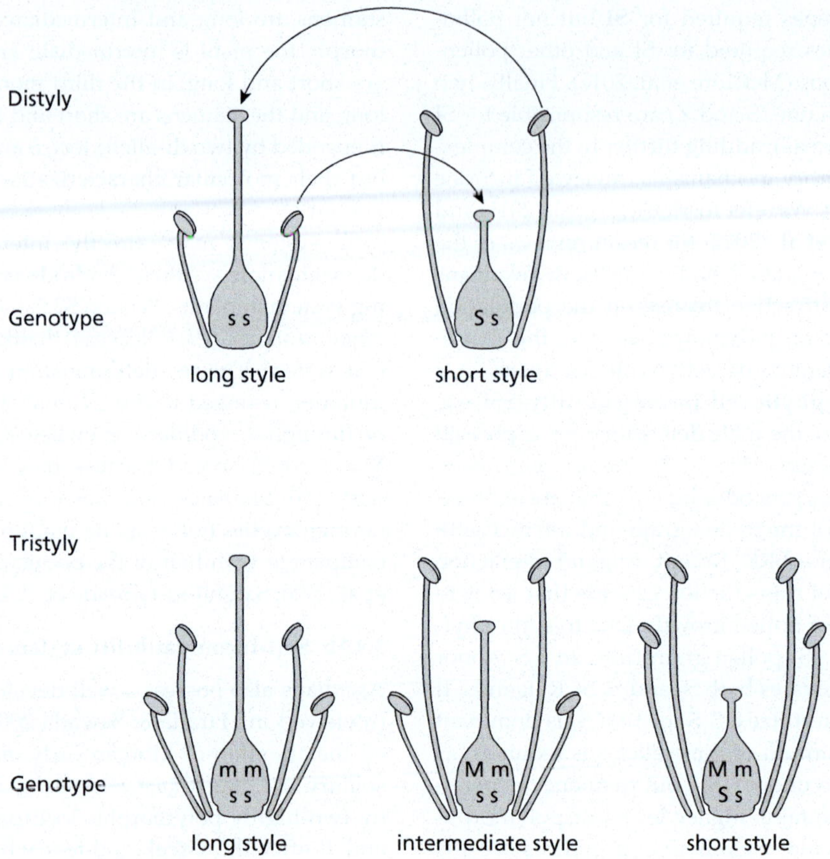


**Figure 3.11** Distyly and tristyly. Distyly is controlled by a single locus S, one type is homozygous ss and the other type is heterozygous Ss. Tristyly is controlled by two loci: one type is homozygous *ss mm*, the intermediate type is partly heterozygous *Mm ss*, and the other type is heterozygous *Mm Ss*. Modified after Stone & Goring (2001).

of non-self. Below we expand on the molecular mechanisms underlying the induction types in ciliates and oomycetes. Similar systems likely exist in Kinetoplastids (*Trypanosoma*) and Bacillariophyceae (Section 2.1.7), but their molecular basis is as yet poorly documented (see Vanstechelman et al. 2013 for recent progress on the mating-type structure of the diatom *Seminavis robusta*).

### 3.4.2a Mating types in ciliates

Mating types have long been known in ciliates (e.g. Sonneborn 1937), and found in almost all species investigated. MAT loci typically encode protein pheromones, with the only exception of *Blepharisma* where one mating type secretes the same non-peptide tryptophan derivative in all species (the alternative mating type uses a 20 kDa protein, homologous but variable between species). Pheromones are either secreted (as in *Euplotes*) or cell-bound (as in *Paramecium*). Cell–cell recognition initiates a cascade of molecular processes that triggers meiosis and nuclear exchange between compatible conjugants.

In *Paramecium*, signal peptides located on ventral surface cilia determine the two mating types O and E (for odd and even). The O peptide is a precursor molecule of E, converted into the E peptide in the cytoplasm of E mating-type cells (Xu et al. 2001). Accordingly, young E cells first display the O phenotype before maturing into an E cell. All main patterns of mating-type inheritance (Box 3.24) occur in this genus. Synclonal inheritance is found in

**Box 3.24 Mating-type inheritance in ciliates**

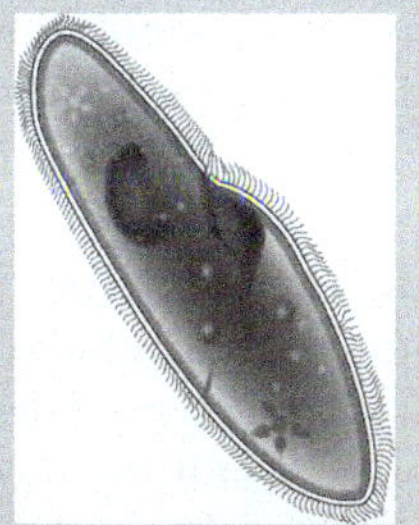

In ciliates, all individuals resulting from a conjugation event are genetically identical (Section S2.2.3c). However, they may present different mating types, because this trait (as all other phenotypic traits) is controlled by the macronucleus. Thus, mating-type inheritance directly depends on the way macronuclei are formed from the micronuclei after conjugation. When macronucleus differentiation is entirely controlled by the newly formed micronucleus, all caryonids resulting from a conjugant pair inherit the same mating type ('synclonal' inheritance). When this differentiation is controlled by the old macronucleus, then mating type will show 'cytoplasmic' inheritance, resulting from maternally inherited epigenetic effects (Duharcourt et al. 1995; Preer 2000). In such a case, all caryonids stemming from one ex-conjugant display the same mating type (inherited from their mother cell), while those born to the partner express a complementary mating type (despite having identical germ line). Finally, when macronucleus differentiation depends on random or environmental factors, all caryonids stemming from the same ex-conjugant pair may present different mating types. This 'caryonidal' inheritance is found, for example, in *Tetrahymena thermophila*, where alleles at the MAT locus determine the distribution of probabilities with which one of the seven mating types is expressed. Cells homozygous for *Mat-B* produce mating types II, III, IV, V, VI, or VII, at frequencies that depend on temperature. Cells homozygous for *Mat-A*, *Mat-C*, or *Mat-D* may all produce mating types I, II, III, V, and VI, but at different frequencies, and also as a function of temperature (Orias 1981; Orias et al. 2011).

In addition, mating type may switch during life, as a result of either maturation of the macronucleus (due to random segregation of MAT alleles during cell fission) or autogamy. In both cases, switches result from the loss of one MAT allele from heterozygous individuals.

early-diverged species (e.g. *Paramecium caudatum*) and is likely ancestral. In this case, the two mating types O and E are determined by a bi-allelic MAT locus; homozygous recessive genotypes (*mm*) produce the O mating type, while heterozygous dominant genotypes (*Mm*) produces E. As conjugation occurs between *Mm* and *mm* individuals, all resulting caryonids present the same type, either E (*Mm*) or O (*mm*) depending on which allele (*M* or *m*) is left and duplicated in conjugant E after the last meiotic division. Note that a type E clone can switch to O by losing its *M* allele during successive macronuclear fissions or during autogamy, but the reverse switch (from O to E) cannot happen, for obvious reasons (similar processes may possibly explain why only the heterogametic mating type $Mt^{+}$ shows homothally in the diatom *S. robusta*; Vanstechelman et al. 2013).

Under **cytoplasmic inheritance** (e.g. *Paramecium biaurelia* and *P. tetraurelia*), the mating-type phenotype is controlled in each ex-conjugant by the parental macronucleus (Box 3.24). All individuals possess both the *M* and the *m* allele, but whether *M* is expressed or silenced is controlled epigenetically, possibly via RNA interference mechanisms. In case of caryonidal inheritance, finally (e.g. in *P. primaurelia* and *P. triaurelia*), epigenetic silencing of the *M* allele is random (also depending on temperature; Sonneborn 1975).

Mating type numbers in ciliates vary from 2 to 12 or more (up to 100 in *Stylonychia mytilus*; Ammermann 1982). Multiple types evolved from an ancestral binary system in several lineages independently (Phadke & Zufall 2009), under selective pressures similar to those applying to primary mating types and SI systems in plants: rare alleles are favoured by frequency-dependent selection. Multiple mating types can be achieved either by multiple loci (as in *P. bursaria*, where a tetrapolar system emerged by duplication of the MAT locus) or multiple alleles at the same locus (as in spirotrichs, which also have synclonal inheritance). The ten mating types of *Euplotes patella* are determined by one locus with four co-dominant alleles (four mating types being homozygous, and six heterozygous; Akada 1985). Each mating type produces specific pheromones and receptors that determine cell identity. Interestingly, both the pheromone and the receptor are produced by the alternative splicing of a single MAT gene (Vallesi et al. 2005). One transcript of the gene translates into a soluble pheromone that is released in the environment. The alternative transcript translates into a trans-membrane receptor, with an extracellular C-terminal sequence, corresponding to the pheromone, and an additional intracellular N-terminal sequence. This receptor acts both as an autocrine growth factor, and as a paracrine mating signal: when binding their own pheromones (constitutively secreted into the extracellular environment), cells grow vegetatively and divide mitotically. When binding a non-self pheromone (produced by cells from a different mating type), they arrest growth and develop competence for mating (Vallesi et al. 2005).

Sex is required in ciliates to prevent aging (by restoring telomeres, progressively erased during asexual cell fissions). This puts a strong selective pressure to regularly undergo the sex cycle. Epigenetic determination of mating types (cytoplasmic or caryonidal; Box 3.24) allows mating compatibility between genetically identical cells resulting from the same conjugation event, thereby limiting the probability of not finding a mating partner (Preer 2000). Under synclonal determination, this risk may be reduced by multiplying mating types (as in *Euplotes*), switching mating types (achieved via random segregation of MAT alleles during successive macronucleus fissions), or autogamy (within-individual fusion of two genetically identical meiotic products from the same nucleus), which evolved in ciliates as an alternative to the multiplication of mating types (Phadke & Zufall 2009). These latter solutions, however, obviously come with a risk of elevated inbreeding.

### 3.4.2b Induction types in oomycetes

Secondary mating types also evolved in oomycetes (Section S2.2.2e), where pheromones from complementary strains are required to reciprocally induce sexual development. Both strains produce male and female gametes, and the resulting zygotes can be either selfed or outcrossed (Judelson 2007, 2009). These 'induction types' (Ko 1988) are *trans*-specific (sexuality can be induced in a given *P. infestans* type by the complementary type from another *Phytophthora* species; Ko 1988). Pheromones in *Phytophthora* are acyclic diterpenes, derived from the phytol supplied from its host through the chlorophyll metabolic pathways. The A2 mating type first converts phytol into the α2 hormone, which is released into environment, and induces gametogenesis in the complementary A1 mating type, which in the process uses α2 as a precursor to synthesize the α1 hormone. In turn, α1 induces sexual differentiation in the A2 mating type (Ojika et al. 2011). The A1 mating type cannot convert phytol directly into α1, and neither does A2 convert α2 into α1.

The MAT locus displays normal Mendelian segregation in *Phytophthora parasitica* (Fabritius & Judelson 1997). The A1 mating type is heterozygous for a dominant allele (*Mm*) and A2 is homozygous for a recessive allele (*mm*). The locus structure, however, is still unknown. It is likely to carry at least two linked genes, one coding for a mating-type-specific pheromone receptor, and another triggering the enzymatic pathway controlling the different steps of phytol conversion. The *M* allele of the latter might encode a transcription factor that either inhibits the A2 developing pathway, or works as heterodimer with its *m* homologue to trigger the A1 pathway (Fabritius & Judelson 1997). Mating-type inheritance seems more complex in *Phytophthora infestans*, being seemingly associated with a balanced-lethal system (Judelson et al. 1995). Recombination within the MAT locus apparently occurs in some species, scrambling the association between pheromones and receptors (Ko 1988). Competence for self-induction ('homothallism') evolved by gene capture, allowing expression of the two induction pathways by the same individual. Transitions to and from 'homothallism' seem frequent (e.g. both homo- and heterothallic species are found in *Pythium* and *Phytophthora*). Mating-type switches also regularly occur in *Phytophthora* strains (Ko 1988).

Some oomycetes may specialize into a predominant male or female role depending on the mating partner (Barksdale 1967), sometimes leading to strictly male or female individuals (such as in the 'dioecious' Saprolegniales *Dictyuchus monosporus*, *Achlya bisexualis*, and *A. ambisexualis*), which obviously enforces outcrossing. The 'female' individual secretes a pheromone A (antheridiol, a C-29 steroid), to which its partner reacts by producing branches on which antheridia are formed and meiosis occurs. The pheromone A simultaneously inhibits the formation of oogonia by the 'male' partner. Meanwhile, this sexually activated male converts antheridiol (taken up from the medium) into pheromone B, to which its female partner reacts by producing oogonia and entering meiosis. Genders seem to only differ regarding pheromone A: males produce less and are more sensitive to it. Hence sex roles are relative, not absolute. A given genotype may act either as male or female, depending on its partner (a nice example of G × E sex determination). This establishes a gradient, from strong females at one end (always producing oogonia) to strong males at the other end (always producing antheridia). The same pheromone A is active in different species, hence allowing the sexual cycle to be triggered by interspecific interactions. As *A. bisexualis* produce more of the pheromone A than *A. ambisexualis*, a male *A. bisexualis* will behave as a female when mated with a male *A. ambisexualis*. Dioecy is thus clearly conditional. If mates are lacking, an individual will produce both oogonia and antheridia and self-fertilize.

CHAPTER 4

# The quantitative genetics of sex determination

## 4.1 Sex as a threshold trait

Although genders occur as a limited number of discrete phenotypes (usually males and females), they are not necessarily inherited along the simple Mendelian mode exemplified by the male or female heterogametic systems of mammals and birds. Inheritance patterns are often much more complex, with no one-to-one correspondence between phenotype and genotype. The concept of **threshold traits** was first proposed by Galton (1889) to account for the occurrence of certain diseases with discontinuous phenotypes. Galton did not know about Mendelian inheritance, but had an obsession with describing any phenotype by using the normal curve; he hypothesized that these traits were underlain by a normally distributed factor (now called the **liability factor**), such that individuals lying beyond a fixed threshold would develop the disease, while individuals below it would not. The genetic basis of threshold traits was later developed by S. Wright to explain the variation in the number of digits in guinea pigs. Many biological traits with discrete phenotypes, such as paedomorphosis in salamanders, or horn and wing development in insects, are best described as threshold traits, being expressed above some critical hormone concentration (Roff 1986).

Sex qualifies as a threshold trait (Bulmer & Bull 1982): individuals develop into one gender if the underlying liability factor exceeds a given threshold, and into the alternative gender otherwise (Fig. 4.1). Even when inheritance patterns appear Mendelian, the underlying processes are quantitative: in *Drosophila melanogaster*, the X signal elements act on their target gene in a dose-sensitive fashion, as already formalized in the **genic balance theory** (Bridges 1925, 1939). In mammals, a minimal threshold of *Sry* expression in precursor cells from the genital ridge is required to trigger proper development of Sertoli cells and tilt the initial balance towards a male fate. The liability factor may be any product of the sex-determining pathway: a transcription factor (as DMRT1 in birds), an RNA-binding protein (as TRA in insects), an enzyme controlling hormone production (as oestrogen-producing aromatase in sea bass (*Dicentrarchus labrax*) or ethylene-producing ACS1 in cucumber), or a pheromone (as antheridiol in oomycetes). Like any other quantitative trait, the phenotypic distribution of liability traits a priori encompasses both genetic and environmental components.

## 4.2 Genetic variance

In the simplest situation, diplo-genotypic sex determination (D-GSD) is achieved by one locus and two alleles, with either male (XY) or female (ZW) heterogamety. Allelic values for the liability trait are such that heterozygotes lie on one side of the threshold and homozygotes on the other side. In medaka fish (*Oryzias latipes*), the duplicated copy of *Dmrt1* on the Y is enough to enhance DMRT1 expression (the liability trait) above the level required for triggering male development. In birds, similarly, the absence of *Dmrt1* on W is enough to lower DMRT1 expression below the level required for male development. Under this conceptual view, whether sex is determined by a dominant allele on the W or by dosage of a gene on the Z is a purely

*The Evolution of Sex Determination*. Leo W. Beukeboom and Nicolas Perrin.

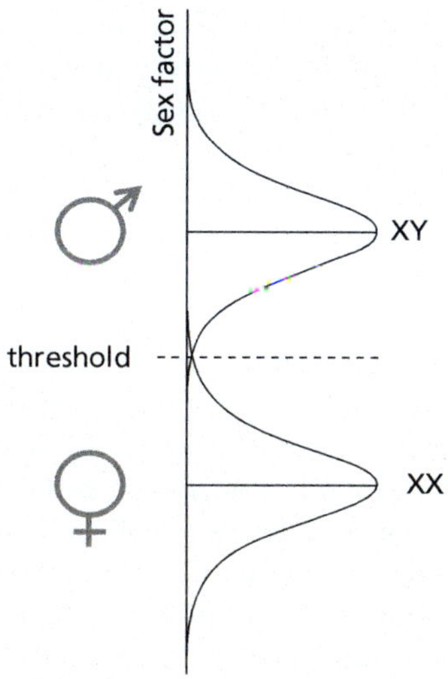


**Figure 4.1** Sex determination in a quantitative genetic framework. Sex is determined by a quantitative sex factor (vertical axis) that depends both on genotype (here XX or XY) and on environment (environmental variance is indicated with a normal distribution around the genotypic mean). Sex reversals (XY females or XX males) are expected when genotypic means are close to the threshold and environmental variance is large.

semantic question: the absence of *Dmrt1* on W is de facto a dominant allele: one copy is enough to trigger female development.

In principle, male and female heterogamety result in simple Mendelian inheritance and 1:1 sex ratios of the progeny; there is genetic variance in the liability trait, but no genetic variance in sex ratios for selection to work on. However, the widespread existence of meiotic-drive sex chromosomes that can distort sex ratios in local populations and autosomal suppressors that restore Mendelian patterns suggests a rich polymorphism of genetic factors that affect sex-chromosome segregation (Atlan et al. 1997; Montchamp-Moreau et al. 2001; Box 6.1). Significant amounts of genetic variance in sex ratios associated with driving Y chromosome have been documented in *Drosophila* (e.g. Varandas et al. 1997; Carvalho et al. 1998). Sex ratios may also vary according to maternal body condition or parturition number in mammals, possibly due to maternal alteration of the vaginal fluids, which may differently favour X or Y sperm. An intriguing claim is that of genetic variation for sex ratio in humans (Gellatly 2009); data collected over half a million individuals from 927 family trees in North America and Europe show that men with more brothers tend to produce more sons, while those with more sisters tend to have more daughters. This trend is not observed in females, consistent with male heterogamety, and possibly explained by autosomal genes affecting X and Y segregation during male spermatogenesis (Gellatly 2009).

The situation also becomes more complex when multiple alleles segregate at the sex locus, such as found in the platyfish (*Xiphophorus maculatus*). The W allele is dominant over Y, itself being dominant over X; in other words, the allelic values for the liability trait of W, X, and Y range from strongly feminizing to mildly masculinizing. This can be further illustrated by assigning allelic values, for example, −3, −1, and +2 for W, X, and Y, respectively. With purely additive effects and a threshold value of 0, the genotypic values for the liability trait become WX = −4, XX = −2, WY = −1, XY = +1, and YY = +4. Genotypes with negative values develop as females, those with positive values as males (note that effects are additive at the level of the liability trait, but dominant at the level of sex expression). Several genotypes thus produce the same phenotype (male or female), and sex ratios can be biased depending on families: XX × YY produce only XY sons, while WX × XY produce three-quarter daughters and one-quarter sons. At the population level, however, sex-ratio selection is expected to maintain the frequencies of X, Y, and W alleles at values producing equal numbers of males and females (the Fisherian sex ratio). This defines a set of neutral solutions, ranging from pure female heterogamety WY/YY to pure male heterogamety XX/XY (Fig. 4.2; Bull 1983).

Further complexity is added by polygenic systems, where different loci simultaneously contribute to phenotypic sex expression. Bull (1983) proposed three criteria as evidence for polygenic sex determination: i) a large between-family variance in sex ratio, which cannot be accounted for by sampling error from a binomial distribution, ii) paternal and maternal effects on family sex ratio, and iii) a sex-ratio response to selection (Section 4.9). Polygenic sex determination has been reported in many plants (e.g. Shannon & Holsinger 2007), fishes (Vandeputte et al. 2007; Ser et al. 2010; Liew et al. 2012), crustaceans (e.g.

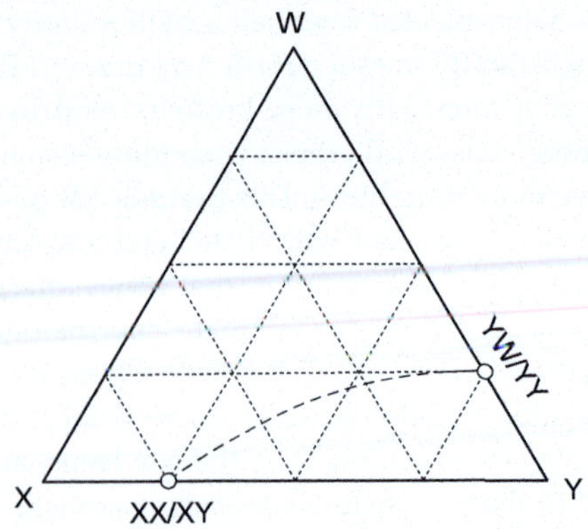


**Figure 4.2** De Finetti diagram of allelic frequencies at a sex-determining locus. The feminizing W is dominant over the masculinizing Y, itself dominant over the feminizing X. Neutral equilibrium values (1:1 sex ratios) are figured by the dotted curve, joining pure male heterogamety (XX/XY) to pure female heterogamety (YW/YY). Sex ratios are female biased above the curve, and male biased below the curve. Dotted lines indicate iso-frequencies of 0.25, 0.50, and 0.75.

Battaglia 1958; Battaglia & Malesani 1959; Voordouw & Anholt 2002), bivalves (Haley 1977; Saavedra et al. 1997), gastropods (Yusa 2007a,b), and polychaetes (Bacci 1965, 1978; Premoli et al. 1996). In the apple snail (*Pomacea canaliculata*), population sex ratios are 1:1, but family sex ratios vary greatly from 0 to 1, with a variance significantly greater than expected from a binomial distribution. Sex-ratio analyses of full-sib and half-sib broods indicate an equal contribution from both parents, with a minimum of four nuclear genes involved in sex determination (Yusa 2007a). In European sea bass, full-factorial crossing experiments have shown a large among-family variance with significant dam and sire effects on progeny sex ratios, with a minimum of two independent bi-allelic loci involved in sex determination (Vandeputte et al. 2007). Sex ratios vary also widely among families in the zebra fish (*Danio rerio*), with the percentage of males ranging from less than 5% to more than 95% (median 51%), and very high repeatability between clutches (Liew et al. 2012). Several sex-linked regions occur on different chromosomes, as revealed by single nucleotide polymorphism (**SNP**) linkage maps (Bradley et al. 2011; Howe et al. 2013). One on chromosome 5 (1.7 cM, 2.2 Mb) contains *Dmrt1*, while another on chromosome 16 (3.5 cM, 4 Mb) contains *Cyp21a2*, which encodes a hydroxylase involved in corticosteroid biosynthesis. Interestingly, various studies on different strains have identified a role for other genes from the sex-determination cascade, including *Cyp19a1*, *Foxl2*, *Sox9*, *Wt1*, and *Amh* (references in Bradley et al. 2011).

Many genes are involved in sex-determination cascades (Chapter 3). Mutations at any step along the cascade have the potential to interfere with the sex-determination pathway, as exemplified by mutation analyses in *Caenorhabditis elegans* (Box 3.15). Mutations segregating within populations form the basis for polygenic sex determination. This mutational variance should be eroded by genetic drift and sex-ratio selection (Section 1.2.3), leading to some equilibrium level of genetic variance in sex determination. Why the equilibrium value is maintained high in some lineages but vanishes in others remains an open and burning question; the dynamics of sexually antagonistic genes might be part of the answer (Chapter 7).

## 4.3 Environmental variance and random effects

Environmental factors may also affect the expression of genes, activity of enzymes, or action of hormones along the sex-determination cascade. These effects will generate an environmental variance in the liability trait. In vertebrates, temperature-dependent sex determination (TSD) was first discovered among lizards (Charnier 1966) and turtles (Pieau 1971, 1972) and later in other reptiles, such as alligators and tuataras (Section S2.2.9f). It has now been found in over 100 species (Valenzuela & Lance 2004; Ezaz et al. 2009; Pokorna & Kratochvil 2009; Gamble 2010), though not randomly regarding phylogeny. TSD might be ancestral to all reptiles (Pokorna & Kratochvil 2009) but disappeared from the avian and mammalian lineages, presumably due to their efficient homeothermy and strongly canalized development. Depending on lineages, the proportion of males may increase or decrease with incubation temperatures (FM or MF pattern, respectively, where F stands for female and M for male), or peak at intermediate temperatures (FMF). The temperature value resulting in equal sex ratios is referred to as the pivotal temperature (Pieau et al. 1994).

In fish, TSD was first reported from southern populations of the Atlantic silverside (*Menidia menidia*), with an FM pattern (Conover & Kynard 1981):

offspring produced early in the season (at low temperature) develop into females, while those produced later in the season (in warmer conditions) develop into males (Conover & Heins 1987a). Since then, thermal influence has been reported from 60 different species of 13 families, several of which are of economic importance, including the Nile tilapia (*Oreochromis niloticus*) (Baroiller et al. 1995), the pejerrey (*Odontesthes bonariensis*) (Strüssmann et al. 1996), the Japanese hirame (*Paralichthys olivaceus*), and the European sea bass, with a predominant FM pattern (Baroiller & D'Cotta 2001; Ospina-Álvarez & Piferrer 2008). Besides temperature, several environmental cues have been identified as sex determinants in fishes, such as pH, oxygen level, density, or food (Devlin & Nagahama 2002). Social cues are also important, particularly among sequential hermaphrodites (Section S2.2.9f). Environmental and social components of sex determination also occur in many invertebrates (Sections 2.2.9c-e), including insects (e.g. aphids; Box 3.22), crustaceans (*Daphnia*, Box 3.16), nematodes, echiurians, and annelids. Outside metazoans, environmental and social cues may also affect the expression of genders or mating types, such as in plants (Section S2.2.6) or ciliates (Box 3.24), respectively.

Part of the non-genetic component of variance in phenotypic sex may be purely random. Such random effects become evident both in GSD and TSD species when the liability trait takes values close to the threshold: turtle eggs incubated at the pivotal temperature have, by definition, a 50% chance of developing into either male or female. Embryos of *Drosophila* with an X to autosome ratio of 2:3 will develop as a mosaic of male and female cells. In ciliates with caryonidal determination of mating types, alleles at the MAT locus determine the distribution of probabilities with which one of several mating types is expressed (Box 3.24).

## 4.4 Mixed systems

In many instances, sex is determined by a combination of genetic and environmental factors. Environmental influence may co-occur with polygenic sex determination as, for example, in the oyster (*Crassostrea virginica*) (Haley 1979) and the apple snail (Yusa 2007a). In zebra fish, the several genomic regions involved in sex determination identified by Bradley et al. (2011) only account for 16% of the variance in phenotypic sex. The largest part of the variance is due to environment (temperature, oxygen, food, or density effects) and/or to random effects. In sea bass, both genetic effects and temperature mediate the differential methylation of the promoter region of *Cyp19a* that drives maleness (Section 3.3.2b; see also Fig 9 in Navarro-Martin et al. 2011). In the Atlantic silverside, environmental and genetic effects coexist in many populations, with evidence for a few genes with major and non-additive effects (Conover & Heins 1987a,b).

Environmental effects may also co-occur with well-defined male or female heterogamety. In medaka fish, maleness is normally due to the extra copy of *Dmrt1* on the Y chromosome (*DmY*). However, the autosomal copy of *Dmrt1* is also upregulated at high temperature (32°C), leading some XX individuals to develop as males, at a rate depending on the strain considered (27–40% 'sex reversal' in the Hd-rR strain, up to 50% in the HNI strain; Sato et al. 2005; Hattori et al. 2007). The Nile tilapia presents a complex sex-determination system combining a major determinant locus on LG1, minor genetic factors on LG3 and LG23, and temperature dependence. At 26°C, the phenotypic sex is mostly under genotypic control, with male heterogamety (XY) due to a major effect gene on LG1. However, XX offspring may develop as males when exposed to temperatures exceeding 32°C during a thermosensitive period (10–20 days post fertilization (dpf); Abucay et al. 1999; Baroiller et al. 2009). Sex reversal is likely to occur in nature: after a period of maternal mouth breeding (from 0 to 10 dpf), the developing fry shoal in shallow water bodies where they encounter masculinizing temperatures (32–34°C; Baroiller et al. 1995, 2009; Bezault 2005). Both XX males and XY females occur in nature; they are viable and fertile, as are YY males (Bezault et al. 2007). In the polychaete *Capitella capitata*, heterogametic individuals develop as females, while homogametics develop as males, females, or hermaphrodites depending on social cues (e.g. population density; Petraitis 1991). Reciprocally, the polychaete *Typosyllis prolifera* is partially protogynous: about half of the individuals differentiate into stable primary males, whereas those that develop into females often change sex at a later stage in the life cycle, depending on population density (Franke 1986).

Environmental effects may also coexist with differentiated sex chromosomes. This has been suggested for *Gekko japonicus* (Yoshida & Itoh 1974; Moritz 1990), although the evidences for chromosomal and temperature effects were gathered from different populations. More conclusive data come from the three-lined skink (*Bassiana duperreyi*). At 22°C, a male heterogametic system prevails, with heteromorphic X and Y chromosomes and balanced sex ratios. At low incubation temperature (16°C), however, clutches produce male-biased sex ratios (Shine et al. 2002). Sex-specific DNA markers show this to be due to the occurrence of sex-reversed XX males, not to differential mortality (Radder et al. 2008). As these treatments mimic the thermal regimes found in natural nests, masculinization by low temperatures is likely to occur in nature. Similar patterns arise in the bearded dragon lizard (*Pogona vitticeps*). Below 32°C, sex is determined by a female heterogametic system (Ezaz et al. 2005). Above 32°C, sex ratios become increasingly female biased; sex-linked markers show this to be due to an increasing proportion of ZZ individuals that develop into females (Quinn et al. 2007). This suggests that sex is determined by the dosage of a gene on the Z chromosome; its downregulation at high temperature prevents male development in ZZ individuals. An intriguing situation exists in the housefly (*Musca domestica*), in which the standard XY male heterogametic system prevails at high latitudes, but female heterogamety at lower latitudes. This clinal distribution occurs at different continents (Franco et al. 1982; Denholm et al. 1986; Tomita & Wada 1989; Çakir & Kence 1996; Hamm et al. 2005; Kozielska et al. 2008) and shows a clear association with temperature (Feldmeyer et al. 2008). The mechanistic details of this temperature dependence are not yet understood (see also Box 3.20).

## 4.5 The classical dichotomous view: genotypic versus environmental sex determination

Environmental (ESD) and genotypic sex-determination (GSD) systems have long been seen as resulting from distinct and mutually exclusive processes, occurring at different timings, and underlain by different mechanisms. According to this classical view, sex is determined either *at fertilization by genes* or *after fertilization by environmental factors*. Since the early days of TSD discovery, however, much evidence has accumulated showing that many systems are mixed, and that TSD and GSD are actually underlain by the same genetic pathways, making this dichotomous view less and less acceptable. Two main lines of arguments have been proposed to maintain it (Valenzuela et al. 2003). The first one holds that some mixed systems are better described as 'GSD modified by environmental effect' (GSD + EE). The idea is that in GSD + EE species, temperature acts on sex *differentiation* (reversing the genetic sex), not on sex *determination*. What matters is the master switch that initiates the cascade of events leading to gonadal differentiation. If this initial switch is environmental, then the system qualifies as ESD. If it is genotypic, then the system qualifies as GSD, even if environmental effects reverse the initial decision later in the cascade.

The second argument is that mixed systems are in fact a combination of GSD and TSD individuals. At the individual level, the process is dichotomic, and the sex of a larva determined *either* by the presence of the sex-determining genes (GSD) *or* by the incubation temperature (TSD). Even in a species where sex is determined by the sum of genetic and environmental effects, if it is temperature that puts a given zygote above or below the threshold, then that individual has TSD. It is the proportion of TSD and GSD individuals that changes as populations adapt to local conditions.

Opposing these arguments, one can first note that the distinction between sex determination and sexual differentiation is mostly semantic (Section 1.4). Sex determination/differentiation is a dynamic process, starting with one or a series of initial cues (genetic, epigenetic, or often a mixture of both) and ending with the commitment of a whole individual or cell lineage into either a male or a female fate. This commitment may be temporary and reversible (as in sequential hermaphrodites), and the initial decision potentially affected by mutations or environmental effects at any step along the cascade. Sex is never determined at conception. In vertebrates, the process starts when the precursor cells from the genital ridge develop into either Sertoli or granulosa cells. The timing and molecular processes are identical between TSD and GSD species. The initial trigger is always transient, and the initial decision must be maintained lifelong by some memory device: the loss of *Dmrt1* in embryonic male mice induces the trans-differentiation of Sertoli cells into granulosa cells, despite the presence of an active *Sry* (Matson et al. 2011). This lability is precisely what allows the evolution of sequential hermaphroditism.

Second, it makes no sense in a quantitative genetic framework to distinguish *at the individual level* whether a particular trait value is due to the genes or to the environment: it is by definition a result of their combined effects. In sea bass, individuals develop as males above a certain methylation threshold of the promoter region of *Cyp19a*, which is affected both by genes and by temperature. Thus, whether an individual develops as male or as female cannot be assigned *either* to its genetic makeup *or* to a specific temperature. Maleness results if the *sum* of genetic and temperature effects exceeds the critical threshold in the liability trait. The only possible distinction between genetic and environmental effects is *at the population level* and concerns the proportion of variance in a phenotypic trait that is due either to the genes or to the environment. If all of the variance in phenotypic sex is genetic, then the system qualifies as GSD; if all of this variance is environmental, then it qualifies as ESD. In between are mixed systems, where a variable proportion of the variance in phenotypic sex is due to genes (defining the trait **heritability**; Section 4.7). ESD and GSD are just the two ends of a continuum (Sarre et al. 2004; Barske & Capel 2008).

## 4.6 Norms of reaction

A particularly insightful view offered by quantitative genetics represents genotypes as **norms of reaction**, defined as n-dimensional structures in the space delimited by environmental and phenotypic variables. Fig 4.3 represents such norms in a two-dimensional space. The horizontal axis is an environmental trait (e.g. temperature), and the vertical axis the liability trait underlying phenotypic sex (e.g. *Dmrt1* expression in medaka fish or aromatase production in sea bass), with the threshold figured by a horizontal line. What phenotype is produced by a given genotype depends on environment (in the case illustrated in Fig 4.3a, both genotypes are more likely to develop as males at higher temperatures). Within-population environmental variance (e.g. differences in incubation temperature among nests or among eggs within nests) translates into a phenotypic variance in the liability trait, potentially affecting the sex ratio produced by a given genotype. Reciprocally, genetic variance among adults (resulting from multiple segregating norms of reaction) will translate into variance in their brood sex ratios, even under a fixed environment. In the general case, the overall resulting phenotypic variance has both an environmental and a genetic component.

The two norms (respectively XX and XY) represented in Fig. 4.3a are both linear and increasing with temperature. If the environment centres around $T_1$ (and the variance is limited), the system will result in pure GSD: the XY genotype develops only in males, and the XX genotype only in females. If however the environment centres around $T_2$ (corresponding to the XX pivotal temperature), then the XX genotype will develop into males and females in equal proportions. The alternative genotype (XY), which would only develop in males at this temperature, is eliminated by sex-ratio selection (XY males have lower fitness than XX males because they produce male-biased sex ratios (3:1), i.e. an excess of the least-wanted sex). Pure TSD results, with a FM pattern (females are produced at low incubation temperatures, and males at higher temperatures, e.g. in *Menidia*). In such a system, a switch from pure GSD to pure TSD may readily result from a climatic change or latitudinal gradient (increase in temperature). Analytical and simulation models have investigated the conditions for transitions from heterogametic or polygenic systems to ESD (Bull 1981; Bulmer & Bull 1982; Grossen et al. 2011; Section 7.2.3b).

Norms can be differently shaped, and be, for example, linear decreasing (resulting in an MF pattern, as found in many turtles; Fig. 4.3b) or curvilinear (resulting in a FMF pattern, as found in many alligators; Fig. 4.3c). The horizontal axis may represent any epigenetic variable, including positional information in simultaneous hermaphrodites. In cucumber (Box 3.8; Fig. 4.3d), the horizontal axis corresponds to the epigenetic gradient of hormone concentration along the shoot, and the vertical axis to ethylene production (controlled by *Acs1*). Little

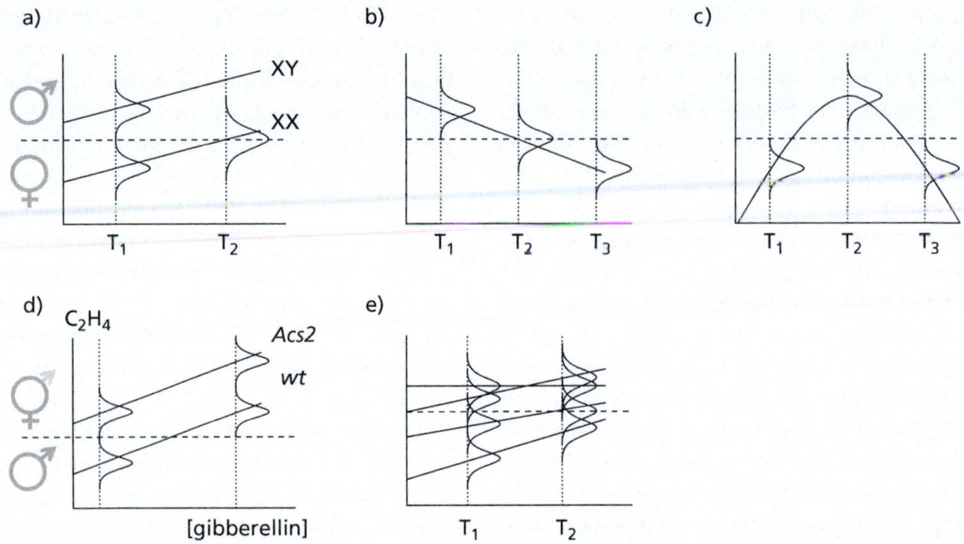


**Figure 4.3** Quantitative genetic models of sex determination. Genotypes are figured as norms of reaction, determining the dependence of the liability trait values (vertical axis) on an environmental or hormonal gradient (horizontal axis). Individuals with liability-trait values above the threshold (horizontal dashed line) develop into one gender, those below the threshold into another gender. (a) Two genotypes with parallel norms of reaction. At temperature $T_1$, the XY genotype produces only males and the XX genotype only females (pure GSD). At $T_2$, the XX genotype produces both male and females in equal proportions, and the XY genotype is eliminated by sex-ratio selection (pure TSD). (b) Pure TSD with a decreasing **norm of reaction** (MF pattern). Males are produced at low temperature ($T_1$), and females at high temperature ($T_3$). (c) Pure TSD with a non-linear norm of reaction, resulting in a FMF pattern of sex determination: females are produced at both low and high temperatures, and males in between. (d) Sex expression in cucumber depends on ethylene production ($C_2H_4$), itself a function of hormone concentration along the shoot. Low $C_2H_4$ values inhibit carpel development, while large $C_2H_4$ values inhibit stamen development. In wild type (*wt*), the overall $C_2H_4$ gradient is such that flowers are male at the bottom of shoots (low GA level) and female at the top of shoots (high GA level). A duplication of the *Acs1* gene (*Acs2*) results in increased overall production of $C_2H_4$, so that flowers always develop as females (gynoecy). In case of loss of function of the M gene, the male function is never turned off, so that the *wt* norm would produce andromonoecious plants (flowers being male below the $C_2H_4$ threshold and hermaphroditic above), while the *Acs2* norm would produce hermaphroditic flowers only. (e) Polygenic system, characterized by a diversity of reaction norms, with different sensitivities to temperature (G × E interactions). The heritability of sex ratios is larger at $T_1$ than at $T_2$ (more of the total phenotypic variance is due to differences among genotypes).

ethylene is produced at low gibberellin (GA) concentration, resulting in male flowers, whereas high ethylene levels (produced at high GA concentration) result in female flowers. Genetic variance was introduced in this system via the *Acs2* duplication at the *F* locus: at the GA concentration that is pivotal for wild genotypes, individuals with the duplicated copy always produce enough ethylene for female development. Growing conditions, which may be under maternal control, often play a similar role: in heterosporous plants, spores larger than a critical threshold size develop into female gametophytes (Korpelainen 1998). In mermithid nematodes, there is a size threshold for sex development above which an individual becomes a female (Petersen 1972, 1977; Blackmore & Charnov 1989). Similarly, a minimum size is required for sea bass juveniles to differentiate as females (Vandeputte et al. 2007).

Polygenic systems result from segregating polymorphism in the norms of reaction (Fig 4.3e). The coexistence of genotypes differing at the sex-determining loci translates into an extra-binomial variance in sex ratios, as largely documented across many lineages, from copepods (Voordouw & Anholt 2002) to gastropods (Yusa 2007b), fish (Liew et al. 2012), and turtles (Bull et al. 1982).

## 4.7 Heritability of sex ratios

The heritability of a quantitative trait measures the proportion of phenotypic variance in this trait that is genetic (broad-sense heritability) or genetically additive (narrow-sense heritability). This can be measured by trait correlations between relatives (e.g. full sibs or parent–offspring). In the case of a threshold trait, however, the underlying liability factor is not directly measurable. The only information available is the class to which individuals belong (male or female in our case). Correlations must thus be derived from discrete observations. Relevant approaches are described, for example, in Bull et al. (1982) and Roff (1986).

Large heritability values for the pivotal temperature have been measured in a variety of systems. In the copepod *Tigriopus californicus*, which shows polygenic sex determination and a significant extra-binomial component of variance in sex ratios among families, heritability was estimated as 0.31 through mother–offspring regressions (Voordouw & Anholt 2002). In sea bass, the heritability of sex ratios was estimated to be 0.62, using families from a full-factorial crossing design reared in a common environment (Vandeputte et al. 2007). A two-generation selection experiment for temperature sensitivity in the Nile tilapia has revealed a heritability of 0.69 (Wessels & Hoerstgen-Schwark 2007). Large inter-clutch differences in sex ratios at the pivotal temperature have also been documented in several species of turtles, such as *Trachemys scripta* (Dodd et al. 2006). Bull et al. (1982) reared clutches from five Ouachita map turtle females (*Graptemys ouachitensis*) at the pivotal temperature (29°C) and found a large component of among-family variance in sex ratios (from 28% to 59% males in the progeny), resulting in a high heritability value of $h^2$ = 0.82. Janzen (1992) similarly collected eggs from 15 females of the common snapping turtle (*Chelydra serpentina*) and randomly assigned them to three temperatures around the pivotal value (27.5, 28, and 28.5°C). The gonadal sex of hatchlings was significantly influenced by both incubation temperature and clutches. Combined over the three temperatures, heritability averaged $h^2$ = 0.56, indicating significant genetic variance for sex ratio: some families produced more males than others; one clutch, for instance, produced no males at any of the three temperatures (Janzen 1992).

Bull et al. (1982) argued that the response of the pivotal temperature to selection also depends on the active choice of thermal environment (e.g. female choice for nest site in turtles, shallow water bodies for shoaling in tilapia fry); the environmentally limited expression of a trait should retard its response to selection. To obtain the *effective heritability*, one should divide the additive genetic variance in pivotal temperature by the total phenotypic variance, comprising both the variance in the pivotal temperature and the variance in nest average temperature, which depends on female choice. Integrating this parameter led to much lower effective heritability values (0.06 for *G. ouachitensis* and 0.05 for *C. serpentina*, respectively), suggesting little scope for the evolution of pivotal temperatures in nature, due to a strong component of female choice. However, McGaugh & Janzen (2011) showed that constant-temperature experiments in the laboratory underestimate the variance in mean temperature that produces mixed-sex clutches in the field. Specifically, fluctuating temperatures induce the production of female offspring at mean temperatures that produce all male offspring in constant-temperature incubation environments. When accounting for this effect, the effective heritability becomes much larger: 0.32 for the *G. ouachitensis* dataset, 0.19 for *C. serpentina*, and around 0.24 for *Chrysemys picta* (McGaugh & Janzen 2011). Thus, heritability of the sex-determination threshold may actually play a significant role in the micro-evolutionary response of TSD.

Female **sex allocation** behaviour also shows some heritability, as documented, for example, in parasitoid wasps (Wajnberg 1994, Kobayashi et al. 2003; Pannebakker et al. 2011). In *Heterospilus prosopidis*, the narrow-sense heritability of sex ratio at a specific host size ranged from 0.19 to 0.32, and that of the sex-changing point (threshold value of the host body size above which females produce daughters rather than sons) ranged from 0.22 to 0.34 (Kobayashi et al. 2003). In *Nasonia vitripennis*, Pannebakker et al. (2011) found a broad-sense heritability of ~0.10 for sex allocation and a weak positive correlation with brood size. They identified one significant and three weak quantitative trait loci for the trait with, however, little variance explained. Although the number of studies is still limited and heritability estimates are weak, the sex allocation behaviour is clearly also expected to respond to selection.

## 4.8 G × E interactions

Many studies on the genetic architecture of sex ratios have revealed significant genotype (G) × environment (E) interactions (e.g. Conover & Heins 1987b; Voordouw & Anholt 2002, Vandeputte et al. 2007; Baroiller & D'Cotta 2001; Baroiller et al. 2009; McGaugh & Janzen 2011). These result from differences among genotypes in the slopes of reaction norms (Fig 4.3e). Sensitivity to temperature may result from a variety of mutations in the genes involved in the sex-determination cascade, with a priori heterogeneous effects. Among the *C. elegans* strains selected by Hodgkin (2002) for their diversity of sex-determination mechanisms, some have pure TSD, and others pure GSD (Box 3.15). Similarly, the Poeciliid fish *Poeciliopsis lucida* displays a large variance in sex-determination mechanisms among laboratory strains. Of the two strains studied by Sullivan & Schultz (1986), one showed strong temperature dependence in sex ratios (from 38% males at 24°C to 92% males at 30°C), while the other did not show any deviation from 1:1 at any temperature. Genotypes with different sensitivities to temperature may coexist in natural populations: Atlantic silverside females from the same populations differ both in the sex ratios of their progeny and in their responsiveness to temperature (Conover & Kynard 1981).

Due to these interactions, heritability values also depend on environment. Rhen et al. (2011), for instance, analysed data from a captive population of Leopard gecko (*Eublepharis macularius*) with a known pedigree. Using an 'animal model' (a general linear mixed model that makes full use of all relatedness relationships) they found a large additive genetic component in progeny sex ratios, but also strong G × E interactions. The heritability of sexual phenotypes was high at 30°C (range 0.37 to 0.44), but much smaller and not different from 0 at 32.5°C (most genotypes producing similar sex ratios at this temperature). These interactions are obviously bound to affect the long-term response of the trait to selection.

## 4.9 Response to selection

Given the high heritability values found in many polygenic systems, it comes as no surprise that sex ratios have been shown to respond quickly to selection in experimental setups (e.g. Ar-Rushdi 1958 on *T. californicus*). Selection experiments were conducted in the Nile tilapia by exposing fry to 36°C during the sensitive period (10–20 dpf) and selecting families that produced either more than 80% or less than 60% males. After three generations of selection, sex ratio reached 92.7% in the high line (with two out of 12 families producing 100% sons), versus 50.4% in the low line, corresponding to a cumulated realized heritability of 0.63 and 0.84, respectively (average 0.73). For both lines, sex ratios did not differ from 1:1 in the controls reared at constant 26°C, and there was no significant correlation between sex ratio at low and high temperature (Wessels & Hoerstgen-Schwark 2007, 2011).

In a selection experiment on Atlantic silverside, five populations from different localities were reared at either 28°C or 17°C. Due to temperature effect on sex determination, the initial proportion of females varied from 5% to 70%, depending on treatment and origin. Frequency-dependent selection for the rarer sex made all populations converge towards 1:1 sex ratios within a few generations (Conover & Van Voorhees 1990). This was accompanied in some populations by a change in TSD, which disappeared from some lineages under high temperature treatment (Conover et al. 1992). The experimental design actually eliminated the adaptive value of TSD (temperature was similar and constant for both sexes), while magnifying its major drawback (skewed sex ratios).

The rainbow trout (*Oncorhynchus mykiss*) also shows large among-family differences in sex ratios after exposure to 18°C. Magerhans & Hörsten-Schwark (2010) selected families with the most male- and female-biased progeny after temperature treatment, to produce high and low sex-ratio lines. After one generation of selection, the female percentage was 58% in the high line versus 44% in the low line, corresponding to a realized heritability of 0.63 and 0.71, respectively. Selection experiments on zebra fishes have also shown a very rapid response to selection (reaching in some families ~ 100% males within three generations), with a concomitant decrease in the coefficient of variation, due to the elimination of genetic variance during selection episodes (Liew et al. 2012). Most sex-ratio selection studies have been performed on fish, but similar responses are expected to occur in other taxa with strong G × E effects on sex determination.

Similar arguments and results apply to sequential hermaphroditism: the threshold trait that determines a sex change is often under polygenic control, with some level of genetic variance within and among populations, allowing for rapid change under selection. In the polychaete *Ophryotrocha puerilis*, individuals start off as male but turn into female later when reaching a certain size (Bacci 1965). Through artificial selection it has been possible to change the size (in terms of number of chaetigerous segments) at which the sex change occurs (Bacci & Bortesi 1961). It seems likely that lineages can be selected towards yielding pure females or males (Bacci 1978; Sella 1980), as also occurred in the transitions from monoecy to dioecy in several angiosperms (Box 3.10). Size-at-sex-change has also drastically diminished in protandrous and protogynous sea bream species, *Lithognathus mormyrus* and *Chrysoblephus puniceus* (Sparidae), following decades of heavy size-selective commercial exploitation (Mariani et al. 2013).

It may come as a surprise that systems relying on specialized sex chromosomes (with expected Mendelian inheritance of sex determination) can also respond to sex-ratio selection. The rapid spread of autosomal repressors of meiotic-drive sex chromosomes reveals a rich polymorphism of genetic factors affecting sex chromosome segregation (Atlan et al. 1997; Montchamp-Moreau et al. 2001; Box 6.1). In experimental populations of *Drosophila mediopunctata* carrying a driving Y chromosome, the proportion of males changed from 16% to 32% within 49 generations (Carvalho et al. 1998); the heritability of sex ratio was $h^2 = 0.41$, indicative of a large additive genetic variance (Varandas et al. 1997).

## 4.10 Local adaptations

Polygenic sex determination is often accompanied by genetic differentiation among populations in the threshold value of the liability trait or in the sensitivity to environmental conditions. This is true whether the sexual commitment is reversible (as in sequential hermaphrodites) or definitive (as in gonochorists): in the partially protogynous polychaete *T. prolifera*, the degree of female lability varies not only among individuals of the same population, but also among local populations (Franke 1986); in the Nile tilapia, the sensitivity of progeny sex ratio to temperature varies greatly among wild populations from different geographic origin and climatic regimes (Tessema et al. 2006; Baroiller et al. 2009).

Genetic differences have been shown to correlate with local selective pressures in several cases, suggesting a role for local adaptation. In the mosquito-parasitizing mermithid nematode *Romanomermis nielseni*, the size threshold above which individuals develop as females varies among populations and increases with average body size, itself a function of host size and quality (Blackmore & Charnov 1989).

Varying climate and seasonality also affect the patterns of ESD. The brackish water shrimp *Gammarus duebeni* shows strong ESD in northern England, where generations do not overlap. Males are produced earlier in the season (in long-day photoperiods), which allows them to benefit from a longer growth period (their mate-guarding behaviour provides high fitness benefits to large males). The longer growing seasons of southern England induce some generation overlap, erasing size differences between early- and late-born animals. The strength of ESD decreases accordingly (Naylor et al. 1988; Watt & Aams 1994). In the snapping turtle *C. serpentina*, populations from different latitudes share the same FMF pattern with a sex ratio culminating at 24.5°C, but the range of temperatures under which males are produced broadens with latitude; females at high latitude also settle their nests in more exposed and warm locations, to hasten embryogenesis (Ewert et al. 2005).

In the Atlantic silverside temperature sensitivity decreases with increasing latitude, reaching in Nova Scotia pure GSD with a few major sex-determination factors (Conover & Heins 1987b; Lagomarsino & Conover 1993). The long growing seasons of southern populations benefit females more than males, selecting for TSD. At higher latitude, however, these fitness benefits decline with the length of growing season. In addition, TSD would cause large inter-annual fluctuations in sex ratio, due to the more variable temperatures during the brief spawning season (Lagomarsino & Conover 1993). Altitude plays a similar role in the short-lived, live-bearing snow skink (*Niveoscincus ocellatus*). Lowland populations display TSD, while highland populations have GSD. Earlier birth is adaptive for females in the long growing seasons of lowland habitats, because they benefit from larger opportunities for growth before maturity: lifetime fitness is about 50% higher in early-born than in late-born females (and selection on body size is stronger in female than in male snow skinks). At higher altitudes the shorter growing season greatly reduces the effect of birth date on female fitness; in addition, the large among-year fluctuations in climate select for GSD because it prevents extreme sex ratios. A simulation model calibrated for *N. ocellatus* life history indeed predicts evolution towards TSD under the parameter values characteristic of lowland populations and towards GSD (with either male or female heterogamety) under values typical of highland populations (Pen et al. 2010).

# Colour Plates

**Plate 1** Phylogenetic distribution of sex determination systems among main eukaryote lineages. Three classification levels are used: (1) whether sex (or mating type) is determined at the haploid or diploid level, indicated by the shape of the symbol, (2) whether sex determination is genotypic or epigenetic, indicated by the central colour of the symbol, and (3) whether individual organisms are unisexual or cosexual, indicated by the colour of the symbol edge. See Figures 2.1–2.5 for further details on sexual cycles.

**Plate 2** Phylogenetic distribution of sex determination systems among insects. The main types are male heterogamety (XX-XY and XX-XO), female heterogamety (ZW-ZZ and ZO-ZZ) and haplodiploidy. The XX-XY category also contains cases of multiple X and Y chromosomes, such as frequent examples in Dermaptera. The miscellaneous category includes paternal genome loss (e.g. in some Hemiptera), X chromosome elimination (e.g. in some Diptera), environmental effects, and other rare sex determination variants. There is one record of an XY system in Strepsiptera. The phylogenetic relationships reflect current knowledge but are not unalterable as the phylogeny of insects is still not completely resolved. Branch lengths have no phylogenetic meaning and are solely chosen for layout purposes.

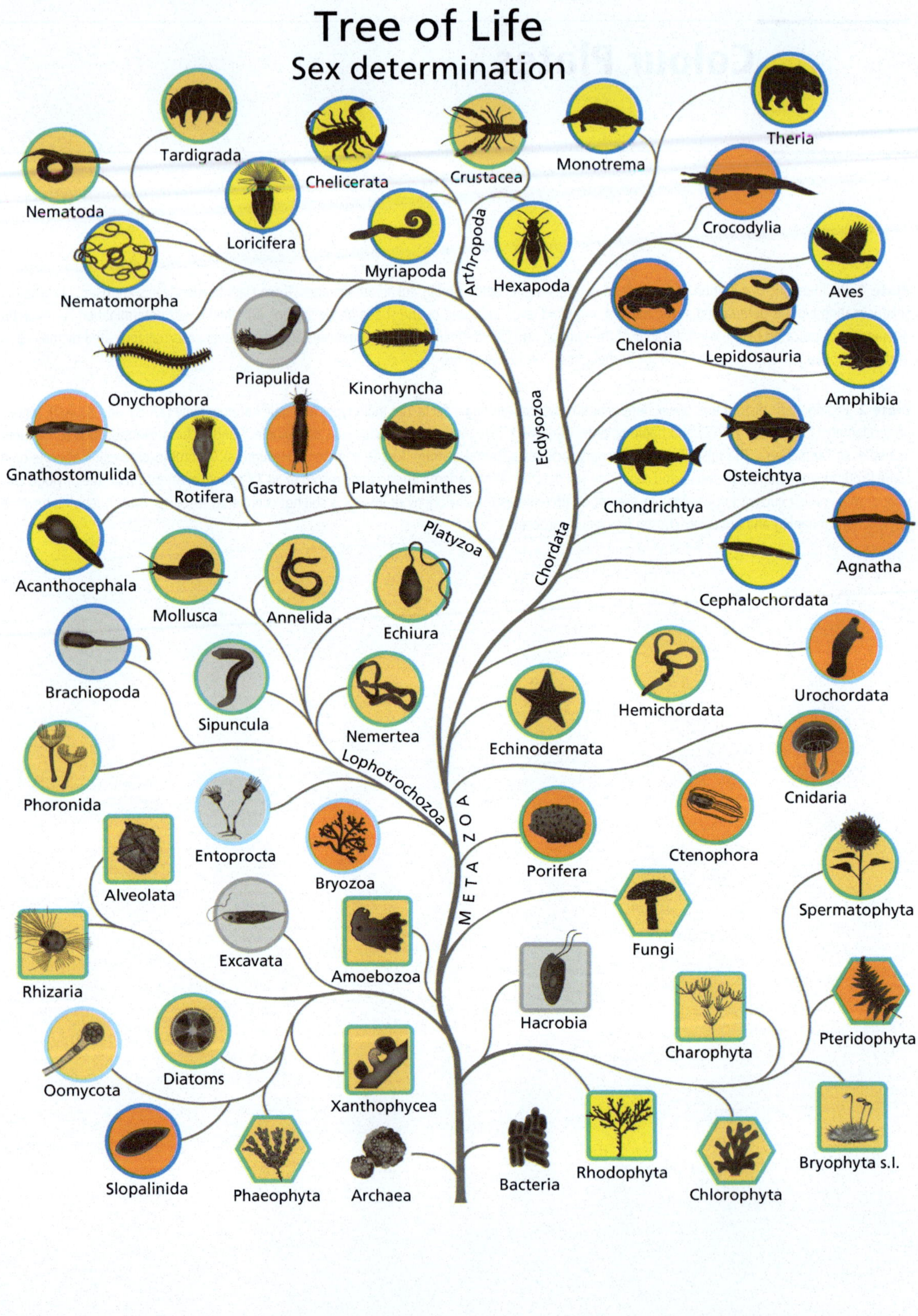

Tree of Life
Sex determination
Theria
Monotrema
Crustacea
Chelicerata
Tardigrada
Nematoda
Loricifera
Myriapoda
Arthropoda
Hexapoda
Crocodylia
Aves
Nematomorpha
Chelonia
Lepidosauria
Amphibia
Onychophora
Priapulida
Kinorhyncha
Ecdysozoa
Osteichtya
Gnathostomulida
Rotifera
Gastrotricha
Platyhelminthes
Chondrichtya
Platyzoa
Chordata
Agnatha
Acanthocephala
Mollusca
Annelida
Echiura
Cephalochordata
Brachiopoda
Urochordata
Hemichordata
Sipuncula
Nemertea
Echinodermata
Lophotrochozoa
Phoronida
Cnidaria
Entoprocta
Ctenophora
Alveolata
Bryozoa
Porifera
METAZOA
Spermatophyta
Fungi
Excavata
Rhizaria
Amoebozoa
Hacrobia
Pteridophyta
Charophyta
Oomycota
Diatoms
Xanthophycea
Bryophyta s.l.
Rhodophyta
Slopalinida
Phaeophyta
Archaea
Bacteria
Chlorophyta

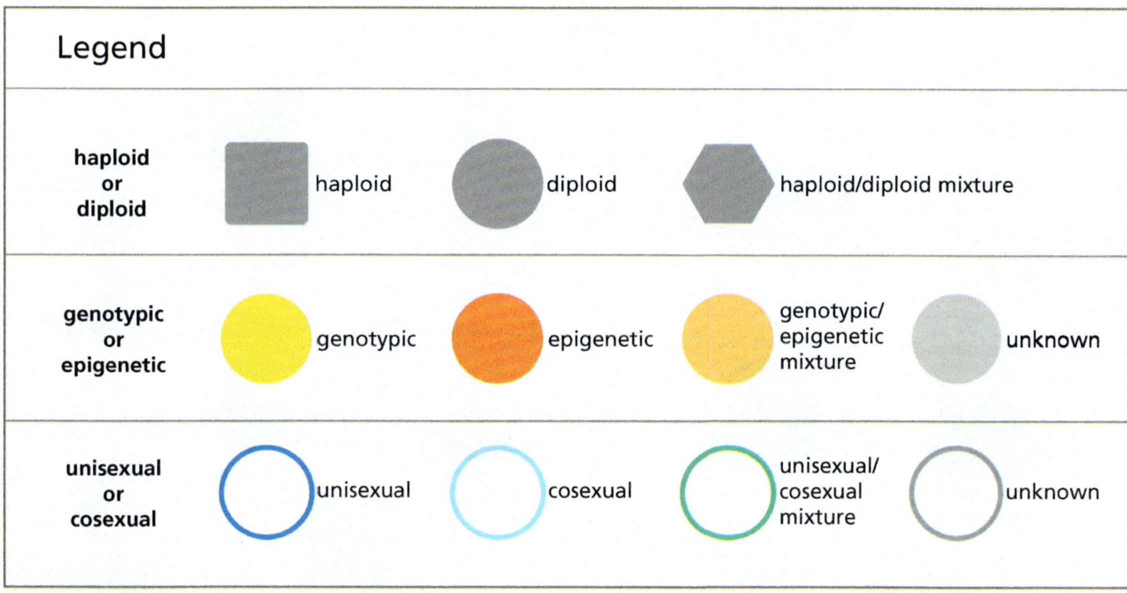
Legend
haploid or diploid
haploid
diploid
haploid/diploid mixture
genotypic or epigenetic
genotypic
epigenetic
genotypic/ epigenetic mixture
unknown
unisexual or cosexual
unisexual
cosexual
unisexual/ cosexual mixture
unknown

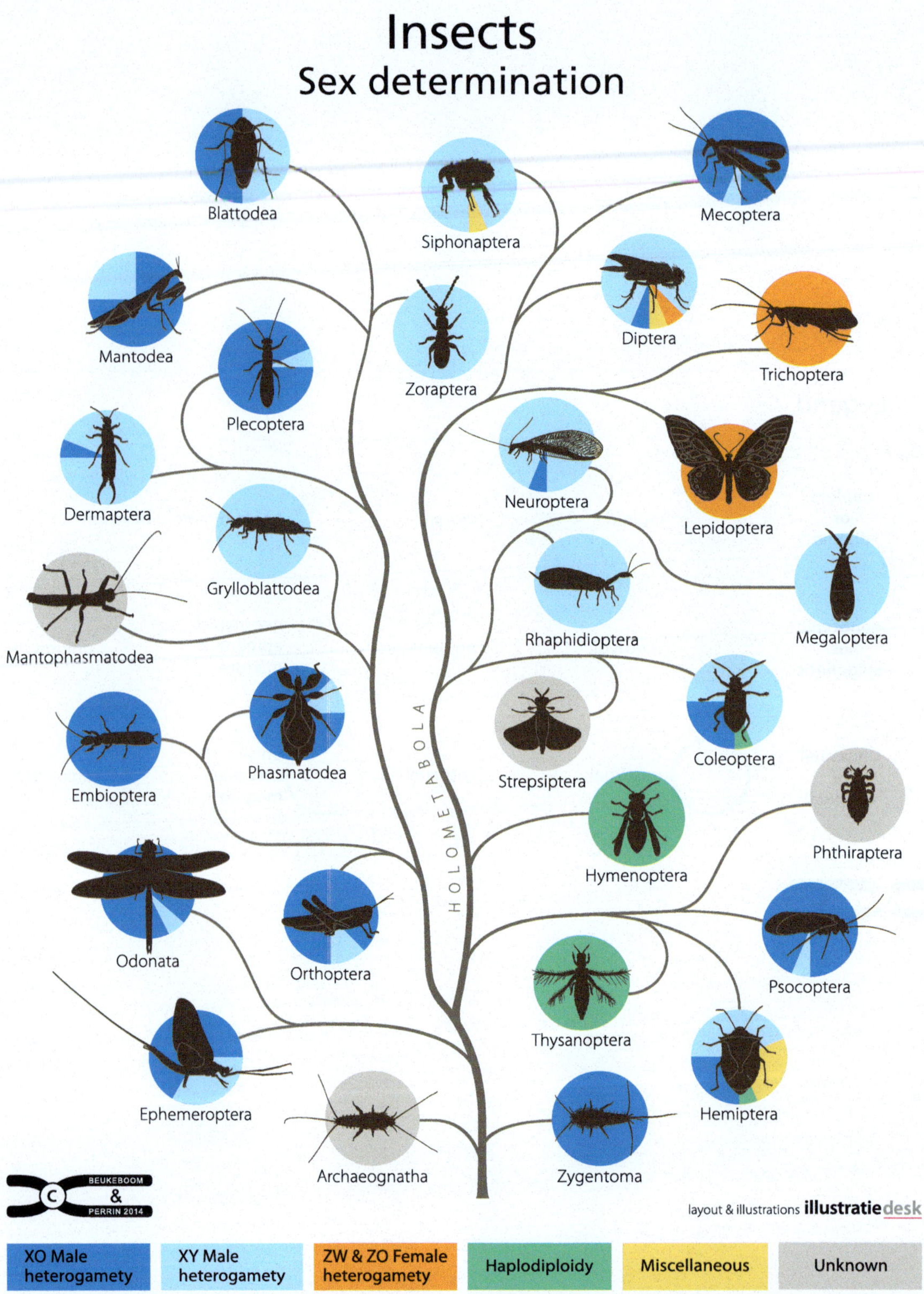
Insects
Sex determination
Blattodea
Siphonaptera
Mecoptera
Mantodea
Diptera
Trichoptera
Plecoptera
Zoraptera
Dermaptera
Neuroptera
Lepidoptera
Grylloblattodea
Mantophasmatodea
Rhaphidioptera
Megaloptera
Phasmatodea
Coleoptera
Embioptera
Strepsiptera
HOLOMETABOLA
Phthiraptera
Hymenoptera
Odonata
Orthoptera
Psocoptera
Thysanoptera
Ephemeroptera
Hemiptera
Archaeognatha
Zygentoma
BEUKEBOOM
&
PERRIN 2014
layout & illustrations illustratiedesk
XO Male heterogamety
XY Male heterogamety
ZW & ZO Female heterogamety
Haplodiploidy
Miscellaneous
Unknown

CHAPTER 5

# The evolution of sex chromosomes

The chromosomes carrying the genes that determine sexes or mating types may show strong morphological differentiation. **Heteromorphic** sex chromosomes were first documented in lineages with diplo-genotypic sex determination (D-GSD), where one member of the pair may even be lacking (as in XX/X0 systems). The German biologist Henking (1891) was the first to note that one element in male meiosis of the firebug (*Pyrrhocoris apterus*) was only transmitted to half of the sperm. He referred to this element as 'X' for unknown. McClung (1902), Sutton (1902), and Stevens (1905) postulated this X element (or 'accessory chromosome') to be connected to sex determination. At about the same time, E. B. Wilson, working with several insect species, noticed that, although male and female cells had equal numbers of chromosomes, one pair was heteromorphic in males, and called this mismatched pair 'idiochromosomes' (Wilson 1905). The involvement of X and Y chromosomes in sex determination was later supported by the work of Bridges (1916, 1925) who analysed *Drosophila* individuals with aberrant constitutions of sex chromosomes (Box 3.18).

Under D-GSD, sex chromosomes are always homomorphic in the so-called **homogametic** sex (e.g. XX females in mammals, ZZ males in birds); differentiated chromosomes are only found in the **heterogametic** sex (XY males in mammals and *Drosophila*; ZW females in birds and snakes). The differential chromosome (Y or W) may be much degenerated and gene poor: in humans, the small Y chromosome only encodes 27 distinct proteins, as compared to > 1000 for the large X chromosome (Skaletsky et al. 2003). Differentiated sex chromosomes also occur under haplo-genotypic sex determination (H-GSD), as in several fungi, algae, and mosses (e.g. Fraser & Heitman 2004). In these systems, the male and female chromosomes (referred to as V and U respectively; Bachtrog et al. 2011) characterize the male and female gametophytes. U and V chromosomes only co-occur in the sporophyte and might be quite divergent: in the liverwort *Marchantia polymorpha*, the female **U chromosome** is twice as large as the male V chromosomes (20 Mb versus 10 Mb), and genetically very distinct (Tanurdzic & Banks 2004). The differential segment of U and V chromosomes may also be variable in size: 1 Mb and ~25% of the sex chromosome in *M. violaceum* (Votintseva & Filatov 2009), at least 6.6 Mb in *Neurospora tetrasperma* (Menkis et al. 2008), and up to 34% of the genome in the moss *Ceratodon purpureus* (McDaniel et al. 2007).

How sex chromosomes came into being captured the attention of the earliest evolutionary biologists. Following Muller (1914, 1918) and Ohno (1967), it is now generally accepted that all sex chromosomes started as an ordinary pair of autosomes, which at some point gained a major sex-determination function by the mutation, translocation, or duplication of gene(s) involved in the sex-determining pathway (Chapter 3). For reasons presented below, these undifferentiated **proto-sex chromosomes** progressively stopped recombining in the heterogametic sex (D-GSD) or in the sporophyte (H-GSD). This arrest had dramatic consequences on genome evolution, inducing the degeneration of the non-recombining parts of chromosomes with ensuing fitness effects on the heterogametic sex, but also affecting the gene content of recombining sex chromosomes. Other indirect consequences include the evolution of dosage compensation in some lineages and high rates of turnovers in others (Chapter 7).

## 5.1 Recombination arrest

Recombination arrest appears as a very general initial step in the evolution of sex chromosomes, taken early after the acquisition of the sex-determining function (Fig. 5.1). At start, it may only affect a very limited segment (the sex-determining region; SDR), but may later expand (or not) over larger portions of the chromosome. In many strongly diverged sex chromosomes, the whole chromosome does not recombine, except for a small **pseudo-autosomal region** (PAR). In so-called **achiasmate** meiosis, recombination is completely aborted in the heterogametic sex (e.g. *Drosophila*, *Bombyx*). Given the dramatic consequences on genome evolution, the reasons why recombination stopped, and the mechanisms involved, are important issues to address.

### 5.1.1 Ultimate causes

The ultimate causes for recombination arrest are generally thought to stem from epistatic interactions between two or more genes involved in sex determination and/or differentiation. Whenever the expression of fully functional male or female phenotypes requires the combined action of different alleles from different genes, then selection will favour linkage of these alleles into a 'supergene', ensuring sex-specific expression of epistatic effects. This explanation for linkage between genes goes back to Dobzhansky's co-adapted gene complexes hypothesis (Dobzhansky & Dobzhansky 1970) and applies to the evolution of non-recombination on any genomic region (not only sex chromosomes), as illustrated by the recombination arrest at the S locus controlling self-incompatibility in plants, or by the parallel evolution of Y-like, non-recombining social chromosomes in fire ants (Wang et al. 2013).

Epistatic interactions may exist between genes directly involved in the determination of sexes or mating types. These genes might, for example, simultaneously affect the liability factor that underlies the development of sex phenotypes (Chapter 4). Their combined action avoids the production of individuals with liability values close to the threshold, likely to develop as **intersexes** or hermaphrodites; linkage thus reinforces the canalized developmental pathways leading to either sex (Nei 1969; Bull 1983). Alternatively, they might have distinct and complementary functions along the cascade, or specify different aspects of sex- or mating-type determination. This occurred recurrently in the evolution of fungal U and V chromosomes. Basidiomycetes originally possess two independent MAT loci, one encoding pheromones and receptors involved in cell–cell recognition, and the other transcription factors controlling the developmental pathways along the sexual cycle (Sections 2.2.8; 3.2.2c). As a result, hyphal fusion may take place between individuals that later reveal incompatible. Linkage of the two MAT loci occurred in several parasitic species, possibly triggered by the limitation of potential partners (50% of matings are compatible in bipolar systems, as opposed to 25%

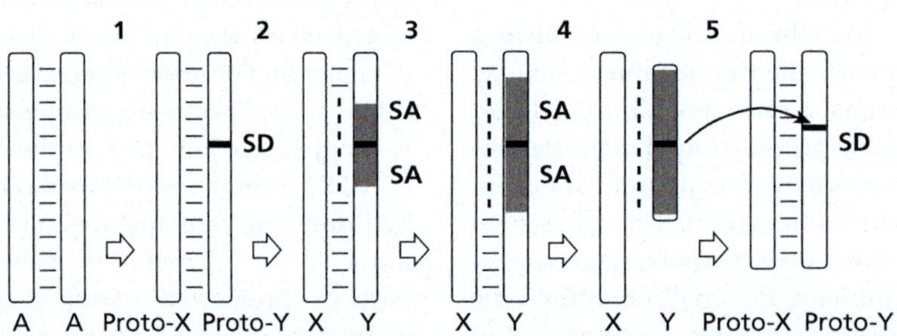


**Figure 5.1** Steps in the evolution of sex chromosomes from a pair of autosomes. (1) Acquisition of a sex-determination gene; (2) acquisition of sexually antagonistic genes and arrest of recombination; (3) extension of the non-recombining region; (4) differentiation and degeneration of the Y through accumulation of deleterious mutations and transposable elements; (5) hijacking of sex determination by a SD mutation on a new autosomal pair; the ancestral Y is lost and the ancestral X is fixed as an autosome. SD, sex-determination locus; SA, sexually antagonistic locus; horizontal dashes indicate occurrence of recombination and vertical dashes absence of recombination.

in tetrapolar systems). The resulting segment spans over 100 kb in *Cryptococcus neoformans*, with more than 20 genes trapped in the non-recombining region (Lengeler et al. 2002). In the dipolar *Ustilago hordei* (a close relative of the tetrapolar *U. maydis*), the non-recombining region encompasses ~500 kb, with the a locus at one end and the b locus at the other end (Lee et al. 1999). In the anther smut fungus (*Microbotryum violaceum*), this event was followed by further expansion of the non-recombining region, which now spans about 1 Mb (~25% of the chromosome). Complementary sex-determination genes may also induce recombination arrest in diplontic lineages: when dioecy evolves from hermaphroditism (as occurred in several lineages of angiosperms), two mutations at least are required, one inducing male-sterility in females, the other female-sterility in males (Charlesworth & Charlesworth 1978; Section 3.3.1 and Box 3.10; Fig. 5.2). Recombination must be stopped to prevent the production of sterile neuter individuals, such as found in *Fragaria virginiana* (Spigler et al. 2008).

Recombination arrest may also foster linkage between the sex locus and other genes involved in the differentiation of sex phenotypes. An example in H-GSD systems would be genes affecting gamete size in *Volvox carteri*, such as *Mat3*, which is autosomal in *Chlamydomonas*. Its linkage to *Mid* contributed to the expansion of the U and V chromosome in *Volvox*, trapping series of other genes in the non-recombining region (Ferris et al. 2010). In D-GSD systems, an example would be the human *Zfy* gene, a *Sry*-linked testis-specific transcription factor, previously thought to be the sex determiner (Decarpentrie et al. 2012). In fact, most of the genes now found on the therian Y chromosome are male-fertility genes; XX individuals that gain *Sry* by XY recombination develop as sterile males. Similarly, XO *Drosophila* individuals develop as sterile males, lacking important male-fertility genes on the Y. These are, however, extreme cases; genes do not need to be essential for the fertility of the heterogametic sex for a recombination arrest to be favoured. The notion of **sexually antagonistic** (SA) genes refers to genes carrying alleles that are beneficial to one sex, but detrimental to the other. A male-beneficial but female-detrimental mutation appearing on autosomes (i.e. transmitted with equal frequencies to sons and daughters) should only spread if the benefits to sons exceed the costs to daughters. If, however, this mutation appears on the Y, close to the sex locus, it will be transmitted to sons more often than to daughters, and thus has a chance to get fixed, even if the costs to females exceed the benefit to sons; at 1 cM from the sex-determining (SD) locus, it will be transmitted 99% of time to sons, and is thus likely to spread even if strongly detrimental to females.

Thus, chromosomal regions located within a few cM of the SDR should constitute hot spots for the accumulation of SA genes, as Fisher (1931) first noticed by commenting on the sex-linked colouration genes in guppies. Bull (1983) further developed the concept, and Rice (1987) formalized the population genetic dynamics of the process, showing that this accumulation depends on the effect strength of SA genes, and can occur under rather unrestrictive linkage requirements. Theoretically, a male-beneficial

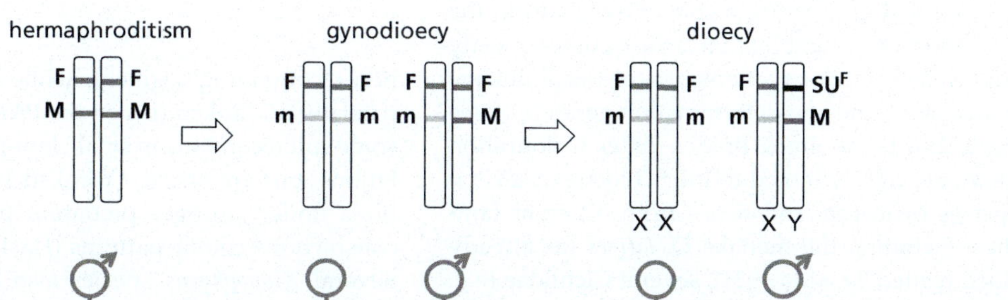


**Figure 5.2** Steps in the evolution of male heterogamety from a hermaphroditic state. A pair of autosomes carries a female (F) and a male (M) fertility locus. A first mutation towards a recessive male-sterility allele (M -> m) results in gynodioecy. Then a mutation towards a dominant female suppressor (F -> $SU^F$) results in dioecy. Recombination between these two genes is suppressed, because mm-$SU^F$ recombinants are sterile.

allele may spread even if lethal to females, provided its recombination with the SD locus is sufficiently rare (Rice 1987). Note that, for the same reason, X chromosomes are also hot spots for the accumulation of sex-antagonistic mutations, but with more complex dynamics (Section 5.2.3).

This accumulation of genes with sex-specific effects generates a selective pressure to suppress recombination in the SDR, and thereby to stop entirely the transmission of male-beneficial alleles to females and vice versa (Rice 1996; Charlesworth & Charlesworth 2000). The process is self-reinforcing: tighter linkage between the SA and sex-determination genes will select for stronger SA alleles, and vice versa. Furthermore, once a non-recombining segment is established, additional SA genes will be later favoured at its border, benefiting from tighter linkage with the SD locus. A 'genetic chain reaction' will thus cause further expansion of the non-recombining segment, which may ultimately encompass the whole chromosome (except for small PARs that might be required for proper segregation during male meiosis; Section 5.1.4). This scenario is supported by the evidence for **evolutionary strata** of genetic divergence between gametologues in mammals (Box 5.1, Fig. 5.3; Lahn & Page 1999), birds (Handley et al. 2004), and plants (Nicolas et al. 2005), testifying to a stepwise expansion. Evolutionary strata have also been identified on U and V chromosomes (Section 5.2.1). In H-GSD systems, SA alleles in either sex select for suppressed recombination, which might favour more rapid expansion of the SDR than in diploid systems, where only alleles that favour the heterogametic sex drive the process (Bachtrog et al. 2011).

Empirical support for a role of SA genes in the evolution of recombination arrest comes mostly from recurrent observations of a physical linkage between SA and sex-determination genes (e.g. see Table 2 in Lindholm & Breden 2002). Colouration genes are tightly linked to the SD locus in several species of insects and fishes from different families, including the medaka *O. latipes* (as already documented by Aida 1921), several Cichlidae (e.g. Roberts et al. 2009), and a number of Poeciliidae, such as swordtails (Kallman 1970, 1975) and guppies (Fisher 1931; Endler 1980; Postma et al. 2011). In guppies, some colouration genes are on the non-recombining segment, while others are only partially sex-linked (i.e. on the **PAR**); furthermore, some still recombine in small, low-predation populations, but are strictly Y-linked in large populations under stronger predation pressure against conspicuous colour patterns (Haskins et al. 1961). Several *Xiphophorus* species also have a P locus (for *Pituitary*) on their Y chromosome that determine age and size at maturity, a trait under strong sex-antagonistic selection (Morris et al. 1992; Ryan et al. 1992).

**Box 5.1 Evolutionary strata on sex chromosomes**

Comparing nucleotide sequences between human X and Y chromosomes, Lahn & Page (1999) identified four discrete regions that differed in their amount of divergence. They proposed that the process of differentiation between gametologues involved four successive events of recombination arrest (Fig. 5.3), each event adding a chromosomal segment to the non-recombining region, resulting in a new evolutionary stratum. The oldest stratum (number 1, at the distal end of the long arm) might have arisen soon after the origin of the therian sex chromosomes, and stratum 2 before the split between eutherians and marsupials. Stratum 3 came into being after the split from marsupials (148 Mya) but before the eutherian radiation, and stratum 4 after the divergence of prosimian and simian primates (30–50 Mya). Strata 1 + 2 correspond to the conserved XCR/YCR region (common to all therians) and strata 3 + 4 to the added XAR/YAR region (autosomal in marsupials; Graves 1995); accordingly, the YAR contains the majority of functional genes on the Y, and XAR is the least dosage compensated. Details of the model have been questioned or altered in subsequent publications (e.g. Skaletsky et al. 2003; Katsura & Satta 2012) and a fifth segment was later identified at the border with the PAR on the short arm (Ross et al. 2005). While the ordering of strata is well preserved on the X chromosome, numerous structural changes and inversions took place over the Y chromosome, scrambling gene order; *Sry*, which was recruited on the oldest stratum (*Sox3* belongs to stratum 1 on the X) is now located at the edge of the PAR (Fig. 5.3). Similar evolutionary strata, with similar interpretation, have been found in other XY, ZW, or UV systems (e.g. Handley et al. 2004; Nicolas et al. 2005; Menkis et al. 2008).

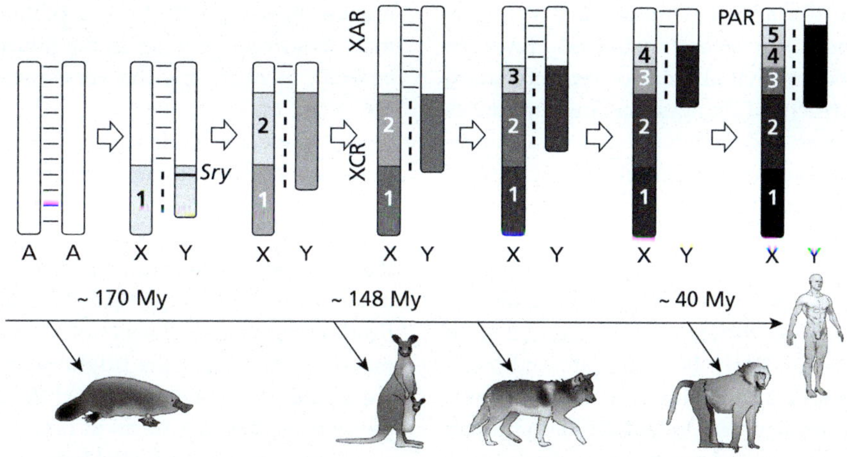


**Figure 5.3** Evolutionary strata on the human Y chromosome. The *Sry* mutation occurred on an autosomal pair shortly after the divergence between monotremes and therians (~170 Mya). Recombination stopped in a stepwise way, creating evolutionary strata 1 and 2. A **neo-sex chromosome** was then formed in eutherians by addition of an autosomal genomic region (the X-added region; XAR) shortly after the divergence from marsupials (~148 Mya). This newly added region then progressively stopped recombining in several steps, the last one having occurred after the split between apes and monkeys (~40 Mya). Recombination is only maintained in the PAR. Horizontal dashes indicate presence of recombination, vertical dashes absence of recombination. Progressive degeneration is indicates with darkening shades. The small PAR on the other Y tip is not shown (after Lahn & Page 1999 and Hughes et al. 2012, modified from Charlesworth & Mank 2010).

A causal role for SA gene accumulation in the recombination loss of the SDR seems widely accepted. However, unequivocal evidence is still limited: genomic evidence often suggests an autosomal location (with steroid receptor regulation) for sexually selected traits (Mank 2009); many sexually dimorphic fishes, for instance, are sequential hermaphrodites, that is, they lack any form of sex chromosomes. Reciprocally, SA genes on sex chromosomes may also have accumulated after recombination arrest. As argued by Ironside (2010), other processes that have been shown to affect recombination in autosomes may also act on sex chromosomes: inversions may spread through genetic drift in small populations or be favoured via heterozygote advantage (Charlesworth & Wall 1999; Hoffmann & Rieseberg 2008). An inversion appearing on an autosome will ultimately get either fixed or eliminated; in any case, recombination arrest will be transient. If however an inversion including the SD locus gets fixed by chance, then the recombination arrest will be definitive in the heterogametic sex. A neutral model might thus suffice to account for the observed patterns. The role of SA genes may also be questioned in species with H-GSD, which usually show little or no differentiation between sexes or mating types. In the liverwort *M. polymorpha*, the male and female gametophytes only differ in sexual organs (antheridia versus **archegonia**), but their sex chromosomes are much diverged (Tanurdzic & Banks 2004). Although there is no doubt that SA genes are key players in the dynamics of sex chromosomes (Chapter 7), their exact role is clearly in need of further empirical investigations.

### 5.1.2 Proximate mechanisms

Highly differentiated sex chromosomes are characterized, both in D-GSD and H-GSD systems, by many structural differences that include **transposons**, repetitive sequences, and inversions. These structural changes, in particular inversions, have often been assumed to constitute the main mechanism that initiates recombination arrest; this idea dates back to Ohno's (1967) suggestion that sex chromosome heteromorphy might be established by a **pericentric inversion**. There is no doubt that an inversion event immediately suppresses recombination in heterozygotes, or, at least, dramatically reduces it in the case of large inversions

(Kirkpatrick 2010), and that inversions were certainly instrumental in several instances. However, the common notion that absence of recombination derives from structural changes may mix up causes and consequences: most of these structural differences may have actually accumulated because of non-recombination.

Structural differences are not required to induce sex differences in recombination. Autosomes, which are shared by both sexes, also show **heterochiasmy,** that is, sex differences in recombination patterns (Burt et al. 1991). Both the rate and localization of crossovers differ between sexes (Lenormand 2003; Lenormand & Dutheil 2005). In human females, for instance, the overall rate is about twice that of males, but males exhibit significantly more recombination in telomeric regions. The ratio of female-to-male autosomal recombination spans close to three orders of magnitude in animals, from 0.14 in the Japanese flounder (Coimbra et al. 2003) to 70 in the common frog (Rodrigues et al. 2013), not to mention **achiasmate** species such as found in Diptera (e.g. *D. melanogaster*; Morgan 1914) or Lepidoptera (e.g. the silkworm *Bombyx mori*; Tanaka 1914), in which one sex does not recombine at all. Which sex recombines more varies among taxa and loci, but heterochiasmy itself is a pervasive feature among gonochoric species, including those with environmental sex determination: females are genetically identical to males in salt-water crocodiles (*Crocodylus porosus*), but display distinct recombination patterns (Isberg et al. 2006).

Sex reversal experiments confirm this dependence of recombination patterns on phenotypic sex rather than on structural genomic differences: sex-reversed XY females in amphibians and fishes display the characteristic recombination patterns of XX females, quite different from those of XY males (e.g. Wallace et al. 1997; Kondo et al. 2001). XX males also display male-specific **synaptonemal complexes** and recombination rates (e.g. Campos-Ramos et al. 2009; Matsuba et al. 2010). The same appears to be true of autosomal recombination in mammals: the rate and patterns of recombination in the oocytes of sex-reversed XY female mice are virtually identical to those of normal XX females; sex, not genotype, is thus the primary determinant of meiotic recombination patterns (Lynn et al. 2005). Sex-reversed XY females have actually been exploited for mapping male-determining genes in the housefly (*Musca domestica*), to overcome the very low recombination rate in XY males (Inoue et al. 1983).

The same dependence on phenotypic sex is found on young sex chromosomes. As already noted, colouration genes are tightly linked to the sex-determination gene in medaka fish; in the d-rR strain, the dominant allele *R* (orange-red) is fixed on the Y chromosome, while the recessive allele *r* (white) is fixed on the X, with absence of XY recombination in males, so that males are bright orange and females are white. In sex-reversed $X^rY^R$ females, however, the two loci recombine as in normal $X^rX^r$ females, resulting in orange-red $X^rX^R$ females and white $X^rY^r$ males among their progeny (Yamamoto 1961). Matsuda et al. (1999) used additional sex-linked markers to confirm that sex-reversed XY females display female-specific recombination patterns on the sex chromosomes, and further showed that sex-reversed XX males (as well as YY males produced by mating sex-reversed XY females with normal XY males) display the male-specific pattern of restricted recombination. Thus, recombination arrest in males does not result from heterogametic sex chromosomes, but from maleness (Matsuda et al. 1999). Many young sex chromosomes actually show strongly reduced or completely arrested recombination in males without any visible structural change (e.g. Schmid et al. 1991). Despite complete homomorphy, XY recombination is absent in European tree frog males (*Hyla arborea* group), which is likely under epigenetic control. Homomorphy is maintained by occasional X-Y recombination, possibly in naturally occurring sex-reversed XY females (Stöck et al. 2011; Guerrero et al. 2012). Thus, the independence of recombination arrest from structural differences in homologous chromosomes argues for sex-specificity of gene expression (rather than pre-existing heteromorphism) as a driving force.

In more differentiated sex chromosomes, structural changes may have accumulated to such a point that recombination is not possible anymore in sex-reversed XY females. However, several detailed studies provide evidence that structural changes followed the cessation of recombination, rather than causing it. From their study of rearrangements on the *Silene latifolia* Y chromosome,

Bergero et al. (2008) concluded that the Y chromosome has been derived through multiple rearrangements, but that none of those detected so far was involved in stopping X-Y recombination. In human sex chromosomes, the *Amelogenin* gene, situated at the boundary between evolutionary strata 3 and 4 (Fig. 5.3), testifies to a progressive expansion of the non-recombining segment. Its 5′ end stopped recombining before the split between primates and ungulates, while its 3′ end stopped after the split between apes and monkeys (Iwase et al. 2003). Inversions were not involved in this process, since none of the species investigated display a truncated Y copy. Thus, the suppression of X-Y recombination proceeded by another, as yet unidentified mechanism (Marais & Galtier 2003).

Recombination rates are known to vary among individuals within populations and to respond rapidly to selection (Chinnici 1971; Kidwell 1972; Charlesworth & Charlesworth 1985; Brooks & Marks 1986; Brooks 1988). Chinnici (1971) selected *Drosophila* lines for the recombination rate between two genes on the X chromosome; after 33 generations, this rate evolved from 15.4% in the initial stock to 8.5% in the low line and 22.1% in the high line. The response was highly site specific (adjacent locations were not affected), polygenic, and not due to inversions or any structural change. Brooks & Marks (1986) showed that different chromosome-2 haplotypes, put in a common genomic background, affect the amount and distribution of recombination differently, both on itself and on other chromosomes. Genetic modifiers of recombination seem thus widespread on the genome, and sometimes very specific in their action. The gene *Prdm9* was recently identified as a major determinant of recombination hotspots in mammals (Baudat et al. 2010; Parvanov et al. 2010). Binding of PRDM9 to specific DNA sequences initiates recombination at specific locations in the genome. These target sequences have high mutation rates and evolve rapidly (Myers et al. 2010), which may account for the rapid response to selection.

Such target sequences may offer a basis for the evolution of sex-specific recombination rates on differentiated sex chromosomes, but not on autosomes or homomorphic sex chromosomes; this requires epigenetic modifiers of recombination. Male and female meioses constitute drastically distinct physiological processes (Tease & Hulten 2004), which creates a large scope for sex differences in recombination patterns. Asynchrony in the replicating pattern of Y or W, for instance, often predates their structural differentiation (e.g. Schempp & Schmid 1981). Comparative work in snakes, where the same pair of ZW chromosomes range from complete homomorphy in Boidae to strong heteromorphy (with deletions and inversions) in Crotalidae and Viperidae, suggests that the development of **allocycly**, rather than the establishment of structural changes, is a likely first step in Z-W differentiation (Ray-Chaudhuri et al. 1971). The patterns of chromatin or DNA methylation and imprinting also differ between sexes: sex-specific configuration of the chromatin may facilitate or prevent the action of genetic modifiers such as PRDM9; imprinted regions are known to display large sex differences in recombination rates (Smalley 1993; Robinson & Lalande 1995; Paldi et al. 1995; Lercher & Hurst 2003).

Structural changes were certainly instrumental in stopping recombination in some nascent sex chromosomes and, clearly, further enforce recombination arrest in differentiated ones. As the appearance of repetitive sequences, transposons, and inversions goes hand in hand with the reduction of recombination, it is often hard to distinguish cause and effects; most likely both processes reinforce each other once the process of degeneration is initiated. However, the potential role of structural changes as initial causes of non-recombination might have been overemphasized. This point certainly deserves greater empirical scrutiny.

### 5.1.3 Neo-sex chromosomes

Neo-sex chromosomes (as opposed to **proto-sex chromosomes**) occur by fusion of autosomes to existing sex chromosomes (Fig. 5.4). Regular autosomal additions contribute to the evolutionary strata documented on sex chromosomes: recently added regions display features of incipient sex chromosomes, being much less differentiated than more ancestral regions. The short arm of the eutherian X chromosomes (the X-added region), for instance, is autosomal in marsupials (Graves 1995); fusion with the long arm (the X-conserved region) occurred

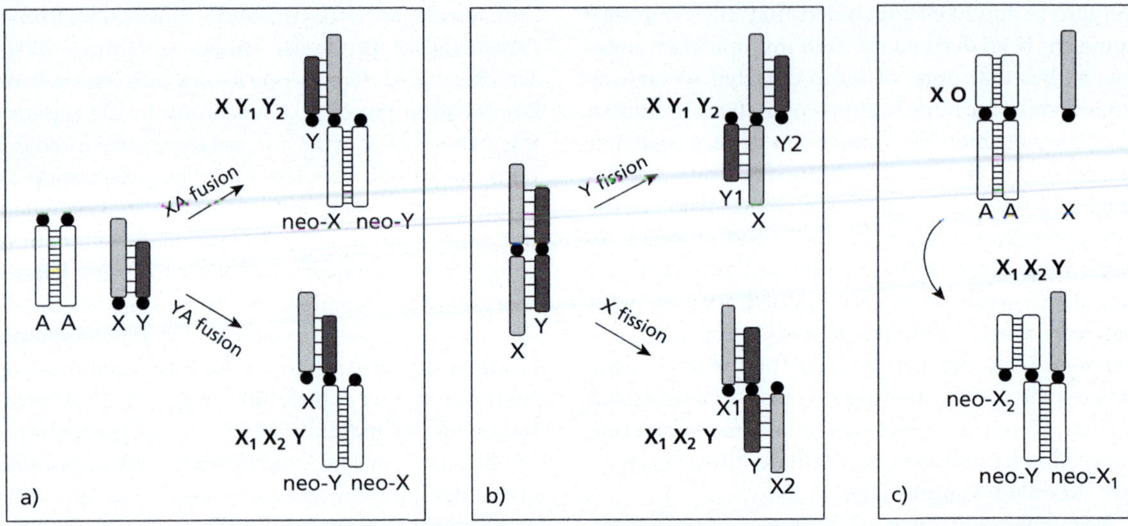


**Figure 5.4** Mechanisms for the formation of neo-sex chromosomes and multiple sex chromosome systems. (a) From a simple XY system, fusion of an X or Y chromosome with an autosome creates an $XY_1Y_2$ or $X_1X_2Y$ system respectively. (b) Centromeric fission of an X or Y chromosome creates a $X_1X_2Y$ or a $XY_1Y_2$ system respectively. (c) A reciprocal translocation or Robertsonian fusion (as shown here) between an X and an autosomal arm in an XO system creates a $X_1X_2Y$ system. Horizontal connections between chromosomal arms represent regions of homology and pairing during meiosis (after White 1973, modified from Yoshida & Kitano 2012 and Kitano & Peichel 2012).

after the split between the two therian clades (148 Mya) but before divergence of the main eutherian lineages (105 Mya; Box 5.1). Its more recent origin translates into lower X-Y differentiation and less advanced dosage compensation (Section 5.2.2e). Autosomal fusions also occurred recurrently throughout *Drosophila* phylogeny, providing useful tools to compare sex-linked regions of different ages. Note that neo-sex chromosomes stop recombining immediately after fusion in achiasmate species such as *Drosophila*.

**Multiple sex chromosomes** result if the fusion only occurs with one of the bivalents, either the X or the Y chromosome. Fusion with the X (but not the Y) produces an $XY_1Y_2$ system, while fusion with the Y (but not the X) produces an $X_1X_2Y$ system. Note that multiple systems may also result from sex chromosome fissions (Fig. 5.4b), but these seem less frequent. Autosomal fusions with the Y chromosome are predominant in fishes (Kitano & Peichel 2012), while fusions with X or Y are equally common in mammals (Yoshida & Kitano 2012). Fusions may involve more than two chromosome pairs, resulting in meiotic **multivalents**, such as are found in some plants (e.g. mistletoe *Viscum*; Wiens & Barlow 1975), invertebrates (e.g. spider *Delena*, Rowell 1991) and vertebrates. The duck-billed platypus (*Ornithorhynchus anatinus*) has five pairs of sex chromosomes that form a 10-chromosome multivalent during male meiosis, with the five X and five Y chromosomes segregating alternatively. The system probably arose by a chain of **Robertsonian translocation** events (Gruetzner et al. 2006).

What selective forces drive chromosome fusions? Charlesworth & Charlesworth (1980) noted that a sexually antagonistic mutation occurring on an autosome would benefit from a linkage with the sex-determination genes; the same selective forces that favour recombination arrest in sex chromosomes might thus be at work. In the Japan Sea populations of sticklebacks (*Gasterosteus aculeatus*), for instance, variation in dorsal spine length (a sexually dimorphic trait) maps to the neo-X chromosome, and thus might have selected for the fusion (Kitano et al. 2009). Similarly, the fusion of a part of chromosome 4 with the Z in Sylvioidea might have been favoured by bringing the *AR* (*androgen receptor*) gene into linkage with *Dmrt1* (Pala et al. 2012). Alternatively, the fusion might be promoted through heterozygote advantage, by enforcing heterozygosity in males (Charlesworth & Wall 1999). Both processes are expected to favour Y- over X-autosomal fusions, and might thus account for the predominant $X_1X_2Y$ systems found in fishes. In the case of mammals, alternative forces must be invoked to account for the high rate of $XY_1Y_2$ systems. Following De Villena & Sapienza (2001a,b), Yoshida & Kitano (2012) proposed a role for female **meiotic drive**: the preferential transmission of fused chromosomes with a single telomere to the egg (versus unfused chromosomes with two telomeres to the polar body) should favour X-autosomal fusions.

### 5.1.4 Pseudo-autosomal regions

The recombining parts of sex chromosomes are referred to as PARs, because they share some features with autosomes (recombination), although being still partly sex linked. The levels of recombination are usually higher in PARs than in the background genome, and higher in males than in females in XY systems (being the only region where X and Y can recombine). PARs vary widely in size; they tend to be large in young sex chromosomes, and small or absent in older ones (such as in therians), in line with a model of progressive expansion of the non-recombining segment; but many exceptions occur, with large PARs in very old sex chromosomes (Otto et al. 2011). Human sex chromosomes have two PARs, at each end of the SDR, which together make up less than 5% of the Y chromosome; PAR2 is a recent addition, specific to humans, while PAR1 is shared with most eutherians, added shortly after the split from marsupials. Its exact boundary is evolutionary labile and differs between lineages, due to differential expansion of the SDR. Mice have one very small PAR, of recent origin, the size and genetic content of which varies among strains. Marsupials have no PAR; their Y chromosome is entirely achiasmate.

The reasons why PARs exist are unclear, and possibly multiple. History may play a role: the SDR might be still expanding and has not yet had enough time to encompass the whole sex chromosome. Alternatively, PARs may persist in a dynamic state, by regular translocation of new genomic regions (neo-sex chromosomes), which then decay (Fig. 5.3). They might also be maintained over evolutionary times by high intrinsic rates of recombination (so that effective linkage with sex is weak enough to stop the genetic chain reaction) or by positive selection. It has for example been suggested that crossovers are required for proper meiotic segregation (but achiasmate sex chromosomes do occur in marsupials, *Drosophila*, and other insects). The specific dynamics of PARs, and the analysis of evolutionary forces acting on them, remain promising areas of research (Otto et al. 2011).

## 5.2 Genomic consequences of recombination arrest

Recombination arrest has manifold consequences for the evolution of sex chromosomes and genomes A first obvious one is a drastic reduction in **effective population size**: the non-recombining parts of U and V chromosomes have their effective sizes reduced to one half, and those of the Y and W chromosomes to one quarter of autosomes (there is only one copy of Y or W per mating pair, as opposed to four copies for autosomes). To a lower extent, the recombining sex chromosomes (X and Z) are also affected, with a number of copies per mating pair reduced to three-quarters of that of autosomes. Lowered effective sizes result in an increased **genetic drift** and a concomitant reduction in the efficiency of selection.

Second, the effective population sizes of non-recombining regions are further lowered (and the action of selection further impeded) by interactions between linked genes, collectively known as Hill–Robertson interferences (Hill & Robertson 1966). As pointed out by Robertson (1961), selection at one locus increases the amount of genetic drift at all other loci on the genome; neutral mutations that occur against a good background will tend to be overrepresented in the next generation, while those occurring against a poor background will tend to be underrepresented. In recombining genomes, this background changes every generation, so that the fixation probabilities of alleles at different loci are still relatively independent. Hill–Robertson effects are an extension of this argument to non-recombining regions: in the case of tight linkage, the random associations of genes will persist for many generations and their effects will be greatly magnified. Linked genes will interfere with each other's fixation by increasing the variance in offspring number, thereby decreasing the effectiveness of selection. This interaction between genetic drift, linkage, and selection will both accelerate the rate of fixation of deleterious mutations and retard that of beneficial ones.

This effect also expands Fisher's (1930) and Muller's (1932) arguments regarding the evolutionary benefits of recombination (Section 1.1.2b), namely that two favourable mutations appearing on different individuals need recombination to get combined; otherwise they will compete against each other, and only one will reach fixation. Hill–Robertson effects actually come in different flavours, which are however conceptually equivalent (Felsenstein 1974). A highly relevant one in the case of sex chromosomes is known as Muller's ratchet: assuming that, as a result of drift, each haplotype in a population contains at least *n* deleterious mutations, natural selection alone will never be able to reduce the minimum number of such mutations below *n*. In the absence of back mutations, the dynamics can only go one way, towards accumulation (Charlesworth & Charlesworth 2000). A click of the ratchet occurs as soon as drift eliminates the category with the lowest *n* value (Box 5.2). The ratchet mechanism is of prime evolutionary importance, because deleterious mutations are far more frequent than beneficial ones. It might actually constitute the major ultimate cause favouring recombination, because such a mechanism allows restoration of categories that are free of mutational load.

Other expressions of Hill–Robertson interferences include **genetic hitchhiking** and background selection. Hitchhiking (or genetic draft) makes neutral or slightly deleterious mutations that appear in a highly fit background increase in frequency, as long as the association persists (Maynard-Smith & Haigh 1974). In addition to fixing potentially deleterious mutations, these **selective sweeps** will erase genetic variance at sites tightly linked to the favourable allele, which, in non-recombining sex chromosomes, include the whole differential segment (Rice 1987). Background selection also induces a loss of genetic diversity, but stems from purifying selection against linked deleterious alleles. Favourable mutations might thus be eliminated because they are linked to unfavourable ones (the 'ruby in the rubbish'; Peck 1994). Both selective sweeps and background selection strongly reduce the effective population size experienced by non-recombining regions. Because only the portion of population

**Box 5.2 Muller's ratchet in Y/W chromosomes**

In infinite populations at mutation–selection equilibrium, the number of deleterious mutations per gamete is expected to follow a Poisson distribution with mean $n = U/2hs$, where $U$ is the mutation rate per diploid genome, $s$ the selection coefficient against homozygotes, and $h$ the dominance coefficient (Kimura & Maruyama 1966). For the non-recombining segment of an incipient Y chromosome, this becomes $U_Y/hs$, with $U_Y$ the mutation rate per segment. From the Poisson distribution, a proportion $f_0 = \exp(-U_Y/hs)$ of the population should be mutation free; in finite populations, this class may only count a few individuals (depending on parameter values and population size) and is thus expected to be eliminated by drift soon or later. The one-mutation class will then become the least loaded, and progressively reach its equilibrium frequency ($f_0$). Successive clicks of the ratchet will progressively increase the number of mutations in the least-loaded class (Charlesworth 1996). Note also that this class is the one that most contributes to the long-term reproduction, which further reduces the effective size of non-recombining sex chromosomes by a factor $f_0$.

Gordo & Charlesworth (2000, 2001) derived analytically the time needed for a click, as a function of population size, mutation rate, and $hs$ values. Assuming for *Drosophila* an effective population size $N_e = 10^6$ (i.e. $5{*}10^5$ Y haplotypes), a mutation rate $U_Y = 0.04$, and a selection coefficient $hs = 0.005$, the number of deleterious mutations on nascent Y chromosomes is expected to average $U_Y/hs = 8$, with $N_e f_0 = 168$ individuals in the least loaded class. Some 1000 generations are needed for a click, so that ~500 mutations should get fixed after 500,000 generations (one tenth the age of *D. miranda* neo-Y chromosome, assuming five generations per year). By this time, male fitness should have dropped to 1/20 of its initial value (Gordo & Charlesworth 2000, 2001). Thus, although Muller's ratchet operates with strongest intensity in lineages with low effective population sizes (like mammals), it can still operate at a biologically relevant rate in very large populations, provided mutations with small fitness effects are sufficiently frequent.

with the fewest mutations will contribute to future generations, the number of loci where an advantageous mutation could arise and be perpetuated in the population is very small (Felsenstein 1974).

As a third consequence of recombination arrest, the Y and W chromosomes, which are limited to the heterogametic sex in D-GSD systems, are sheltered by permanent heterozygosity from the effects of deleterious recessive mutations. Loss-of-function mutations and gene deletions may accumulate even when incompletely recessive, due to the strong genetic drift affecting these chromosomes. This effect, fully expressed in purely diplontic lineages (e.g. animals), is likely mitigated in sub-diplontic lineages (e.g. angiosperms), because some sex-linked genes may be expressed during the haploid phase, purifying recessive deleterious mutations (Haldane 1933). In haplo-diplontic lineage with H-GSD (e.g. fungi, algae, bryophytes), many genes on the U and V chromosomes are likely expressed in the gametophytic phase, which strongly limits such effects.

Fourth, male-limited sex chromosomes (V and Y) are more exposed than female-limited ones (U and W), because males typically have higher mutation rates than females (due to a higher number of cell divisions during gametogenesis) and are under stronger sexual selection (which further contributes to lowering effective populations sizes, and offers more scope for selective sweeps). Note that this last argument does not apply to the mating-type chromosomes of fungi, because sexual selection is not expected to operate in isogamous lineages.

Finally, non-recombining sex chromosomes are particularly prone to evolve meiotic-drive elements, because any gene on the U, X, or W chromosome can drive against homologous genes on the V, Y, or Z (and vice versa; Section 6.3.2). Driving U, X, or W chromosomes will bias sex ratios towards females, triggering an evolutionary response to suppress the drive, which may include a turnover in sex determination or silencing of the driver. In mice, spermatid-expressed genes seem to have been amplified in response to an X-Y driver–suppressor interaction (Cocquet et al. 2012). Selection to silence drivers on the permanently heterozygous Y or W chromosome may also contribute to their degeneration. Meiotic sex chromosome inactivation (**MSCI**), which induces a transcriptional silencing of most genes on the X or Z chromosome during meiosis in the heterogametic sex, might be part of the genomic defence against drivers (Section 5.2.3a).

As a consequence of all these effects, non-recombining genomic regions are exposed to a strong drift, which will translate into the accumulation of non-coding DNA sequences such as transposons and other repeated elements, structural rearrangements including inversions, deletions or translocations, as well as silencing and loss of functional genes. The strength of these effects depends on the size of the non-recombining region: larger segments include many genes, which creates much scope for Hill–Robertson interferences; the larger the non-recombining segment, the quicker its decay. Their strength is also expected to vary greatly between categories of sex chromosomes, because they are differently exposed to the manifold consequences of recombination arrest.

### 5.2.1 U and V chromosomes

Similar to Y and W chromosomes in D-GSD systems, the V and U chromosomes in H-GSD systems are sex limited, being present only in the male and female gametophytes respectively. They differ from the formers by a less extensive drop in effective population size (one half of autosomes, compared to one quarter for Y and W), and by not being sheltered by permanent heterozygosity; gene expression in the haploid state allows purifying selection against deleterious recessive mutations. For genes that are only expressed in the diploid stage (sporophyte), detrimental mutations can accumulate, but only insofar as they are compensated by the other chromosome (Bull 1978). Thus, neither the U nor the V chromosome is expected to experience extensive degeneration; evolutionary changes should be mostly limited to inversions, translocations, and accumulation of genetic material (transposable elements, or genes related to sex-specific functions), rather than gene losses. In addition, many H-GSD lineages are isogamous or present poorly differentiated gametophytes, reducing the scope for sexual selection; the two chromosomes should thus suffer from similar and limited decay.

Nevertheless, the non-recombining region of sex chromosomes can be quite large in some fungi. It spans >100 kb in *C. neoformans*, including some 20 genes, with extensive rearrangements and inversions, as well as many transposons, transposon remnants, and repetitive sequences. Recombination arrest seems very old: silent divergence between the two allelic forms reached saturation in the most ancient region (Lengeler et al. 2002; Fraser et al. 2004). In *U. hordei*, this non-recombining region covers about one-sixth of the mating-type chromosome (~500 kb for MAT-1, ~430 kb for MAT-2), with the *a* locus at one end and *b* at the other end. In between these two loci several rearrangements are found together with many transposable and repetitive elements. In *N. tetrasperma*, it covers 75% of the chromosome (almost 7 Mb), being flanked by distal regions with obligate crossovers (PARs). Two evolutionary strata are recognized, a large and older one (6.6 Mb) and a small one where recombination ceased more recently (Menkis et al. 2008). In *M. violaceum*, the non-recombining region encompasses 90% of the chromosome, and is also flanked at each end by a PAR (Hood et al. 2013). The region in between is also rich in transposable elements, translocations and inversions. Interestingly, the two chromosomes are dimorphic in size (3.3 and 4.0 Mb respectively), and carry unequal densities of functional genes. Three evolutionary strata have been suggested by Votintseva & Filatov (2009), but not confirmed by Hood et al. (2013). The ultimate causes favouring the expansion of non-recombining regions in some fungi remain an enigma, given the unlikely role of SA genes in these lineages.

In the isogamous, haplo-diplontic brown alga *Ectocarpus siliculosus*, the SDR stopped recombining some 80–100 Mya, and encompasses ~900 kb (i.e. one-fifth of the U and V chromosomes). The low gene density (~20 genes) is mostly due to very long introns and high densities of transposable elements. Evidence for decay is limited to a few pseudogenes, a suboptimal codon usage, and less abundant transcripts, possibly due to the degradation of promoter regions. The ratio of synonymous to non-synonymous substitutions is particularly high, likely due to purifying selection during the gametophytic phase, where genes from the SDR tend to be particularly expressed (Ahmed et al. submitted). In the isogamous, haplontic green alga *Chlamydomonas*, the 200–300 kb multigenic MAT locus includes several inversions, translocations, gene duplications, and mutations, and also contains many transposable elements and repetitive sequences. The same MAT locus is functional in the anisogamous *Volvox carteri* (which diverged from *Chlamydomonas* some 200 Mya), but greatly expanded, with signs of evolution towards fully-fledged U and V chromosomes. The non-recombining region is ~500% larger (with U slightly larger than V) and contains over 70 genes, including some coding for anisogamy. It is repeat-rich, with lower gene density and more intronic sequences than genomic average. Interestingly, the divergence between allelic pairs is up to two orders of magnitude larger than in *Chlamydomonas*, which might relate to sexually antagonistic selection, or to stronger Hill–Robertson interferences on a larger non-recombining segment. Evolutionary strata are also recognized, with two internal **syntenic** blocks that are relatively similar between U and V, suggesting they were acquired more recently (Ferris et al. 2010).

Sexual selection is also expected to operate to some extent in bryophytes: males should display a greater variance in reproductive success, due to their many small and dispersing gametes (as opposed to the few larger ones retained by female gametophytes). In addition, sexually antagonistic alleles in either sex may select for suppressed recombination between U and V chromosomes, which might favour more rapid expansion of the SDR than in diploid systems, where recombination arrest is only driven by SA alleles favouring the heterogametic sex (Bachtrog et al. 2011). Accordingly, the U and V chromosomes are often markedly differentiated (with repeated elements and low gene density) and heteromorphic (the U chromosome being generally larger; Bull 1978). In *M. polymorpha*, the male V chromosome is 10 Mb and harbours thousands of copies of repetitive elements. Of the 64 genes detected, 14 are unique to the male genome and show sexual organ-specific expression. The U chromosome is twice as large (20 Mb), and carries large clusters of ribosomal DNA. Female homologues have been detected for five male genes, with 70.6 to 93.6% of nucleotide identity. **Synonymous substitutions** have reached saturation for some of them, suggesting that recombination has been suppressed for a long time (Okada et al. 2001; Ishizaki et al. 2002; Tanurdzic & Banks 2004).

## 5.2.2 Y and W chromosomes

### 5.2.2a Evolutionary forces acting on Y and W

Many features of sex chromosome evolution converge to drive strong differentiation and decay in Y and W chromosomes (Fig. 5.5). First, contrasting with X and Z, Y and W chromosomes do not recombine over the SDR. Second, their effective population size is only one quarter that of autosomes, magnifying the effect of genetic drift. Third, they are strictly sex limited, making them hotspots for sexually antagonistic genes benefiting the heterogametic sex, and thereby for selective sweeps (Rice 1987). Fourth, being sex limited, they are also hotspots for meiotic-drive genes; selection to restore balanced sex ratios can favour silencing of parts of the Y and W chromosomes, contributing to their degeneration (Jaenike 2001). Fifth, they are sheltered from selection by permanent heterozygosity, so that deleterious recessive mutations may freely accumulate over time (Muller 1918). Even loss-of-function mutations of housekeeping genes might have only minor effects on fitness when heterozygous: experiments on *Drosophila* suggest that recessive lethals reduce the fitness of heterozygotes by only ~2% on average (Crow 1993). Sheltering might be mitigated in sub-diplontic lineages (e.g. angiosperms), if sex-limited genes are expressed in the haploid phase, and thereby exposed to purifying selection (Haldane 1933). Y chromosomes are also expected to degenerate more rapidly than W chromosomes, because males usually have both higher mutation rates and lower effective population sizes than females (Caballero 1995). Sexual selection is also stronger in males, with more rapid shifts, so that Y chromosomes should carry more sexually antagonistic genes, and undergo more frequent selective sweeps, than W chromosomes.

### 5.2.2b Empirical evidence

As a result of all these effects, Y and (possibly to a lesser extent) W chromosomes are expected to display high rates of gene erosion (Rice 1996; Charlesworth & Charlesworth 2000; Charlesworth et al. 2005). Empirical studies are indeed documenting a

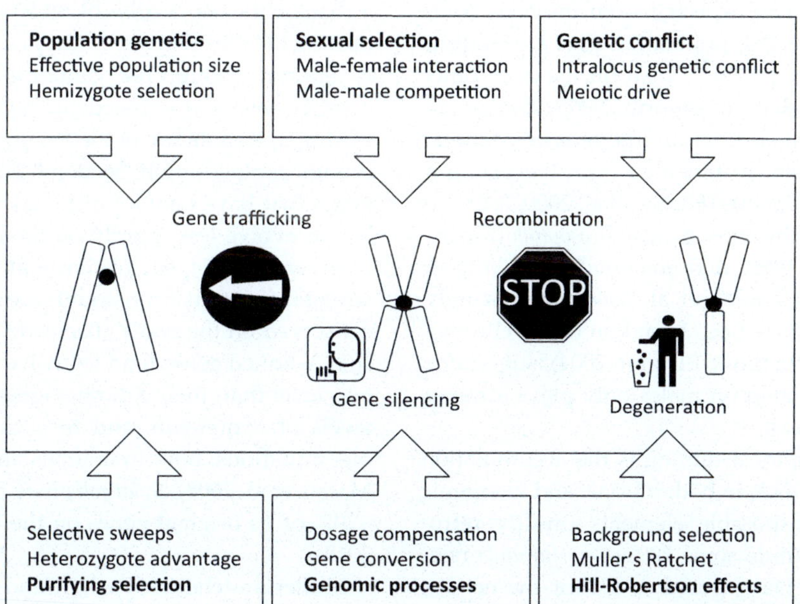


**Figure 5.5** Evolutionary forces that act on the gene content of sex chromosomes. Middle panel, from left to right: autosomes, the X (or Z) and the Y (or W) chromosome.

strong decay in several male and female heterogametic systems such as in mammals, birds, snakes, and *Drosophila*. However, these systems, being relatively old (ca 160 and 130 My in mammals and birds) are not suitable for examining the dynamics of processes, and disentangling the evolutionary forces involved in the early stages of degeneration. Several young Y chromosomes (i.e. less than 10 My), from both angiosperms and animals, are now being characterized at the molecular level, shedding some light on these early stages. Neo-sex chromosomes are especially useful in this context, allowing comparison between non-recombining regions of different ages. *Drosophila* in particular turns out to be a useful model, due to its male **achiasmy** (recombination is immediately arrested after fusion to a Y chromosome) and the several independent events of fusion that occurred in different lineages.

The most extensively studied young Y chromosome to date is the neo-Y of *D. miranda*, which resulted some 1–2 Mya from the Y chromosome fusion with an autosomal element harbouring more than 2000 genes (about 4.3 Mb of coding sequences). A first feature to note is the drastic loss of diversity at neutral markers on the neo-Y chromosome (both microsatellites and sequence data), much stronger than expected from the mere fourfold drop in effective population size. Diversity is ~25 times lower on the Y than on the X chromosome (Bachtrog & Charlesworth 2000, 2002), pointing to strong Hill–Robertson interferences. Similar deficits have been documented from the young Y chromosomes of plants (Filatov et al. 2000, 2001), as well as older Y chromosomes of *Drosophila* (Zurovcova & Eanes 1999) and mammals (Hellborg & Ellegren 2004; Handley et al. 2006). Interestingly, similar deficits have been found on the W chromosomes of birds (Berlin & Ellegren 2004), suggesting that sexual selection on males only plays a minor role in this context.

A second empirical finding is the accumulation of non-coding DNA, in both intronic and intergenic sequences. Transposable elements (mostly **retroelements**) constitute some 20% of *D. miranda* neo-Y chromosome (as opposed to 1% for the neo-X); in total, more than half of the neo-Y appears to be made of repeated sequences and junk DNA, including transposable elements and tandem replications (Bachtrog 2008). Similar analyses of young Y chromosomes in sticklebacks (Peichel et al. 2004), medaka (Kondo et al. 2006), *Silene* (Marais et al. 2008) and papaya (Liu et al. 2004) show that repetitive elements are accumulating faster on the Y than in homologous X-linked regions. This accumulation largely contributes to the tendency for plant Y chromosomes to be larger than their X homologues (Charlesworth 2013). Transposable elements are clearly early invaders of nascent Y chromosomes, and may play an important role in the degeneration process (Steinemann & Steinemann 2005; Bachtrog et al. 2008); however, the potential contribution of repetitive elements and non-coding RNAs to gene regulation might have been underestimated (Section 5.2.2f).

A third prominent feature is the decay of coding regions. About half of the > 2000 genes on the neo-Y of *D. miranda* have become non-functional, due to frame-shift mutations or premature stop codons; some are completely lost (Bachtrog et al. 2008; Zhou & Bachtrog 2012a,b; Bachtrog 2013). Other mutations have accumulated in genes that are still transcribed, or in their regulatory regions, which should negatively affect fitness. Amino acid substitutions have accumulated about two-fold faster on the neo-Y than on the neo-X (**dN/dS** = 0.50 versus 0.24), supporting the idea that the efficacy of natural selection is reduced (though one cannot exclude a role for positive selection to inactivate genes on the Y, following up-regulation of the X copy in the process of dosage compensation; Section 5.2.2e). Interestingly, genes that have kept a functional copy on the neo-Y also evolve less rapidly on the neo-X (dN/dS = 0.19 versus 0.31), suggesting that genes subject to stronger functional constraints are more likely to be conserved on the neo-Y (Bachtrog et al. 2008). As a rule, Y-linked genes tend to evolve faster at the protein level than their X-homologues, to show lower levels of expression, and to accumulate deleterious mutations, both in animals and in plants (e.g. Marais et al. 2008). *S. latifolia* may have already lost ~20% of its original genes on the Y (Charlesworth 2013).

Muller's ratchet and background selection are likely to be the main selective forces driving gene erosion on nascent sex chromosomes, due to Hill–Robertson interferences between the many functional genes

(Bachtrog 2008). Analytical and simulation studies indeed confirm that Muller's ratchet alone constitutes a significant force, even at large population sizes, provided the non-recombining segment is large enough (Box 5.2). Its effects will be further amplified by background selection. Empirical estimates of the loss of diversity and accumulation of deleterious mutations on the neo-Y of *D. miranda* are indeed compatible with the action of Muller's ratchet combined with background selection (Kaiser & Charlesworth 2009, 2010). Although selective sweeps may also act on specific genes at these early stages (Bachtrog 2004), only at later stages are they expected to become a prominent cause of decay, as more and more genes are inactivated (Bachtrog 2008).

Other features that become apparent at later stages of decay, besides a generalized gene erosion, is a marked gene trafficking between autosomes and sex chromosomes (many coding sequences being translocated to and from sex chromosomes; Skaletsky et al. 2003; Koerich et al. 2008) and a masculinization of the Y chromosomes: the few functional genes left often have a specialized male function. These two features are partly linked; genes that are important for both sexes are better located on autosomes (where they avoid decay) and those that specifically bear male benefits (at some cost to females) are better located on the Y.

The Y chromosome of *D. pseudoobscura* is non-homologous to that of *D. melanogaster* (the ancestral Y was lost from this lineage, and several of its genes translocated to autosomes). The neo-Y chromosome, which arose some 15 Mya from the fusion between an autosome and the X chromosome, now only retains some 15 genes of its originally > 2000 genes, most of which with a male specific function. Of these, ten have homology with the neo-X, while others were translocated from autosomes (Carvalho & Clark 2005). The Y chromosome of *D. melanogaster*, estimated to be at least 60 My old, retained some 20 genes, most of which involved in spermatogenesis (e.g. encoding sperm flagella motor proteins). All of them apparently arose by duplication from autosomal genes (Carvalho et al. 2000, 2001). Their location on the Y reflects male-specific advantage, with possibly deleterious effect for females. Overall, the Y chromosomes of *Drosophila* show a preponderance of gene gains over losses (i.e. transferred to Y rather than the reverse; Bernardo & Carvalho 2009).

Genome sequence analyses of the human Y have also uncovered seven gene families whose closest paralogues are autosomal (hence transposed from autosome to Y). Most are predominantly expressed in testes. Two additional gene families with testis-specific expression have evolved from X-linked homologues (Skaletsky et al. 2003). Reciprocally, some genes originally on the ancestral Y chromosome are now autosomal (and have purged much of the **repetitive DNA** accumulated during their tenure on the Y chromosome).

#### 5.2.2c Fitness costs to the heterogametic sex

Many genes are involved in regulatory networks, where the balance of products is critical. Although networks might be robust against small dose differences, larger imbalance will generate defects in the cascade, so that the whole network may fail to recover (Gupta et al. 2006; Oliver 2007). This is the reason why **aneuploidy** (e.g. trisomy) is much more deleterious than **polyploidy**; the only viable monosomic condition in humans is Turner syndrome (XO females). There is large empirical evidence for fitness costs in heterozygotes from single loss-of-function mutations: recessive lethal mutants in *Drosophila* are partially dominant, with selection coefficient against heterozygotes (*hs*) averaging ~2% (Crow & Simmons 1983). As the genes occurring on sex chromosomes are also involved in such networks, the accumulation of loss-of-function mutations on Y and W chromosomes is expected to severely impact fitness in the heterogametic sex. From analytical and simulation studies, the accumulation of detrimental effects mostly depends on the product of *hs* with effective population size *N*, and fitness costs are most severe at intermediate *Nhs* values (around 0.4 for autosomal mutations, and 1.6 for Y or W chromosomes, with effective sizes one quarter those of autosomal genes; Lande 1994; Blaser et al. 2013). Strongly deleterious mutations are eliminated, while very mild mutations, though accumulating freely, are not deleterious enough to significantly impact fitness. Muller's ratchet alone has the potential to lower male fitness by several orders of magnitude in *Drosophila*, under biological relevant time scales and mutation parameters

(Box 5.2). Still larger effects are expected in vertebrates (with much lower effective population sizes), with the potential to drive small populations to extinction (Grossen et al. 2012). Additional Hill–Robertson interferences (e.g. background selection) are expected to further accelerate the process, particularly for large non-recombining segments (Kaiser & Charlesworth 2010). This dramatic selective pressure has important evolutionary consequences; it may eventually result in a sex-chromosome turnover, during which the decaying Y is replaced by a new masculinizing mutation occurring on an autosome (Blaser et al. 2013; Section 7.2.2b). Alternatively, it may elicit evolutionary responses to prevent or slow down the decay, or to accommodate it via dosage compensation.

#### 5.2.2d Opposing the decay: gene conversion and X-Y recombination

In this category are to be listed processes favouring X-Y and Y-Y recombination or gene conversion. X-Y recombination bears costs by reshuffling SA alleles (Section 5.1.1), but also brings benefits by counteracting Y degeneration; a non-zero evolutionarily stable rate should thus result. However, the mechanisms controlling male recombination also matter: epigenetic control certainly allows much better fine-tuning than inversions (Section 5.1.2). Epigenetic modifiers may allow occasional X-Y recombination either in males, or in sex-reversed XY females (the fountain-of-youth hypothesis; Perrin 2009). Individual-based simulations by Grossen et al. (2012) allowing evolution of the liability factor (Chapter 4) under the opposing forces of deleterious mutation load and SA selection indeed predict a non-zero equilibrium rate of sex reversal and ensuing X-Y recombination. Empirical support for the fountain of youth was obtained from species of European tree frogs (*Hyla arborea* group) that diverged > 5 Mya. All species from this group share the same pair of sex chromosome, with no sign of Y decay; despite complete absence of male recombination, X and Y alleles cluster by species, not by gametologue, pointing to rare events of X-Y recombination (Stöck et al. 2011). Using approximate Bayesian computations, Guerrero et al. (2012) estimated this rate to be $< 10^{-5}$ lower than X-X recombination, but highly significantly different from zero.

Rare events of X to Y gene conversion may also occur. Empirical evidence was provided by the sequencing of mammalian X and Y chromosomes (Slattery et al. 2000; Rosser et al. 2009; Trombetta et al. 2010). The punctual replacement of degenerated Y genes by fully functional copies from the X should oppose the decay of Y. Sequencing work has also provided evidence for Y-Y gene conversion, with similar consequences. Most gene families on the mammalian Y chromosome are organized in **palindromic** sequences, with similar repeats pointing in opposite directions, connected by spacers, and showing high levels of gene conversion (Skaletsky et al. 2003; Rozen et al. 2003). Simulation work by Marais et al. (2010) support the view that these large-scale duplication events and high rates of Y-Y gene conversion have been selected to oppose the Y degeneration. Interestingly, the U and V chromosomes of fungi also show high densities of inverted palindromic sequences, enabling repair via intra-chromosomal recombination (Fraser et al. 2004; Fraser & Heitman 2004, 2005).

#### 5.2.2e Accommodating the decay: dosage compensation

The dosage imbalance resulting from the loss of function of Y or W-linked genes selects for an increased activity of X or Z homologues. Deleterious mutations occurring in genes that are still transcribed may also select for an inactivation of the maladapted genes, and concomitant dosage compensation by X or Z homologues. Note that imbalance occurs at two levels: between sexes, and between sex chromosomes and autosomes in the heterogametic sex (Marin et al. 2000; Heard & Carrel 2009; Vicoso & Bachtrog 2009). Dosage compensation (DC) was first described in *Drosophila* by Muller (1932), and subsequently mostly studied in this and a few other model organisms including *C. elegans* and mice, all male heterogametic. Interestingly, different mechanisms occur in different taxa; see reviews by Gendrel & Heard (2011), Conrad & Akhtar (2012), Gribnau & Grootegoed (2012) and Gschwend et al. (2012). The simplest and seemingly most efficient one is found in *Drosophila* and sciarid flies, where DC is achieved by up-regulation of the single X in males, simultaneously solving the sex- and X-autosome imbalance (Box 5.3).

**Box 5.3 X up-regulation in *Drosophila* males**

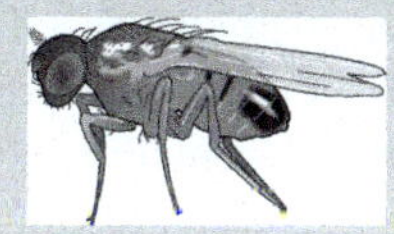

X up-regulation in *Drosophila* males is mediated by a dosage compensation complex (DCC) made of proteins from five *Male-specific lethal* genes (*Msl*-1, *Msl*-2, *Msl*-3, *Mle*, and *Mof*) and non-coding RNA products from two additional genes, *RoX1* and *RoX2* (*RNA on the X*). The DCC binds to hundreds of sites along the male X chromosome, doubling its transcription by altering chromatin structure (acetylation of lysine 16 of H4, a histone required for the folding of **nucleosomal arrays**), thereby making the genetic material more readily accessible to the transcription machinery (Park & Kuroda 2001; Taipale & Akhtar 2005; Alekseyenko et al. 2008). *RoX1* and *RoX2* may act as chromatin-entry sites, from which epigenetic spreading of the **compensasome** proceeds along the chromosome. A functional compensasome is not produced in females, because *Msl*-2 transcripts are repressed by the female-specific, master sex-determination gene *Sex-lethal* (Marin et al. 2000; Box 3.18). This mechanism is shared by all *Drosophila* species and thus at least 60 My old. The degree to which the new arms of neo-X chromosomes are dosage compensated correlates with the amount of decay on the corresponding neo-Y (Bone & Kuroda 1996; Marin et al. 1996, 2000). In *D. americana*, most genes are still active and not dosage compensated on arm B, which fused with the X less than 0.2 Mya. On the arm C of *D. miranda*, which fused 1–2 Mya with the Y, many loci are still expressed, but a substantial fraction have become non-functional or are lost; homologues on the X have become dosage compensated. Finally, arm D has fused 13 Mya with the X in the *pseudoobscura* subgroup; the homologous neo-Y has completely lost its activity, and full DC is found on the X-linked D arm.

In mammals and *C. elegans*, the sex imbalance is solved instead by a down-regulation of X expression in females, achieved in different ways. In *C. elegans*, X is first hyper-transcribed in both sexes (equalizing expression between X and autosomes in males), then both Xs are secondarily hypotranscribed in hermaphrodites (Gupta et al. 2006; Box 5.4).

In mammals, sex imbalance is solved by inactivating one X in females (Lyon 1961; Box 5.5).

**Box 5.4 X repression in C. *elegans* hermaphrodites**

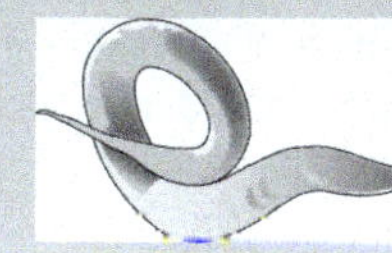

X repression in *C. elegans* hermaphrodites results from an increased condensation of the X chromosomes by a dosage-compensation complex (DCC) made of nine proteins at least, some of which also involved in sex determination (the sex-determination and dosage-compensation genes, or SDC genes; Box 3.15). All are supplied maternally to both XX and XO embryos, except for SDC-2, expressed only in XX embryos where it both triggers the development of a hermaphroditic phenotype and recruits the DCC complex. Two types of DCC-binding sites coexist on the X: the high-affinity *Rex* sites (recruitment elements on the X) act as primary binding sites, and recruit DCC independent of their location. The low-affinity *Dox* sites (*dependent on X*) only bind the DCC when located on the X, and are responsible for the spreading of the DCC to coat the entire chromosome (Meyer 2000, 2005).

However, the way X-autosomal imbalance is solved is still controversial: some studies report an up-regulation of the active X (Nguyen & Disteche 2006) while others oppose this conclusion (Xiong et al. 2010; Julien et al. 2012). Alternatively, the protein products of the X homologues might have evolved improved efficiency, to function at half the gene dose, selecting in turn for X-inactivation in females.

As a possible evolutionary scenario, the transcription (or functional efficiency) of X gametologues was first increased in all three groups following Y decay, to restore functional gene networks in males. In *Drosophila*, the male-specific up-regulation solved by the same token both the sex-specific and X-autosomal dosage problems. In *C. elegans* and mammals, up-regulation or hyper-functionality extended to females, driving the evolution of complementary mechanisms to reduce transcription in females (Charlesworth 1978b; 1996). Engelstädter & Haig (2008) formalized this scenario, showing that it may work in principle, but requires precise quantitative relations between the costs of under-expression in males and those of over-expression in

**Box 5.5 X silencing in female mammals**

In eutherians, imprinted inactivation of the paternal X occurs transiently during early embryonic stages (2 to 4 cells), followed by random X inactivation at the blastocyst stage; the inactive state is then stably inherited, giving rise to adults that are mosaics for two cell types, expressing either the paternal or maternal X chromosome (Payer & Lee 2008). However, paternal X inactivation is maintained lifelong in marsupials, as well as in the extra-embryonic tissues of some placental mammals (including mice and cattle); it possibly constitutes the ancestral state, which originated with live-bearing habits. Gene inactivation is only limited and partial in platypus (Deakin et al. 2008); male-to-female ratios of expression for genes on the non-recombining regions of X chromosomes average 0.6 (Julien et al. 2012).

Random X-inactivation in eutherians involves a recognition step in which the number of X chromosomes is counted relative to cell ploidy. Silencing of the supernumerary X is controlled by two non-coding **antisense RNAs,** *Xist* and *Tsix*, transcribed from the same X-locus (the X-inactivation centre, or XIC), but in opposite directions. *Xist* and *Tsix* function as mutual repressors of transcription: on the active X, *Tsix* is expressed and *Xist* repressed, while on the inactive X, *Tsix* is repressed, so that *Xist* is transcribed; it spreads in *cis* and coats the X chromosome, recruiting two polycomb repressive complexes (PRC1 and PRC2) that hypoacetylate histones, leading to a heterochromatic state and stably inactivated X (Marin et al. 2000; Avner & Heard 2001; Heard & Disteche 2006; Straub & Becker 2007). **LINE** elements on the X may play a role in the spreading (Lyon 1998).

A few genes escape X inactivation, some of which without active Y homologues, suggesting a gene-by-gene mechanism of silencing. Their localization correlates with evolutionary strata (Fig. 5.3): dosage compensation is most effective in the oldest stratum, and least effective in the most recent ones (Graves & Foster 1994; Marin et al. 2000). In humans, the 15% or so X-linked genes that escape inactivation cluster in the X-added region, which is autosomal in marsupials (Carrel & Willard 2005).

females. As such costs are expected to differ from gene to gene, it might not be valid as a general mechanism. In addition, it seems difficult to see why up-regulation of X in males would also affect females, given that gene expression in mammals is sex-biased at many loci (e.g. Yang et al. 2006). Alternatively, X inactivation might have resulted from parental conflict: males have interest to up-regulate the X-linked foetal growth genes in their daughters, even at the expense of mothers; X inactivation would constitute the female response (Haig 2006; Engelstädter & Haig 2008).

Dosage compensation was long thought to be a necessary consequence of sex chromosome decay, and thus a widespread feature of all sex chromosomes. However, evidence is now accumulating that many or most groups lack any global mechanism; DC is either absent, as in sticklebacks (Leder et al. 2010), or only partial, as in monotremes (Deakin et al. 2008; Julien et al. 2012), birds (Itoh et al. 2007, 2010; Ellegren and Parsch 2007), and Lepidoptera (Zha et al. 2009; Arunkumar et al. 2009). Female birds show lower expression of Z-linked genes than males, but the drop is only ~ 0.8-fold on average (i.e. less than the gene dosage; Itoh et al. 2007). Most Z genes seem not or under-compensated, while a few ones show the same level of expression as in males. Levels of expression also vary by tissue and ontogenetically (Mank & Ellegren 2009; Itoh et al. 2010). Compensation in birds thus seems to proceed in a subtle way, occurring locally, on a gene-by-gene basis, when and where balanced transcription is needed. Similar patterns occur in silkworms (Zha et al. 2009).

It appears therefore that a global mechanism of DC is not always required. For many gene networks, product doses are buffered by complex feedback loops that regulate gene expression; despite lowered Z-linked gene expression, female birds apparently achieve enough dosage compensation to balance their gene networks (Naurin et al. 2010). In addition, a global mechanism affecting whole chromosomal regions is expected to further accelerate the degeneration process: genes on the Y that are still functional, but trapped in a compensated region, should be soon silenced as an adaptive response to DC, inducing their subsequent decay (Orr & Kim 1998; Bachtrog 2006). This raises the question of why some groups evolved global mechanisms, given the feasibility of a more parsimonious approach, which compensates only genes with critical expression titres.

A possible hint comes from noting that global DC was only documented in male heterogametic systems; the female heterogametic birds and Lepidoptera operate on a gene-by-gene basis (Mank 2009). One might speculate a role for the differential mutation rates and strengths of sexual selection in male versus female heterogametic systems: Y chromosomes are male limited, and thus expected to experience faster degradation than W chromosomes, because of higher mutation rates in males, and reduced effective size due to stronger sexual selection. In addition, sexual selection on males favours the accumulation of male-beneficial alleles on the Z in female heterogametic systems (Section 5.2.3); under-expressing these might be beneficial to females (Ellegren 2009; Mank & Ellegren 2009). The mode of expression of Hill–Robertson interferences may also matter: if the decay occurs mostly by occasional fixation of strongly deleterious mutations via hitchhiking (selective sweeps), then a gene-by-gene compensation might be favoured. But if it occurs by the massive accumulation of mildly deleterious mutations, due to inefficient purifying selection, then a global mechanism, that compensates large blocks of genes at once, might be favoured (Charlesworth 1978b; Rice 1987; Vicoso & Bachtrog 2009). All of this remains highly speculative, however, given the very low number of independent cases properly studied. Information on the dynamic history of the process is also lacking; it appears crucial to document DC evolution in more groups, with both highly differentiated and incipient sex chromosomes (e.g. snakes and frogs, respectively).

#### 5.2.2f Are Y and W chromosomes doomed to extinction?

The patterns documented in several well-studied lineages might give the impression of an ineluctable decay; as dramatically expressed by Steinemann & Steinemann (2005), Y chromosomes might be 'born to be destroyed'. In line with this expectation, the ancestral Y chromosome has disappeared from several *Drosophila* and rodent lineages (e.g. the mole vole *Ellobius*), and many groups of arthropods lack a Y, being characterized by XX/XO systems (Chapter 2). Aitken & Marshall Graves (2002) estimated the rate of gene loss since the origin of the mammalian Y chromosome to be around five genes per My, and from this number predicted 'self-destruction' of the human Y within 10 My (see also Graves 2005, 2006; Sun & Heitman 2012).

Several arguments, however, run counter this bold prediction. First, the assumption of linearity in the rate of decay is unrealistic; an exponential negative would be a more appropriate null model. Second, the intensity of Hill–Robertson interferences are expected to decline as erosion progresses, lowering the rate of decay (Bachtrog 2008). Third, the genes left on highly differentiated sex chromosomes are often essential for male fertility, and thus subject to selective forces strong enough to oppose drift. In addition, these genes are often protected from erosion by YY recombination between palindromic sequences (Section 5.2.2d). Thus, the exponential decay is expected to strongly decelerate in mature sex chromosomes.

A comparison between the Y chromosomes of humans and Rhesus monkeys (with 25 My divergence) shows that losses are limited to the last evolutionary stratum (established 30 Mya, that is, predating the split by 5 My only). No single additional gene has been lost from strata 1 to 4 (Fig. 5.3), neither in the human nor in the Rhesus lineage (Hughes et al. 2012). When including chimpanzees in the comparison (6 My divergence from humans), the decay is best fitted by an exponential decline with a non-zero baseline, implying that a series of genes are protected from erosion. The decay is rapid at first, with a gene half-life shorter than 5 My, then markedly decelerated when approaching the baseline. Human and chimp Y chromosomes also differ radically in sequence structure and gene content, indicating extensive rearrangements over the past 6 My. About 30% of sequences on the Y are non-homologous (as opposed to 2% for the remainder of the genome), mainly in coding sequences. This mostly results from gene losses in chimps, and gene acquisition (including an X-to-Y transposition) in humans. Given that many genes on the Y are involved in sperm production, the higher rate of erosion in chimpanzees possibly results from the strong sperm competition in males, which may induce important selective sweeps (Hughes et al. 2010). Ongoing research on ancient sex chromosomes are thus revealing much more dynamic processes than envisaged previously, involving losses and gains of individual genes or large genomic regions, and no strong evidence for an imminent extinction.

Recent developments in the longstanding 'junk' DNA controversy are also relevant to the preservation of Y and W chromosomes (Maher 2012; Pennisi 2012). Highly heterochromatic regions of sex chromosome are traditionally considered as genetically inert, but evidence is growing that repetitive DNAs may be a source of non-coding RNAs that mediate gene expression, for example, by regulating transcription factors (Barrett et al. 2012). Lemos et al. (2008) introgressed Y chromosomes from geographically distinct *D. melanogaster* populations into a uniform genetic background, and found a strong effect on genome-wide patterns of gene expression; see Piergentili (2010) for a more in depth coverage of the genetic effects of the *Drosophila* Y chromosome. Although still confined to fruit flies, these results suggest that the Y chromosome has become a major regulator of gene expression in males (Rice & Friberg 2008). Interestingly, the introgressed Y chromosomes also affected genes that interact with the mitochondria, suggesting that these may compensate for mitochondrial or cytoplasmic effects in the context of genomic conflict. The current surge of studies into potential transgenerational epigenetic effects may add further to the importance of paternal or maternal effects mediated by the Y or W respectively (e.g. Friberg et al. 2012).

## 5.2.3 X and Z chromosomes

### 5.2.3a Evolutionary forces on X and Z

Contrasting with other sex chromosomes, X and Z are recombining over their whole length, and for this reason were initially assumed to have diverged little from their autosomal progenitors (Ohno 1967). However, comparative genomic and expression studies are now challenging this view, revealing discreet but important features that underline distinct evolutionary fates; X and Z chromosome are not passive players, but actively respond both to their sex-biased transmission and to the degeneration of their homologues (Bachtrog 2006; Vicoso & Charlesworth 2006; Gurbich & Bachtrog 2008; Bachtrog et al. 2011). Table 5.1 and Fig. 5.5 depict a number of evolutionary forces that act in concert on the homogametic sex chromosomes.

X and Z chromosomes first differ from autosomes by their lower effective population size (three quarters that of autosomes) and absence of recombination in the heterogametic sex (implying less recombination in average). Both features should induce more drift and less efficient selection; sexual selection on males is expected to amplify this effect in ZW systems, but restrain it in XY systems. Opposing drift, however, hemizygous exposure in the heterogametic sex induces purifying selection against deleterious recessive mutations, and more efficient positive selection for recessive beneficial mutations. Both drift and positive selection are expected to fasten the evolution of X and Z chromosomes (faster-X processes).

X and Z also have sex-biased patterns of transmission and expression, spending two thirds of their time in the homogametic sex, but being exposed in hemizygous state in the heterogametic sex. This generates complex patterns of selection on sexually antagonistic genes (Rice 1984). When occurring on autosomes, SA mutations are unlikely to be maintained in stable polymorphism: they will spread to fixation if the net benefits to one sex exceed the costs to the other sex, and disappear otherwise. On the X chromosome, in contrast, a dominant mutation will at start be expressed twice as often in females than in males, and thus be initially disfavoured if female detrimental, but favoured if female beneficial. If the cost to males is low, it may spread to fixation, contributing to a 'feminization' of the X; if the cost is high, a stable polymorphism may result, because costs increase linearly with mutation frequency $p$ in males, but benefits saturate in females (specifically, the phenotypic expression increases as $p$ in males and $1-(1-p)^2$ in females). Reciprocally, a recessive mutation will first be expressed only in males, and will thus be initially disfavoured if male detrimental, but favoured if male beneficial. If the costs to females are low, it may spread to fixation, resulting in a 'masculinization' of the X; if costs are high enough, it will be counter-selected as soon as homozygotes appear in females, resulting again in a stable polymorphism. Similar processes occur, *mutatis mutandis*, on the Z chromosome. Thus, X and Z

**Table 5.1** The evolutionary forces that act on X (or Z) chromosomes

| Theory | Mechanism | Model | Predictions | Supportive evidence | References |
|---|---|---|---|---|---|
| Population genetics | Hemizygous selection (positive selection through rounds of haploidy) | Charlesworth et al. 1987, Vicoso & Charlesworth 2006, 2009 | More recessive beneficial male-expressed alleles on the X (faster X)<br>More recessive beneficial female-expressed alleles on the Z (faster Z) | See sexual selection below[1] | Presgraves (2008)[2] |
| | Drift (effective population size is reduced for X and Z) | | Accumulation of mildly deleterious alleles on X | primates | Xu et al. 2012 |
| | | | Accumulation of mildly deleterious alleles on Z | Chicken (*Gallus gallus*) and zebra finch (*Taeniopygia guttata*) | Mank et al. 2010 |
| | | | Monogamous species have more variation than polygamous (Z) | Shorebirds | Corl & Ellegren 2012 |
| Sexual selection | Fisherian runaway (on male-male competition and female choice) | Reeve & Pfennig (2003), Kirkpatrick & Hall (2004) | Male sperm competition genes on the X | Chimpanzee (*Pan troglodytes*) and human | Torgerson & Singh 2003, Khaitovich et al. 2005, Xu et al. 2012 |
| | | | Female mate preference genes on the Z | Moth (*Utetheisa ornatrix*) | Iyengar & Reeve 2010 |
| | Good genes | Kirkpatrick & Hall (2004) | Female mate preference genes on the X | | |
| Conflict | Intragenomic conflict<br>Meiotic drive and X-Y conflict | Hamilton (1967), Haig and Grafen (1991) | Gene amplification of distorter and suppressor genes (e.g. spermatid expressed genes) | House mouse (*Mus musculus*) | Cocquet et al. 2012 |
| | Intralocus sexual conflict (sexual antagonism) | Rice (1987), Bonduriansky & Chenoweth 2009 | Genes that increase male mating success and decrease female fitness are X-linked | *Drosophila melanogaster* | Morrow et al. 2008, Dean et al. 2012 |
| | Zygotic drive | Rice & Friberg (2008) | Accumulation of male harmful epigenetic mutations on the X<br>Accumulation of female harmful epigenetic mutations on the Z | Egg-to-adult survival in *Drosophila melanogaster* | Friberg et al. 2012 |

*continued*

**Table 5.1** Continued

| Theory | Mechanism | Model | Predictions | Supportive evidence | References |
|---|---|---|---|---|---|
| Dosage compensation | Dosage compensation | Charlesworth 1978b | Underrepresentation of male-biased genes on X (X feminization) | *Drosophila melanogaster* | Parisi et al. (2003), Ranz et al. (2003) |
| | | | | *Caenorhabditis elegans* | Reinke et al. 2004 |
| | | | | NOT in mouse (*Mus musculus*) | Wang et al. 2001, Yang et al. 2006 |
| | | | | Human | Lercher et al. 2003 |
| | | | | *Tribolium castaneum* | Prince et al. (2010) |
| | Meiotic sex chromosome inactivation (MSCI) | Wu & Xu 2003 | Retrotransposition of genes from X to autosomes | Testis-specific genes in *Drosophila* | Betran et al. (2002), Sturgill et al. (2007), Vibranovski et al. (2009) |
| | | | | Spermatogenesis genes in human | Emerson et al. (2004), Vinckenbosch et al. (2006), Makino & McLysaght (2008), Potrzebowski et al. (2008), Sin et al. (2012) |
| | | | Retrotransposition of genes from Z to autosomes | Ovary-specific genes in silkmoth (*Bombyx mori*) | Wang et al. 2012, but see Toups et al. (2011) for a counter example |

[1]No independent evidence for hemizygous selection is available as its predictions are indistinguishable from the sexual selection theories.
[2]For a discussion on evidence in Drosophila.

chromosomes should be enriched both in SA genes and in SA variation, but whether masculinization or feminization occurs depends on the nature of underlying genetic variation (Rice 1984).

An additional feature of X and Z chromosomes is their transcriptional repression during meiosis in the heterogametic sex. This process, known as meiotic sex-chromosome inactivation (MSCI), relates to a general silencing of unpaired chromatin during meiosis, possibly a defence against transposon invasion. However, the accumulation of female-beneficial SA alleles on the X (or male-beneficial on the Z) might also select against their transcription during meiosis in the heterogametic sex; alternatively, MSCI might be favoured because it silences meiotic-drive genes during heterogametic meiosis. Whatever its ultimate causes, however, MSCI is expected to affect the dynamics of SA alleles on X and Z chromosomes (Wu & Xu 2003).

Finally, the evolution of dosage compensation (Section 5.2.2e) resulting from Y inactivation necessarily affects the evolution of X chromosomes. Positive selection for the possibly numerous entry sites of the DC complex may contribute to the faster-X effect. X inactivation also affects the patterns of expression: hemizygous expression is extended to the homogametic sex (only the maternal X is expressed in female marsupials), amplifying positive and purifying selective pressures.

### 5.2.3b Empirical evidence

Empirical investigations reveal an accumulation of repetitive elements (e.g. Bellott et al. 2010), as well as retrogenes and pseudoretrogenes (Potrzebowski et al. 2010), on X and Z chromosomes above genome average, presumably resulting from enhanced drift (but repetitive elements may also play a role in DC processes, including X up-regulation in male *Drosophila* and X inactivation in therians). In chicken, the density of interspersed repeats is 60% higher in the Z chromosome than in autosomes, resulting in large expansions of intergenic regions and lower gene density (11 genes per Mb, versus 25 for autosomes; Bellott et al. 2010). X and Z chromosomes also show a reduced level of standing genetic variation at neutral markers relative to autosomes (e.g. Moriyama & Powell 1996; Begun & Whitley 2000) and more rapid evolution (faster-X; e.g. Counterman et al. 2004 for *Drosophila*). Both features may stem from several causes, including larger drift, purifying selection, or positive selection due to hemizygous exposure in the heterogametic sex or to the acquisition of DCC binding sites (Section 5.2.2e). Potential causes can only be disentangled in few specific instances; due to the particular life cycle of aphids, for instance, X chromosomes have the same effective population size as autosomes and similar levels of allelic richness, but show significantly higher dN and dN/dS ratios, supporting a role for positive selection on recessive genes expressed at hemizygous state in males (Jaquiéry et al. 2012). Comparisons of homologous sequences of sperm proteins in human, mouse, and rat also show more positive selection on the X than on autosomes, suggesting a role for hemizygous expression in males (Torgerson & Singh 2003, 2006). Positive selection for X-linked genes is similarly suggested by comparisons between humans and chimps, with 30% more frequent non-synonymous substitutions than on the autosomes (Nielsen et al. 2005; Lu & Wu 2005).

In birds, dN/dS ratios are significantly higher for Z-linked than for autosomal genes, despite similar ratios of non-synonymous to synonymous polymorphism, and even accounting for the fact that the Z chromosome harbours more male-biased genes, with higher rates of evolution (Mank et al. 2007, 2010). Fast-Z effects are pronounced: the mean divergence between chicken and turkey (based on intron sequences) is 12% for the Z, as compared to 8% for autosomes, and 5% for the W chromosome. Comparing more than 2500 orthologues shared between zebra finch, chicken, human, mouse, and opossum, Ellegren (2009) found a large overrepresentation of fast evolving genes on the bird Z chromosomes relative to autosomes; accelerated genes had higher dN/dS values, due to both higher dN and lower dS. Surprisingly, no such patterns were found in the three mammal species involved, suggesting that faster-Z effects exceed faster-X effects. This might stem from the confounding effects of stronger sexual selection and lower effective population size in males, as well as over-representation of male-biased genes on the avian Z chromosome. As there seem to be no differences whether genes are expressed predominantly in males or in females, the strong faster-Z effects in birds might stem more from drift than from positive selection (Mank et al. 2010).

Selective episodes seem to happen rapidly following recombination arrest. Chromosome-wide diversity in the 1-My-old neo-X of *D. miranda* is two-fold lower than on the ancestral X chromosome, and 10–15% of neo-X genes appear to have been targets of recent adaptive evolution, versus only 1–3% of genes on the ancestral X chromosome (Bachtrog et al. 2009). The rate of adaptation is also one order of magnitude higher than background levels in the genome, an acceleration that goes in parallel with the evolution of dosage compensation (Section 5.2.2e).

As expected from sex-biased inheritance and hemizygous expression in the heterogametic sex, X and Z chromosomes are enriched in genes associated with sex and reproduction. Table 5.2 compares the abundance of a number of gene classes in the light of these predicted 'sex roles'. Both masculinization and feminization have occurred, mostly depending on the patterns of heterogamety. The Z chromosomes of birds are depauperate in feminizing (e.g. ovary-specific) genes and enriched in male-biased genes (Kaiser & Ellegren 2006; Storchova & Divina 2006). The chicken Z chromosome, for instance, contains massive tandem arrays of duplicated genes expressed in testes, expanding

**Table 5.2** Predictions and evidence for the distribution of specific gene classes on the sex chromosomes of animals in the context of their 'sex roles'. Data and references relating to female heterogamety are given in brackets.

| Gene class | Y (or W) chromosome | X (or Z) chromosome | Species | Reference |
|---|---|---|---|---|
| Sex specifically expressed genes | Enriched for male (female)–biased genes, depleted for female (male)-biased genes | Depleted for male (female)-biased genes. Enriched for female (male) biased genes | *C. elegans*, several *Drosophila* species, mosquito, mouse, human | Reinke et al. 2000, Parisi et al. 2003, Ranz et al. 2003, Skaletsky et al. 2003, Khil et al. 2004, Sturgill et al. 2007, Kaiser et al. 2011, & Bachtrog 2012b, Magnusson et al. 2012, Meisel et al. 2012 |
| | | | NOT in medaka, mouse, human | Kondo et al. 2006, Wang et al. 2001, Yang et al. 2006 |
| | | | (chicken, flycatcher, zebra finch, common whitethroat, crow) | (Saether et al. 2007, Kaiser & Ellegren 2006, Storchova & Divina 2006, Wolf & Bryk 2011, Naurin et al. 2012) |
| Fertility genes | Male (or female) fertility genes | Female (male) fertility genes | Several *Drosophila* species, human | Kennison 1981, Hackstein & Hochstenbach 1995, Carvalho et al. 2000, Skaletsky et al. 2003; Lange et al. 2009 |
| | | | (chicken) | (Hori et al. 2000) |
| Gonad specific genes | Enriched for testes, depleted for ovaries | Depleted for testes, enriched for ovaries | *D. melanogaster*, *D. miranda*, mouse, human | Betrán et al. 2002, Skaletsky et al. 2003, Sturgill et al. 2007, Mueller et al. 2008, Zhou & Bachtrog 2012b, Meisel et al. 2012 |
| Sexual selection genes | Male (female) specific | Female (male) specific | Sticklebacks, medaka fish, poeciliids, cichlids | Kallman 1970, Wada et al. 1998, Lande et al. 2001, Lindholm & Breden 2002, Streelman et al. 2003, Fernandez & Morris 2008, Kitano et al. 2009, Roberts et al. 2009 |
| | | | (warbler) | (Pala et al. 2012) |
| Sexually antagonistic genes | Beneficial for males (females) | Depleted for male (female) beneficial when recessive | *D. melanogaster*, *D. miranda* | Innocenti & Morrow 2010, Zhou & Bachtrog 2012b |
| Growth genes | Promoting male (female) growth | Promoting female (male) growth | Mammals | Mittwoch 1986, 1989, Kraak & De Looze 1993 |

over 15% of the chromosome (Bellott et al. 2010). This masculinization of Z may suggest a role for dominant male-beneficial mutations; however, the lower dosage compensation in birds, together with Z inactivation during female meiosis (MSCI; Schoenmakers et al. 2009) certainly contribute to favour Z linkage of sexually selected genes, with direct consequences on the sex-specific heritability of male fitness (Chapter 6).

Conversely, the X chromosomes of flies and nematodes are generally enriched in genes that are female-biased in expression or female-specific in activity, and impoverished in male-biased genes (Reinke et al. 2000, 2004; Parisi et al. 2003; Oliver & Parisi 2004; Sturgill et al. 2007). Demasculinization mostly affects genes that are highly expressed in the male germ line and during meiosis, pointing to a strong effect of MSCI (e.g. Hense et al. 2007; Vibranovski et al. 2009), but some genes with male-biased expression in somatic cells are also depleted from the X, suggesting an additional role for sex antagonisms (Bachtrog & Charlesworth 2002; Bachtrog 2003; Oliver & Parisi 2004; Sturgill et al. 2007). Male-biased genes usually show a high turnover rate; demasculinization of X chromosomes occurs through both less frequent emergences and more frequent losses, occurring through retroposition of male-specific genes out of the X (new autosomal copies are intronless with polyA tracts; Betrán et al. 2002; Schrider et al. 2011). The same patterns are documented on the neo-X chromosomes of *D. pseudoobscura* and *D. willistoni*, which show a depletion of male-biased genes on the neo-X, and an excess of testis-expressed genes duplicated off the neo-X chromosome (Meisel et al. 2009). By cytogenetically cloning series of *D. melanogaster* X chromosomes in different genomic backgrounds, Gibson et al. (2002) also confirmed the expected strong X enrichment in SA variation: at the juvenile stage (where gender roles are most similar), the X chromosome made no detectable contribution to genome-wide fitness variation, while at the adult stage (in which gender roles strongly diverge), the X chromosome harboured ~45% of the genome-wide fitness variation and 97% of the negative between-sex covariance for adult fitness (compared to 20% expected from its relative size).

The mammalian X is similarly enriched in female-biased and female-specific genes (Khil et al. 2004; Zhang et al. 2010), but male-biased genes display a more complex pattern, apparently moulded by the action of MSCI, which occurs later during germline development than in *Drosophila* or *C. elegans*. Namely, male-specific genes expressed in the mitotic stage of spermatogenesis (before X is inactivated) are over-represented on the X, those expressed at X inactivation are depleted, and those expressed later in meiotic stages are completely absent (Khil et al. 2004). Retroposition was also instrumental in relocalizing these male-biased genes to- and from the X chromosome (Emerson et al. 2004; Potrzebowski et al. 2008, 2010). Interestingly, the X is also enriched in genes with post-meiotic testis expression, which are present in multiple copies, possibly to compensate for the repressive action of chromatin in post-meiotic stages (Turner 2007; Mueller et al. 2008). Similar to selective episodes in *Drosophila* (Bachtrog et al. 2009), much of the gene trafficking to- and from the X chromosome seem to have occurred in early stages after recombination arrest (Potrzebowski et al. 2010).

## 5.3 Heterogeneity between lineages

One of the most enigmatic features of sex-chromosome evolution is the striking heterogeneity found among lineages in the patterns of differentiation and rates of decay of Y or W chromosomes. Such heterogeneity may occur at all phylogenetic levels. Unlike animals, for instance, angiosperms show no strongly degenerated sex chromosomes. The contrast is marked, even accounting for the generally more recent origin of dioecious lineages in angiosperms; the non-recombining region is still very limited on the sex chromosomes of Salicaceae, despite a very ancient origin of dioecy in this group; the sex chromosomes of *S. latifolia* show much less decay than the neo-sex chromosome of *D. miranda*, despite being older (10 Mya versus 1 Mya; though the number of generations might be similar). A possible cause lies in the lower level of sexual dimorphism and sexual selection in plants. Sex chromosomes may facilitate the evolution of sexual dimorphism (Rice 1984; Chapter 6) and reciprocally, sexual conflicts might trigger stronger sex chromosome differentiation. However, there are many other differences between campions and flies, which confuse interpretation. Purifying selection may operate in the haploid gametophytic stage of plants (Bergero & Charlesworth 2011; Chibalina & Filatov 2011), which is lacking in animals; the non-recombining segment expanded stepwise in *Silene*, but encompassed the whole neo-sex chromosome at once in *Drosophila*, inducing stronger Hill–Robertson interferences. In addition, dosage compensation was already in place in *Drosophila*, which might have accelerated gene loss on the Y chromosome.

Within vertebrates, sex chromosomes are usually strongly heteromorphic in the warm-blooded birds and mammals, but homomorphic in the cold-blooded fishes and amphibians. Sexual selection cannot be invoked here (many fishes are under strong sexual selection). The less canalized ontogeny of sex determination may play a role, by facilitating sex reversal and X-Y recombination; turnovers are also more frequent in cold-blooded vertebrates, which possibly stems from a lack of proper dosage compensation (Section 7.2.2b). Even more surprising are differences within vertebrate classes: among snakes, the Boidae share with Viperidae the same pair of ZW sex chromosomes, which are however homomorphic in the former and highly heteromorphic in the latter. Among birds, similarly, the ratites share with neognaths the same sex-determination system and sex chromosomes, but with no Z-W differentiation despite being the same evolutionary age. Ratites might have adopted alternative strategies to solve sexual conflicts, relying on sex-biased expression rather than recombination arrest (Vicoso et al. 2013). Why and how these early-diverged lineages maintain sex-chromosome homomorphy remain open and fascinating questions.

CHAPTER 6

# Evolutionary correlates of sex-determination systems

Sex-determination systems have profound and manifold consequences on the biology of organisms, affecting many aspects of their life histories. The mere existence of two sexes with alternative ways of obtaining reproductive fitness drives the evolution of disparate sexual traits, manifested as sex differences in morphology, behaviour, and reproduction: organisms with separate genders often show strong sexual dimorphism in size, colouration, and mating behaviour. Sex-determination systems also impact on the evolution of reproductive modes (such as constraining the establishment of parthenogenesis), on life history (such as sex allocation patterns), and on speciation processes. From the variety of potential evolutionary consequences, we restrict ourselves to a limited number of traits and processes that have been addressed in some detail in the scientific literature, that is, we will consider how sex-determination systems interact with the evolution of polyploidy and parthenogenesis, reproductive modes and sex allocation, sexual selection, and speciation. Readers should keep in mind that such interactions are often bidirectional: the mode of sex determination may dictate the evolution of sex-specific traits, but differences in selective forces between the sexes may also feedback on sex-determination systems. Such two-way interactions are well illustrated by the processes of sexual selection; sex-determination systems provide the basis for the evolution of the sex roles, but sexual selection in turn drives changes in the genome, and particularly in the genetic composition of sex chromosomes (Chapter 5).

## 6.1 Sex determination and polyploidy

Certain sex-determination systems may facilitate the evolution of polyploidy. As already pointed out by Muller (1925), one reason why polyploidy is much more prevalent in plants than in animals might reside in their widespread hermaphroditism: polyploidy likely disrupts sex determination in gonochoric animals. Dioecious plants, furthermore, often have **multifactorial sex determination**, which might more easily accommodate changes in chromosome copy number than systems with distinct sex chromosomes. Nonetheless, there are some reports of a disruption of sex determination in plants following polyploidization (Pannell et al. 2004). Polyploids are thus expected to be rare in genetic sex determination (GSD) animals, but their occurrence would be less constrained in taxa with environmental sex determination (ESD), hermaphroditic, or asexual reproduction (Orr 1990; Otto & Whitton 2000). Consistently, polyploidy appears more frequent in taxa with hermaphroditic reproduction and epigenetic sex determination (Mable 2004) and often associates with parthenogenesis (Suomalainen et al. 1987; Stenberg & Saura 2009).

The sex-determination disruption argument of Muller (1925) was based on crosses with *Drosophila melanogaster*, in which the X:A ratio system of sex determination is indeed bound to be disrupted under a ploidy increase. Is the more prevalent dominant-Y system (or W under female heterogamety) more prone to lead to polyploidy? In such a system, the sex ratio might be biased towards the heterogametic sex in a newly arisen population of

*The Evolution of Sex Determination*. Leo W. Beukeboom and Nicolas Perrin.

tetraploids (as XXXY, XXYY, XYYY individuals will all be male and only XXXX individuals will be female), but sex-ratio selection should quickly restore the balance (Otto & Whitton 2000). Although the predicted association between a Y-dominant sex determiner and polyploidy in GSD taxa has not been formally tested, a simple analysis of the predicted sexes of polyploids under Y or W dominance versus X or Z chromosome counting systems reveals that the latter are more prone to produce intersexes (Wertheim et al. 2013).

Another obstacle for the evolution of polyploidy in gonochorists is the disruption of dosage compensation (Orr 1990; Otto & Whitton 2000). The X-counting mechanism that initiates X-silencing in female mammals, or up-regulation of the single X chromosome in the heterogametic sex, are likely to be disrupted by polyploidization, leading to an imbalance between the sex-chromosome and autosome gene dose. As a result, polyploidy should be more frequent in animal groups with neither degenerate Y (or W) chromosomes nor compensatory gene dosage mechanisms. In support of this prediction, polyploidy is relatively common in groups such as amphibians where heteromorphic sex chromosomes are rare and dosage compensation is likely absent (Schmid et al. 1991; Eggert 2004; Schartl 2004).

## 6.2 Sex determination and parthenogenesis

One major distinction in the mode of reproduction is between sexual and asexual reproduction. Asexual (i.e. clonal) reproduction is widespread among all major clades of eukaryotes; many organisms reproduce by budding or **fission**. In some lineages (e.g. euglenids, bdelloid rotifers), it is the only documented mode of reproduction. Parthenogenesis represents a specific mode of asexual reproduction, secondarily derived from sexual reproduction, and refers to the development of a multicellular organism from an unfertilized egg. It thus requires a functional female (and sometimes also male) apparatus. It has evolved multiple times independently in a wide variety of groups. In some haplo-diplontic lineages (as widespread among algae), unfertilized gametes may enter a parthenogenetic asexual cycle by germinating without fusion to produce either a gametophyte or a partheno-sporophyte; the latter produces mitospores that develop as gametophytes, completing the asexual cycle (Peters et al. 2008; Coelho et al. 2012). Parthenogenesis is also found among diplontic or sub-diplontic cycles; it has evolved multiple times in insects and angiosperms. Different modes can be recognized, with distinct consequences for the homozygosity of the progeny (Suomalainen et al. 1987). In apomictic (= mitotic) parthenogenesis, no reductional division occurs, so that offspring are clonal copies of their mother (i.e. the patterns of heterozygosity are not affected). In automictic (= meiotic) parthenogenesis, a reductional division first occurs before diploidy is restored. This most frequently occurs by fusion with a polar body, so that heterozygosity is halved at each generation (equivalent to selfing); alternatively, the meiotic product can be duplicated, followed by fusion (as in the autogamy of ciliates), resulting in complete homozygosity. Note that automictic parthenogenesis can also be considered as a mode of sexual reproduction, insofar as it involves meiosis, even though no genetic material is exchanged between individuals.

Certain taxa may be more prone than others to evolve parthenogenesis, due to a number of developmental constraints including the mechanism of sex determination (Engelstädter 2008). Parthenogenesis can only sustainably evolve if it leads to the development of females. Female heterogamety is thus ill-adapted to automictic parthenogenesis: polar-body fusion would result in a mixture of ZZ males, ZW females, and likely unviable WW offspring. Parthenogenesis has been shown to occur in some birds (turkey) from the duplication of a haploid post-meiotic set of chromosomes, resulting in males only (WW are not viable). Male heterogamety is a priori better suited, because XX females only produce XX daughters. In mammals, however, parthenogenesis is prevented because of parental imprinting: both a paternal and a maternal genome are required for proper embryonic development.

**Haplodiploidy** seems less constrained in this respect, because it already involves egg development without fertilization (haploid males) in the absence of specialized sex chromosomes. Parthenogenesis has indeed evolved multiple times independently in haplodiploid insects (Normark 2003). However, the specific genetics of sex determination and cytological mode of parthenogenesis may hamper its evolution in some groups. In the complementary sex-determination (CSD) system prevalent in Hymenoptera, femaleness depends on heterozygosity at the sex-determination locus; homozygosity yields diploid males (Box 3.21). Parthenogenesis in CSD species can therefore only evolve if the cytological mechanism retains heterozygosity at the sex locus, as in apomictic species (Van Wilgenburg et al. 2006; Heimpel & De Boer 2008). Automictic parthenogenesis may be compatible with CSD if the sex locus lies in a non-recombining region, such as close to a centromere or within an inversion (Beukeboom & Pijnacker 2000). Parthenogenesis-inducing endosymbionts, such as *Wolbachia* and *Cardinium*, occur widely among Hymenoptera and are only able to invade species that lack CSD, because the bacteria predominantly restore diploidy by gamete duplication resulting in complete homozygosity, and hence diploid males.

ESD may also seem less restrictive in the evolution of parthenogenesis (because there are no obstructive genetic mechanisms) but requires additional adaptations to produce females only; otherwise, environmental factors will operate in the same way as in sexually produced offspring, resulting in both males and females (Engelstädter 2008). Parthenogenesis in cladocerans (e.g. *Daphnia pulex*, *D. magna*) is apomictic and cyclical: mothers clonally produce only daughters during the growing season, and sons when conditions deteriorate (Box 3.16). Some asexual clones have lost the ability to produce males, apparently due to a dominant allele at a single locus (Galimov et al. 2011). In the pea aphid (*Acyrthosiphon pisum*), XX females apomictically produce all-daughter generations during the summer, and a mixture of XO males and XX females at the onset of winter; XO males arise from the random loss of one of the two maternal X chromosomes in the apomictic egg. Note that cyclical parthenogenesis is prevented in XY systems (because XX females cannot restore XY males) but is theoretically feasible with a ZW system: dropping the W and duplicating the Z chromosome would restore males.

Although parthenogenetic reproduction might seem facilitated in hermaphrodites (because all offspring possess the female function), the number of documented cases in animals is remarkably low. One possible reason is that hermaphrodites may alternatively rely on selfing as a reproductive insurance. Another factor may be the difficulty of detecting it; given that hermaphrodites retain the male function, apomictic parthenogenesis can only be identified by showing that oogenesis is mitotic rather than meiotic, and automictic parthenogenesis by showing that sperm does not contribute to egg fertilization. Caution needs to be exerted here, as some parthenogenetic hermaphroditic Turbellaria and Oligochaeta are sperm-dependent (pseudogamous); sperm is required for egg activation but subsequently expelled from the zygote (Beukeboom & Vrijenhoek 1998). A particularly intriguing case is that of the flatworm *Schmidtea polychroa* in which sexual and sperm-dependent parthenogenetic hermaphrodites co-occur and exchange sperm by sperm-trading (Michiels & Kuhl 2003). Further study of hermaphroditic invertebrates is expected to yield more examples of parthenogenesis.

The best illustration of the accommodation of epigenetic sex determination in the evolution of parthenogenesis comes from plants. The vast majority of plants are hermaphroditic, and **apomixis** has evolved in many groups, such as in Poaceae (grasses), Liliaceae, Asteraceae, and Brassicaceae (Ozias-Akins & Van Dijk 2007). Apomixis in plants encompasses the development of a diploid female gamete without meiosis (**apomeiosis**), development of the embryo without fertilization (parthenogenesis), and developmental alterations to ensure endosperm development. The (epi)genetic regulation of apomictic egg development is still poorly known, despite great efforts in the study of *Arabidopsis* and maize, in particular, from the perspective of agricultural applications (e.g. Tsanev et al. 2006). The evolutionary dynamics of parthenogenetic hermaphrodites are discussed in Joshi & Moody (1998), Weinzierl et al. (1998), and Meirmans et al. (2006). Whether dioecy in plants has restricted the evolution of apomixis has not been addressed as far as we know.

## 6.3 Sex determination and sex allocation

The theory of sex allocation is rightly considered as one of the success stories of evolutionary biology, supported with both qualitative and quantitative evidence (West 2009). Sex allocation refers to the relative investment of resources to the male versus the female function (Charnov 1982). In unisexual species (dioicous or dioecious), differential investment is measured in terms of the number of male and female offspring born to mothers; in simultaneous hermaphrodites, it refers to the relative investment into testes and ovaries; in sequential hermaphrodites, to the switching time between the male and female role; and under social sex determination, to the decision to develop as male or female. The concept implies flexible control over the production of gametes or offspring of different sexes, or over sex change, which is obviously constrained by the sex-determination system. In fact, sex determination and sex allocation cannot be viewed as independent processes. Changes in sex determination that are manifested as a shift from males to females (or vice versa) have immediate consequences for sex allocation; a gene that can cause a change in sex is therefore also a sex allocation gene. The reverse is not always true: genes that affect sex allocation can operate without affecting the mechanisms of sex determination. An example is a meiotic-drive X chromosome that results in the overproduction of females, but only does so by altering the proportion of X versus Y-bearing sperm produced by the heterogametic sex (Hamilton 1967).

### 6.3.1 Parental control and genetic conflicts over sex allocation

A good understanding of the evolutionary dynamics of sex determination and sex allocation requires further specification of their genetic regulation and the levels of selection involved. Werren & Beukeboom (1998) distinguished three categories of genetic elements with distinct selection levels (Fig. 6.1). The first one comprises parental sex-ratio (or sex-allocation) genes, which act in the parents to affect the relative proportion of males and females in the progeny. In the second category are **parental-effect sex determiners**, that is, genes that are expressed in the parents but act in the developing zygote to affect its sex. The third category comprises the 'standard' zygotic sex-determination genes that are expressed in the developing zygote and affect its sex. Because genes from the three categories are expressed at different stages, they also undergo different selection regimes; genes from the first two categories are selected with respect to parental fitness, while zygotic sex determiners evolve according to offspring fitness. Many opportunities for genetic conflict arise as a result of these differential selection levels, both between male (paternal) and female (maternal) determiners within categories, but also between male and female determining elements of different categories. Examples include parent–offspring conflict over sex determination (Trivers 1974) and **nucleo-cytoplasmic conflicts**. It follows directly from this conceptual view that sex ratios result from the interplay between sex-allocation and sex-determination genes. In extreme cases, the effects of sex-allocation genes may be counteracted by sex-determination genes, as in the case of meiotic-drive sex chromosomes or cytoplasmic male-sterility genes on mitochondria, counteracted by autosomal suppressors (Section 6.3.2); sex-ratio selection induced by sex-allocation genes may select for turnovers in sex-determination systems (Bull 1983; Section 7.2.3).

Sex chromosomes play an important role in the control of sex allocation because they can both constrain adaptive sex allocation and protect against invasion of sex-ratio distorters. A fair meiosis ensures Mendelian segregation of male- and female-determining chromosomes, resulting in a 1:1 offspring sex ratio. Heterogametic systems rarely display strong biases in **primary sex ratios**, except for cases of meiotic drive (Section 6.3.2). Nevertheless, a number of studies have reported slight but significant deviations in insects, birds, and mammals (Clutton-Brock et al. 1984, 1986; Pike & Petrie 2003; Rosenfeld & Roberts 2004; Sheldon & West 2004; reviewed in West 2009), including some contentious claims in humans (James 2010; Dixson et al. 2011). Biases are generally in line with Trivers & Willard's (1973) predictions (Section 2.1.6e). Evidently, sex chromosomes do not put absolute constraints on sex allocation. Some spectacular examples of adaptive shifts in primary sex ratios are known from birds (Dijkstra et al. 1990; Komdeur et al. 1997; Nager et al. 1999), where hormones and

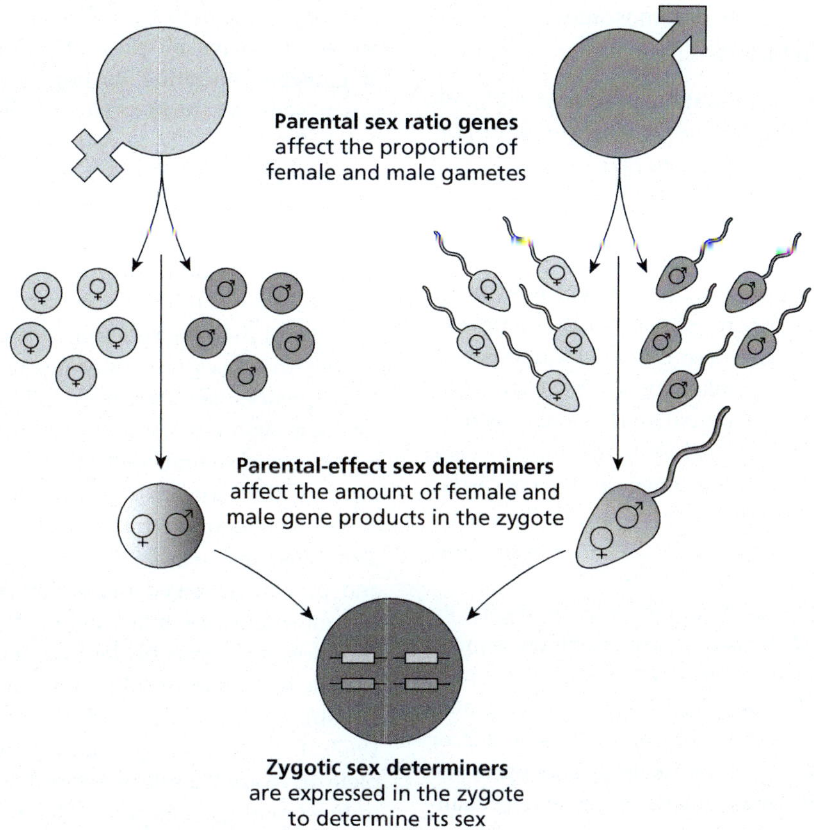


**Figure 6.1** Sex-ratio and sex-determination genes. In diplo-genotypic sex-determination systems, genes affecting sex determination or sex allocation can be classified in three categories: parental **sex-ratio genes**, parental-effect sex determiners, and zygotic sex determiners. These categories differ in the generation in which they are expressed (parents versus offspring) and between the two sexes. Adapted from Werren & Beukeboom (1998).

meiotic mechanisms possibly affect the segregation of Z and W chromosomes in females (reviewed in Rutkowska & Badyaev 2008).

Haplodiploid sex determination depends on egg fertilization (typically under maternal control), which allows fine-tuning of sex allocation to the benefits of mothers. Haplodiploid species have been extensively used for testing sex-allocation theory (West 2009), most spectacularly in fig wasps (Cinipidae) and parasitoid wasps (Chalcidoidea) (Werren 1983; Herre et al. 1997). Interestingly, both groups lack CSD, a mode of sex determination that may constrain sex allocation (as inbreeding and homozygosity leads to the production of diploid males; Cook & Crozier 1995). Sex allocation is generally expected to be the least constrained under diverse forms of epigenetic sex determination. In simultaneous hermaphrodites (Section 2.1.6a), examples are known where the investment is drastically shifted towards the male or female function (reviewed in Goldman & Willson 1986; Charlesworth & Morgan 1991; Sakai 2000 and Burd 2011 for plants; Schärer 2009 for animals). Similarly, sequential hermaphroditism and environmental or social sex determination have provided nice examples where sex determination evolves to maximize reproductive output through differential investment in the two sexes, whenever male and female reproductive values vary differently with the environment or the social context (Ghiselin 1969; Charnov & Bull 1977; Sections 2.1.6a–d).

### 6.3.2 Meiotic-drive sex chromosomes and sex-ratio distorters

The flexibility in sex allocation associated with epigenetic sex determination might increase the opportunities for genetic elements to take over control of sex allocation or sex determination. Sex-ratio distorters are genetic elements that enhance their own transmission by biasing Mendelian segregation during meiosis, by altering the sex of their host, or by affecting their host's investment in the male and female functions. They are found ubiquitously in plants and animals (Jaenike 2001; Werren et al. 2008). A plethora of such sex-ratio distorters have been reported in the literature, including meiotic-drive sex chromosomes, **B chromosomes**, and cytoplasmic elements, such as mitochondria and microorganisms (reviewed in Jaenike 2001; Werren et al. 2008; Duron et al. 2008).

Meiotic-drive sex chromosomes segregate to more than half of the gametes in the heterogametic sex. Female heterogamety is vulnerable to true meiotic drive, which relies on the asymmetry of meiosis in females; sex-ratio biases will result if Z or W gametologues are more likely to end up in the polar body. Under male heterogamety, the mechanisms have to be post-meiotic, relying normally on two closely linked genes, the driver (or killer) locus and the responder (or target) locus (Montchamp-Moreau et al. 2006; Tao et al. 2007a,b). A chromosome carrying both the killer and resistant alleles accrues its drive by preventing the development of gametes inheriting its homologue, which carries a sensitive (non-resistant) allele (Meiklejohn & Tao 2010, Fig. 6.2). Meiotic drive can plague any chromosome (e.g. the *t*-haplotype in mice), but sex chromosomes are particularly vulnerable, because they provide more opportunity for linkage disequilibrium and carry many heterochromatic regions, two conditions favouring the evolution of a linked driver-responder complex. In addition, meiotic-drive sex chromosomes have conspicuous consequences (sex-ratio biases) and may segregate for long evolutionary times under the counteracting effects of sex-ratio selection (autosomal drivers often readily spread unnoticed to fixation). They have been reported from insects (including male dipterans and female lepidopterans), lemmings, and plants (reviewed in Jaenike 2001). Driving X chromosomes are more frequent than driving Y chromosomes, probably because the harmful effects of male bias, in terms of population extinction rate, are stronger than those of female bias (Hamilton 1967). For the same reasons, elements with complete drive are at a selective disadvantage compared to those with partial drive. Meiotic drive elements can also have negative effects on host fitness, such as reduced fertility (Jaenike 1996), which may further counteract their spread. Meiotic-drive sex chromosomes seem to occur widely in populations,

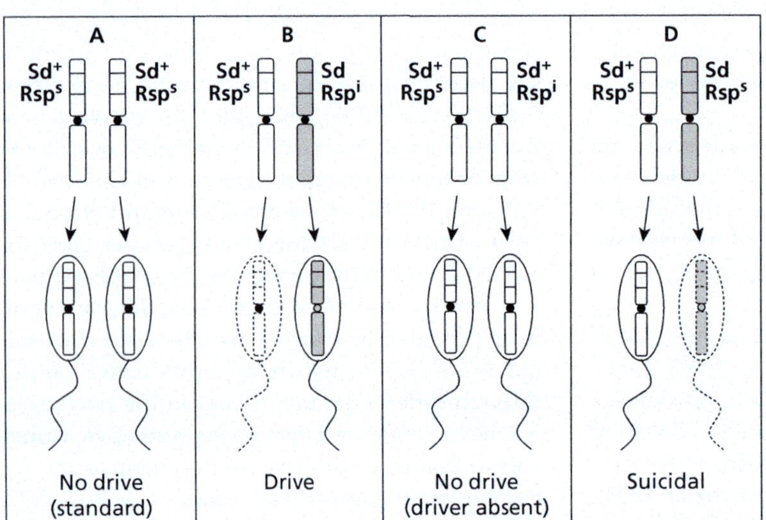


**Figure 6.2** Meiotic drive. A **segregation distorter** (Sd) is linked with a responder locus (Rsp) and causes loss of the non-carrier homologue. (A) The Sd wild-type allele (Sd $^{+}$) does not cause drive, (B) The driver allele (Sd) causes death of the homologue that carries a sensitive allele of the responder (Rsp$^{s}$). It is immune itself because of a linked insensitive responder allele (Rsp$^{i}$). Recombination between these two loci would result either in the loss of the drive (C) or in a suicidal behaviour (D).

but are controlled locally by suppressor genes and, therefore, often not detected (Box 6.1). They may induce transitions in sex-determination systems, as recent theoretical work suggests (Kozielska et al. 2010; Section 7.2.3c).

B chromosomes are supernumerary chromosomes, not essential for the life of their host, and usually mostly or entirely heterochromatic. They offer thus similar opportunities for the spread of selfish elements, including sex-ratio distorters. B chromosomes have been found to affect sex ratios in wasps, shrimps, frogs, and fishes (Green et al. 1993; Beladjal et al. 2002; Yoshida et al. 2011; reviewed in Camacho et al. 2011; Box 6.2). Being born

**Box 6.1 How abundant are meiotic-drive sex chromosomes?**

The prevalence of meiotic-drive sex chromosomes in nature and their potential as an evolutionary force have long been subject of debate. The work of C. Montchamp-Moreau and colleagues on *Drosophila simulans* is shedding important light on this issue. A driving X chromosome called '*sex-ratio*' ($X^{SR}$) recently appeared in this species, from a 37.5 kb tandem duplication spanning six genes (Fouvry et al. 2011). Crossing $X^{SR}$ into a standard (ST) laboratory strain showed the SR trait to be fully expressed (brood sex ratios with > 90% females) in individuals with a complete set of ST chromosomes, partly expressed in individuals that carried one of the original SR strain autosomes or the Y chromosome, and entirely suppressed in individuals with a complete set of SR chromosomes (Atlan et al. 1997, 2003, Montchamp-Moreau et al. 2001; Fig. 6.3). This means that suppressors of $X^{SR}$ have evolved in the original SR strain on all major autosomes plus the Y chromosome, which together are able to completely suppress the X drive. The recent invasion and spread of $X^{SR}$ in East Africa and Madagascar was quickly followed by complete drive suppression, then by the replacement of $X^{SR}$ by $X^{ST}$ due to a fitness cost of the segmental duplication involved in the drive. This supports a scenario of rapid cycling of X chromosomes: once suppression has evolved, standard $X^{ST}$ chromosomes locally replace costly $X^{SR}$ chromosomes within a few decades (Bastide et al. 2011).

These studies suggest that meiotic-drive sex chromosomes are abundant, but remain cryptic (due to locally evolved suppressor genotypes) and are short lived. Meiotic drive is, presumably, an important force in shaping genomic processes and the genetic composition of sex chromosomes. However, extensive studies have not been performed outside dipterans and rodents, which leaves some doubt as to the frequency and taxonomic distribution of driving sex chromosomes and their ubiquity as a motor of sex-chromosome evolutionary change.

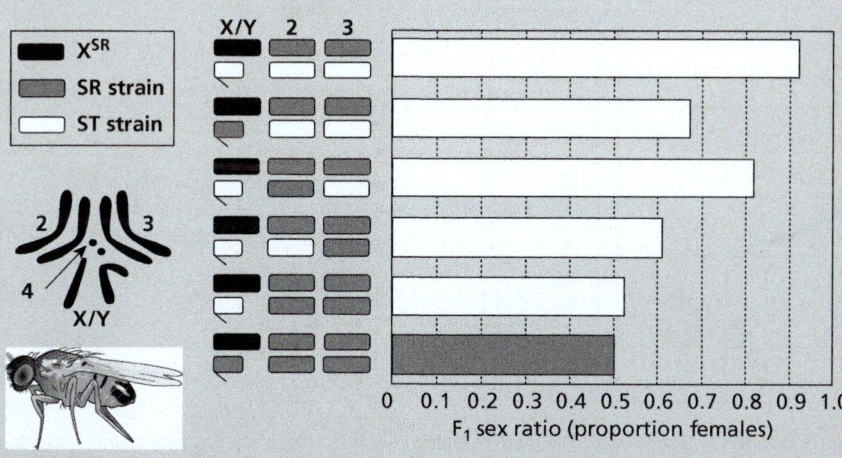


**Figure 6.3** Suppression of X chromosome drive in *Drosophila simulans*. Suppression of drive depends on the whole genetic background. Drive (expressed as proportion of females in the progeny) is strong when $X^{SR}$ is introduced in a standard strain, partly suppressed when counteracted by either the Y chromosome or autosomes #2 and #3 from the SR strain, and fully suppressed in its original SR genetic background. ST = standard strain, SR = sex ratio strain, $X^{SR}$ = driving X. Adapted from Cazemajor et al. (1997).

### Box 6.2 Sex-ratio distorters in *Nasonia vitripennis*

Four different types of sex-ratio distorters have been described from *Nasonia*. Their action is depicted in Fig. 6.4, and their evolutionary dynamics described in Werren (1983). Most *Nasonia* species are doubly infected with a representative from each of two major *Wolbachia* groups (A and B); due to haplodiploidy, the mechanisms of

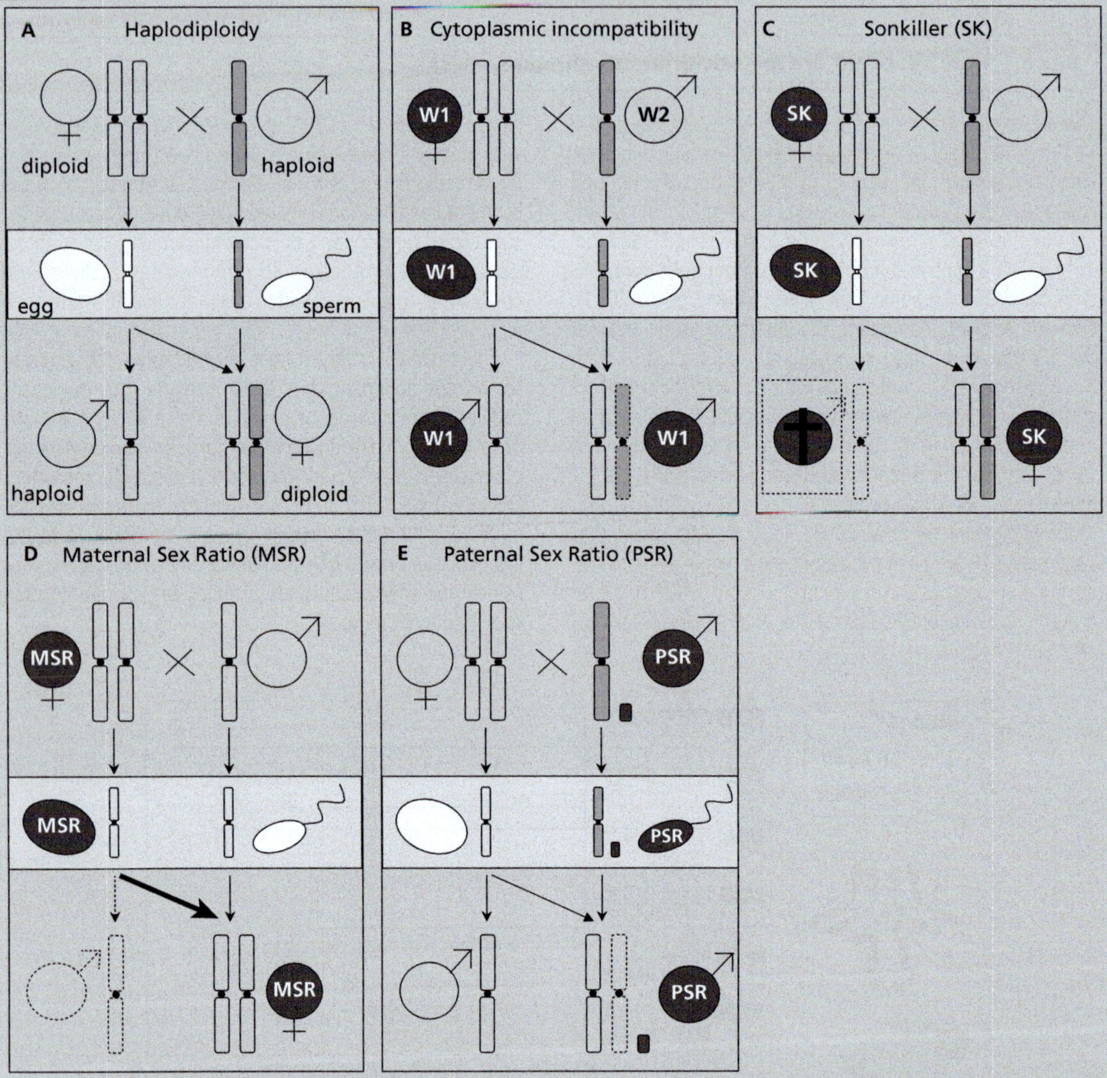


**Figure 6.4** Sex ratio distorters in *Nasonia vitripennis*. (A) Haplodiploidy: Females have control over fertilization; diploid daughters develop from fertilized eggs, and haploid sons from unfertilized eggs. (B) Cytoplasmic incompatibility: males and females harbour different *Wolbachia* bacteria (W1 and W2) resulting in paternal chromosome elimination in the fertilized egg, which dies or develops into a male. (C) **Sonkiller**: females that carry the sonkiller bacterium produce only daughters because haploid eggs die. (D) Maternal Sex Ratio: infected females produce only daughters because they fertilize all of their eggs. (E) Paternal Sex Ratio: males that carry the PSR chromosome produce only sons because PSR destroys the paternal chromosomes in the fertilized egg.

*continued*

**Box 6.2** *Continued*

cytoplasmic incompatibility via paternal genome loss may result in viable males, rather than complete zygote failure (Breeuwer & Werren 1990; Raychoudhury & Werren 2012). The maternally inherited bacterium *Arsenophonus nasoniae* selectively kills haploid (male) embryos (Skinner 1985; Werren et al. 1986, Gherna et al. 1991); son-killer elements are known from many arthropods (Hurst et al. 1997) and thought to benefit from the better survival of daughters in the brood. Maternal Sex Ratio (MSR) causes carrier females to produce all-female or strongly female-biased progeny (Skinner 1982). MSR is maternally transmitted but has not been identified yet; as it is insensitive to antibiotic treatment, it is likely a mitochondrial variant that enhances its own transmission by somehow inducing females to fertilize all of their eggs. The most remarkable sex-ratio distorter is Paternal Sex Ratio (PSR), a supernumerary B chromosome that is paternally inherited and causes all-male progenies (Nur et al. 1988, Werren 1991). PSR destroys the paternal chromosomes of fertilized eggs in crosses between females and carrier males. It effectively turns diploid (female) eggs into haploid males that inherit the element. Being the sole survivor of the male genome, it may be considered the most selfish genetic element ever described.

as selfish elements, B chromosomes might gain by retroposition genes that benefit their host and thus evolve towards normal A chromosomes. This scenario was actually suggested for the emergence of *Drosophila* Y chromosomes, which show little or no homology with the X chromosome (Hackstein et al. 1996; Carvalho 2002).

Many sex-ratio distorters reside in the cytoplasm of their host's cells, including eggs, and gain a transmission advantage by inducing their host to invest more energy in the production of the female sex, through which they are transmitted. As sperm do not contain cytoplasm, males constitute a dead-end for these elements. Many hermaphroditic plants show cytoplasmic male sterility, due to mitochondria that induce male sterility by disrupting pollen development (reviewed in Burt & Trivers 2009). The occurrence of male-sterile individuals in populations selects for both a biased investment toward the male function in hermaphrodites and genetic suppressors of the sterilizing mitochondria. Although cytoplasmic male sterility is also expected to evolve in animals, very few examples have been reported (e.g. Clancy et al. 2011).

Intracellular endosymbionts, such as bacteria (*Wolbachia, Cardinium, Rickettsia, Spiroplasma*) and fungi (microsporidia) occur widely in arthropods, where they manipulate their host to produce more female offspring (reviewed in Werren et al. 2008; Duron et al. 2008; Cordaux et al. 2011). Being vertically inherited through the egg's cytoplasm, they can gain a transmission advantage by inducing parthenogenesis in their host, by killing or feminizing males or by generating **cytoplasmic incompatibility**; the latter is accomplished by modifying the sperm of infected males in such a way that it becomes compatible only with eggs infected by the same strain. *Gammarus duebeni* females produce an excess of daughters when infected by intracellular microsporidian parasites, which convert genotypic males into fully functional phenotypic females (Dunn et al. 1995; Terry et al. 1998). Feminization is likely achieved by directly interfering with the host sex-determination pathway; *Wolbachia* affects the sex-specific splicing of *doublesex* in the moth *Ostrinia scapulalis*, so that ZZ individuals produce the female isoform (Sugimoto & Ishikawa 2012; Beukeboom 2012). Haplodiploids seem particularly vulnerable to such parasites that take over their sex allocation control; the jewel wasp *Nasonia vitripennis* provides an illustrative example for the prevalence and diversity of sex-ratio distorters, including different types of bacteria, mitochondrial variants that skew sex ratios toward females, as well as a supernumerary B chromosome that distorts sex ratios toward males (Box 6.2).

The sex-ratio biases induced by microorganisms are also expected to drive turnovers in sex-determining mechanisms (Section 7.2.3c); in the female heterogametic woodlice *Armadillidium vulgare*, the feminization of genotypic males by *Wolbachia* selected for a progressive loss of the W chromosome, that resulted in a switch towards male heterogamety via a dominant autosomal masculinizing mutation (Rigaud et al. 1997; Bouchon et al. 2008; Cordaux et al. 2011; Box 7.1).

## 6.4 Sexual dimorphism and sexual selection

### 6.4.1 Are sex chromosomes attractors for sexually selected traits?

Sexual dimorphism is a defining feature of dioicous/dioecious species. It usually extends far beyond the mere differentiation of gonads and includes many sex specificities in morphology, physiology, behaviour, or life histories; males and females may differ to the point that they have sometimes been assigned to different species. Sexual dimorphism reflects an adaptive response to selection favouring different optimal trait values in the two sexes. Sexual selection, stemming from the contrasted reproductive roles of males and females, plays a central role in this context (Section 1.3). Strong male–male competition for access to females often selects for elaborated secondary sexual characters, aimed at outcompeting rivals or attracting females; this puts females under selection to choose partners that potentially bring most direct or indirect benefits, leading to a co-evolution between male signalling and female preference (Box 6.3).

Divergent optimal trait values are likely to generate conflicts between sexes: males and females share essentially the same genome, which necessarily creates genetic correlations (Rice & Chippindale 2001). A first way of solving such conflicts relies on the sex-specific expression of autosomal genes. The complex and elaborate pathways of gene regulation in multicellular lineages certainly facilitate the epigenetic control of autosomal genes. Steroid hormone signalling and receptors underlie the strong dimorphism of many sequential hermaphrodites (e.g. many reef-dwelling fishes) and ESD lineages (e.g. the dwarf males of *Bonellia*). Sexually dimorphic expression of autosomal genes is also widespread among GSD lineages, including insects, nematodes, birds, and mammals: more than half of the genes in *D. melanogaster* and *Mus musculus* show sex-biased expression (Ellegren & Parsch 2007).

**Box 6.3 Coupling of male traits and female preferences**

Female preference for specific male traits evolves under direct selection if the female's offspring directly benefit from their father's good condition (e.g. via paternal care); selection is said to be indirect if offspring benefit from the genes inherited from their father. Under good-genes models, the male trait bears intrinsic costs (handicap), but is genetically correlated with a trait under positive selection, such as a good body condition; hence, females preferring ornamented males benefit from healthy offspring. Under Fisherian (or runaway) models, females benefit from sexy sons: males inherit their father's attractiveness, and females their mother's preferences. Under both processes, the co-evolution between male trait and female preference depends on their genetic covariance; however, the linkage disequilibrium generated by female mating preference is independent of whether or not the preference and male trait genes are physically linked.

In a survey of existing literature at that time, Bakker & Pomiankowski (1995) found abundant evidence for genetic variation in mate choice, and good support for a genetic, but not a physical, coupling of male-trait and female-preference genes. Genetic coupling, however, is not always found (e.g. Ritchie et al. 2005; Allison et al. 2008; Van der Sluijs et al. 2010). Almost 20 years later, there still are surprisingly few studies that have simultaneously mapped male-signal and female-preference genes. A quantitative trait locus (**QTL**) study by Limousin et al. (2012) found no evidence for physical linkage of male ornament and female preference genes in the acoustic moth *Achroia*. In contrast, Kronforst et al. (2006) found coinciding QTLs for wing colour and female choice in the butterfly *Heliconius*, but this was autosomal and not on the Z chromosome. Wiley et al. (2012) found that four out of five QTLs for male song pulse rate were physically linked with QTLs for song preference in female Hawaiian cricket *Laupala*. Interestingly, all four were located on autosomes, whereas the fifth was X-linked. Thus, at present, there is supportive evidence for linkage disequilibrium of sexual selection genes, as well as for their sex-linked inheritance, in a subset of studied systems, but the few mapping studies into their physical linkage and preferential location on the sex chromosomes do not yet yield consistent results.

An alternative way of solving conflicts, specific to organisms with GSD, is achieved by physically linking sexually antagonistic (SA) genes with the sex-determination locus. **Sexual antagonism** is indeed regularly invoked as the driving force behind recombination arrest in sex chromosomes (Section 5.1.1). Thus, one might, a priori, expect sexual dimorphism to evolve more readily if SA genes occur on sex chromosomes or can be translocated there (Fairbairn & Roff 2006). Sex-limited chromosomes (U, V, Y, W) might seem to constitute ideal places to locate such genes; there is indeed some empirical evidence for male-beneficial genes on Y chromosomes (e.g. Lindholm & Breden 2002; Chapter 5). However, many Y chromosomes still recombine over large regions (even sporadically), which weakens sex linkage and thus attractiveness for SA genes; non-recombining regions degenerate (Section 5.2.2b), restricting the scope for the long-term maintenance of SA genes. Furthermore, sex-limited chromosomes are lacking in systems with XO, ZO, or haplodiploid sex determination. Whether sexual antagonism and sexual dimorphism are less prevalent in these lineages seems doubtful; scale insects, for instance, in which maleness is determined by paternal genome elimination, display extreme sex dimorphism.

The homogametic X and Z sex chromosomes are also expected to constitute hot spots for SA genes and sexual antagonistic variation (Rice 1984; Patten & Haig 2009; Section 5.2.3). Patterns of inheritance and expression of X chromosomes should facilitate the initial spread of recessive male-beneficial and dominant female-beneficial mutations. Then, from a state of segregating alleles, modifiers that restrict gene expression to the benefiting sex are expected to evolve, driving eventual fixation of male- or female-benefiting alleles (Rice 1984). Several approaches have been used to investigate the sex linkage of SA genes, including reciprocal crosses between divergent populations, within-population quantitative genetic studies, correlation of sexual dimorphism with sex-determination systems, and genomic mapping of genes with sex-biased expression. These studies have yielded contrasting results. Crossing divergent populations has produced some evidence of sex linkage for sexually selected behavioural and morphological traits (reviewed in Reinhold 1998). Sex linkage of either male traits or female preferences has been gathered from a variety of taxa (Qvarnström & Bailey 2009), including birds (Saether et al. 2007; Pryke 2010), butterflies (Grula & Taylor 1980; Iyengar et al. 2002), and fish (Lindholm & Breden 2002; Roberts et al. 2009; Kitano et al. 2011). The situation in *Drosophila* seems controversial (Heisler 1984; Reinhold 1998; Fitzpatrick 2004); Bailey et al. (2011) reported significant clustering of candidate mate choice genes on the X chromosome of *D. melanogaster*, but the genomic location of the male signal traits is unknown. Other studies failed to find any bias towards X linkage for genes affecting sexually selected traits (Gleason et al. 2002; Fitzpatrick 2004).

Overall, the evidence for a preferential **sex linkage** of sexually dimorphic traits is profoundly mixed (Fairbairn & Roff 2006). Quantitative genetic analyses suggest that sexually selected traits are not more likely than others to be sex linked, and that the degree of sex-linked genetic variance does not correlate with sexual dimorphism of the trait (e.g. Husby et al. 2013). Once corrected for phylogeny, turtles show the same levels of sexual dimorphism in GSD and temperature sex determination species (Ceballos et al. 2013). Genomic surveys find no strong tendency for genes with sex-biased expression to be on the X or Z chromosome; the autosomes harbour the lion's share of sex-biased genes in mice (Yang et al. 2006), birds (Mank et al. 2008), and flies (Parisi et al. 2004). Given the scarcity of empirical studies and the limited and inconsistent evidence, it seems largely premature to conclude that there is preferential sex linkage of sexually selected traits.

### 6.4.2 Is female heterogamety more conducive to flashy males?

Theoretical studies also suggest that the patterns of heterogamety may affect the outcome of sexual selection. Hastings (1994) argued that ZW systems are more likely than XY to evolve ornate male sexual characters via **handicap selection**: female-preference genes located on the W chromosome are only inherited by daughters, and thus unconditionally selected, being associated with the high-fitness alleles inherited from the father, but not with the handicap (expressed only in males). As suggested by Hastings, this might explain why elaborate male ornaments relying on female choice are more developed in birds and butterflies (which are ZW) than in mammals, where male traits predominantly result from inter-male rivalry. Rice & Chippindale (2001) also noted that in female heterogametic systems the father-to-son transmission of the Z chromosome should increase the heritability of Z-linked traits, so that sexual selection should be intrinsically more effective. Simulations of a Fisherian process by Reeve & Pfennig (2003) suggested that the co-evolution of male ornaments and female preference is easier in ZW/ZZ than XX/XY systems, haplodiploidy constituting the least favourable settings, though these results appear to stem from implicit assumptions in the model rather than from the effects of sex linkage (Kirkpatrick & Hall 2004). Analytical models by Kirkpatrick & Hall (2004) predicted Z-linkage of female preference to be especially conducive to a Fisherian runaway, whereas X-linkage of female preference, coupled with autosomal inheritance of male displays should favour the operation of a good-genes process. The W-linkage of either trait or preference should make the runaway effect impossible, but the W-linkage of preference would be prone to the evolution via good genes (in line with Hastings 1994). Albert & Otto (2005), finally, proposed that, for an SA trait on the X chromosome, females should evolve to prefer mates carrying alleles beneficial to daughters, while for a Z-linked trait, females should evolve mating preferences for mates carrying alleles beneficial to sons (that is, flashy displays).

Empirical evidence is, however, also controversial. Data gathered by Reeve & Pfennig (2003) suggest that more exaggerated male secondary sexual characters occur in ZW than in XY systems (and more in diploid than in haplodiploid species), but a broad scale comparative analysis of teleost fishes, correcting for phylogenetic non-independence, failed to find any significant effect of male or female heterogamety on male ornaments (Mank et al. 2006). Here again, empirical support seems insufficient to conclude that there is an effect of heterogamety on the outcome of sexual selection.

## 6.5 Sex determination and speciation

Sex-determination systems are expected to interfere with reproductive isolation and speciation processes. Sex chromosomes in particular are likely to accelerate speciation on several accounts. One way is through **chromosomal speciation**, which refers to changes in chromosomal composition or structure (polyploidization, fissions, fusions, inversions, or translocations) that creates partial or complete reproductive isolation between populations, owing to a failure of homologous chromosomes to pair adequately during meiosis. Inversion polymorphisms contribute to speciation not only by perturbing meiosis, but also by preventing recombination and sheltering genomic regions from introgression (see Hoffmann & Rieseberg 2008 for a discussion of the population dynamics of inversions and their possible role in speciation). In the fruitfly *Rhagoletis pomonella*, for instance, inversions contain genes for hatching time and diapause, which cause **allochronic** isolation between populations (Feder et al. 2003). Due to their highly dynamic nature, sex chromosomes are especially prone to undergo structural changes, and thus likely to contribute significantly to chromosomal speciation.

Other features stemming from the recombination arrest in the heterogametic sex (including faster X evolution and accumulation of SA or sex-biased genes) is expected to facilitate the emergence and spread of speciation genes. Speciation genes may act either at pre-mating (female choice) or post-mating stages (hybrid sterility or unviability). **Post-mating isolation** is thought to result from the disruption of co-adapted gene complexes: alleles that are fit in one genetic background may be unfit in others. Negative epistatic interactions between genes with different evolutionary histories are collectively known as **Dobzhansky–Muller (DM) incompatibilities** (Box 6.4). Strong support for the DM model has been gathered mostly from *Drosophila* studies, but also from other animals (e.g. the melanoma genes in *Xiphophorus*) and from plants (e.g. *Mimulus*; Fishman & Willis 2006; reviewed in Coyne 1992; Wu & Ting 2004; Presgraves 2010).

Sex chromosomes play a disproportionate role in **hybrid dysgenesis**, largely contributing to the several 'rules of speciation' (Orr & Coyne 1989), namely **Haldane's rule**, Darwin's corollary, and the large X effect (or **Coyne's rule**).

**Box 6.4 Dobzhansky–Muller incompatibilities**

Bateson (1909), Dobzhansky (1937), and Muller (1940, 1942) independently proposed that hybrid sterility or unviability stem from incompatible interactions between alleles from two or more genes. Consider two such loci with alleles *a* and *b* fixed in an ancestral species; all individuals are *aabb* (Fig. 6.5); allopatric speciation makes one descendant population fix allele *A* at locus 1, while another population fixes allele *B* at locus 2 (e.g. *Ab* and *aB* interactions might work better than *ab* in these populations contexts). Upon secondary contact, crosses between *AAbb* and *aaBB* individuals produce *AaBb* hybrids, inducing interactions between alleles *A* and *B*. As these alleles have never been tested previously in the same genome, they might be revealed as incompatible. The outcome also depends on the patterns of dominance: problems will only appear if both *A* and *B* are expressed, but not if either one is recessive. DM incompatibilities also include cytonuclear incompatibilities, where organelle alleles are always expressed.

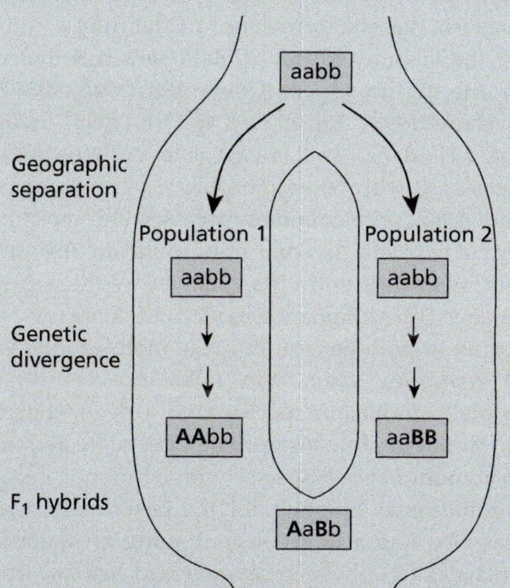


**Figure 6.5** Dobzhansky–Muller incompatibilities. A two-locus-two-allele model for the evolution of negative epistatic interactions. A population of *aabb* individuals becomes split. Allele *a* is substituted by *A* in population 1 and allele *b* by *B* in population 2. The new alleles are beneficial in their respective backgrounds. Upon reunion, the derived alleles *A* and *B* do not interact favourably and cause hybrid dysgenesis. Adapted from Presgraves (2010). Reproduced with permission from Nature Publishing Group.

### 6.5.1 Haldane's rule

Upon studying hybridization patterns over a range of taxa, Haldane (1922) observed that 'when in the $F_1$ offspring of two different animal races one sex is absent, rare or sterile, that sex is the heterozygous [heterogametic] sex'. In other words, male hybrids are more susceptible to hybrid incompatibility under male heterogamety (XY or XO males), while female hybrids are more susceptible under female heterogamety (ZW or ZO females). This observation was later known as Haldane's rule, as it holds for the vast majority of hybridizations studied to date, covering hundreds of species pairs in nine phyla (Orr & Coyne 1989; Schilthuizen et al. 2011). Haldane's rule is remarkably well obeyed: recent surveys suggest that it holds in all animals known to possess sex chromosomes, and it was recently extended to plants (Brothers & Delph 2010). Haldane's rule captures an early stage in speciation, where the homogametic sex is still fertile and viable, and its fitness not yet compromised by hybridization. Hybrid sterility typically evolves before hybrid unviability, as shown for *Drosophila* (Wu & Davis 1993) and Lepidoptera (Presgraves 2002).

Several models have attempted to explain Haldane's rule; all assume some form of incompatibility leading to the disruption of co-adapted gene networks, but evolving under different evolutionary causes. Muller (1940) first proposed a role for X-A imbalance in the heterogametic sex: in XY systems, $F_1$ males have copies of autosomes from both of the parental species, but not of the X chromosome from one of the parental species. Coyne (1985) later adjusted Muller's model to accommodate differences in hybrid effects on sterility and unviability; Turelli & Orr (1995) noted that sex-differences in hybrid fitness crucially depend on the patterns of dominance of genes involved in incompatibilities (Box 6.5), and formalized Muller's idea into an explicit theoretical framework, known as the dominance model. A twist to the model is added by faster-X effects, stemming from positive selection for recessive beneficial alleles in hemizygous males (Section 5.2.3a); fast-evolving X alleles are more likely to generate incompatibilities with autosomal genes in a hybrid context. The dominance model generally applies to both sterility and viability incompatibilities, under both male and

**Box 6.5 The dominance model of Haldane's rule**

Consider DM incompatibilities between an autosomal and an X-linked locus, for which males are hemizygous. Crossing an *Abb* male with an *aaBB* female produces *ABb* $F_1$-males and *AaBb* $F_1$-females. *AB* incompatibilities will be differently expressed depending on the patterns of dominance: if both *A* and *B* are dominant, then both sexes will suffer; if *B* is recessive, neither sex will be affected; but if *B* is dominant and *A* recessive, only males will suffer. In other words, when one sex is affected, that is the heterogametic one, in accordance with Haldane's rule.

female heterogamety. Haplodiploids are expected to be particularly susceptible, because males are hemizygous for all chromosomes, not only for the X chromosome (Koevoets & Beukeboom 2009).

The faster-male model notes that, due to stronger sexual selection, male-biased genes evolve faster, and show greater divergence between species, than female-biased or unbiased genes (e.g. Civetta & Singh 1995; Meiklejohn et al. 2003; Parisi et al. 2003). Gene incompatibilities causing hybrid male sterility should thus accumulate faster than those causing female sterility (spermatogenesis is also known to be more easily disrupted by mutations than oogenesis). Faster-male effects cannot account for hybrid unviability (male-biased genes should only affect fecundity); neither can they account for Haldane's rule under female heterogamety (but see Tao & Hartl 2003 for a 'fast-heterogametic sex' alternative). They may, however, also apply to organisms with homomorphic sex chromosomes (contrary to the dominance model, which depends on X:A imbalance).

Alternative models underline the potential for interactions and **genetic conflicts** between sex chromosomes. Y chromosomes interact with all chromosomes, including the X; absence of the conspecific X chromosome in $F_1$ males should thus disrupt co-adaptations. Sex-ratio distorters might play an important role in this context: a meiotic-drive X chromosome counteracted by a Y-specific repressor will be unleashed in hybrid males, causing distortion and partial sterility (Frank 1991; Hurst & Pomiankowski 1991). Note that this argument applies neither to viability genes (unaffected by meiotic-drive sex chromosomes), nor to XO or haplodiploid systems. Finally, the disruption of dosage compensation or meiotic sex chromosome inactivation (MSCI) mechanisms, as proposed by Lifschytz & Lindsley (1972), might contribute to hybrid male sterility in XY systems.

Empirical data support both the dominance and the faster-male models, but in different contexts. Orr (1993) crossed *Drosophila teisseri* males with *D. simulans* females that had their two X chromosomes attached; hybrid females received two X chromosomes from *simulans* and were thus as imbalanced as hybrid males; they also displayed the same patterns of mortality, supporting the dominance model for unviability. Similar experiments, performed with other species pairs of *Drosophila*, show that imbalanced females tend to be unviable in crosses for which $F_1$ males are unviable, but fertile in crosses for which the $F_1$ males are sterile; this suggests that the dominance model fully accounts for the viability aspect of Haldane's rule, but not for the sterility aspect (Coyne 1985; Orr 1993; Wu & Davis 1993). Presgraves & Orr (1998) investigated Haldane's rule in two genera of mosquitoes: *Anopheles*, with heteromorphic sex chromosomes, and *Aedes*, with homomorphic sex chromosomes. Unviability effects were only found in *Anopheles*, and seem thus entirely explained by the X:A imbalance (the dominance model). Male sterility occurred in both genera, but was more pronounced in *Anopheles*. Thus, X:A imbalance contributes to male sterility in species with a degenerated Y chromosome, but faster-male effects are required to account for it in species with homomorphic sex chromosomes. Support for the faster-male model was also found in an elegant study of Malone & Michalak (2008); using sex-reversed *Xenopus* frogs, they showed male sterility to be determined by the phenotypic sex (male or female), not the genotypic sex (ZZ or ZW). The situation in marsupials is of special interest: because females inactivate the paternal X chromosome, they show the same X:A imbalance as males; in accordance with the faster male theory, hybrids show no sex difference in unviability, but sterility affects males more than females (Watson & Demuth 2012). Faster evolution of

X-linked loci may also contribute to the dominance model, though the evidence is weaker (Orr 1997; Orr & Coyne 1989).

Models based on X-Y interactions and meiotic drive seem less well supported (Coyne & Orr 1993; but see McDermott & Noor 2010; Presgraves 2010; Johnson 2010). The effect of the heterogametic chromosome (Y or W) on reproductive isolation is usually less pronounced than that of the homogametic chromosome. Haldane's rule (both sterility and unviability aspects) also applies to systems lacking Y or W chromosomes (such as XO in grasshoppers; Wu & Davis 1993; Coyne & Orr 2004). The Y chromosome seems responsible for sterility in a number of *Drosophila* hybrid species pairs (Turelli & Orr 2000; Coyne & Orr 2004; Sweigart 2010; Sackton et al. 2011), but incompatibilities may stem from interactions with autosomes, not with the X chromosome (Johnson et al. 1992, 1993). X-Y incompatibilities have been suggested to play a role in causing sterility in house mice hybridization (Campbell et al. 2012, Box 6.6) but Z-W incompatibilities have not been reported from female heterogametic systems. Disruption of MSCI has recently received support from male hybrid-sterility analyses in mice (Campbell et al. 2013).

**BOX 6.6 Hybrid zone and sex chromosomes in *Mus musculus***

The two subspecies of house mouse *M. m. musculus* and *M. m. domesticus*, which diverged ~0.5 Mya, are morphologically distinguished by tail length and craniofacial shape (Macholán 1996). *Mus m. domesticus* ranges across Western Europe and the Mediterranean area, whereas *Mus m. musculus* occurs in Eastern Europe and Northern Asia; they form a **hybrid zone** from the Jutland Peninsula over Eastern Germany and the Balkan to the Bulgarian coast of the Black Sea (Boursot et al. 1993; Payseur et al. 2004). Crosses between the two subspecies result in partial hybrid male sterility consistent with Haldane's rule. Intensive studies on the hybrid zone have shown that different parts of the genome introgress to different degrees in the two subspecies (references in Payseur et al. 2004). X-linked loci introgress much less than autosomal loci (consistent with the large X-effects), while Y chromosomes typically do not spread at all within the other subspecies' range (suggesting it carries one or more important speciation genes). Introgression is also asymmetrical: *M. m. domesticus* alleles spread less far into the *M. m. musculus* range than vice versa, an illustration of Darwin's corollary (also evidenced by asymmetric male sterility: only $F_1$ males with a *musculus* mother are sterile; Good et al. 2008; Campbell et al. 2013).

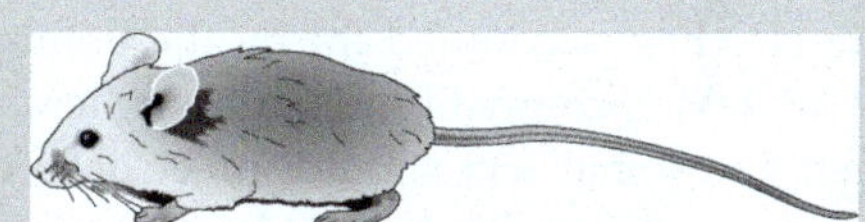

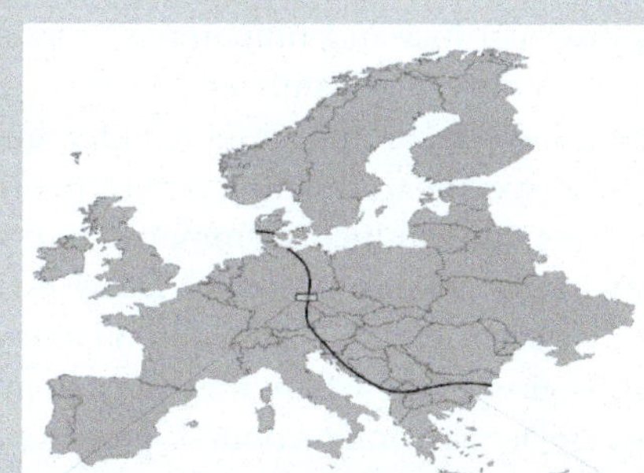

Using a genomic cline approach, Janoušek et al. (2012) identified contiguous regions on the X chromosome likely involved in epistatic interactions, with substantial overlap with QTLs for hybrid sterility identified in laboratory crosses. Analysis of gene content of these regions suggests a key role for sexual conflict and sexual selection operating at post-mating stages. The central region of the X chromosome seems particularly enriched in male sterility factors, and shows exceptionally low levels of gene flow across the hybrid zone. Good et al. (2011) found evidence for rapid evolution and recurrent positive selection in several genes from this region, responsible for their reduced introgression. One of these genes, *testis-specific gene a8* (*Tsga8*), showed extreme nucleotide divergence between the two subspecies; although its specific function remains unknown, *Tsga8* is highly expressed during spermatogenesis and likely involved in sperm chromatin remodelling. This gene is misexpressed in sterile males, and might contribute to the disruption of MSCI found in $F_1$ males (Campbell et al. 2013). Identification of the genes responsible for incompatibilities and elucidation of their interactions in future studies will further reveal the nature of DM interactions between autosomes and sex chromosomes.

Altogether, empirical evidence suggests that Haldane's rule is a composite phenomenon, resulting from several causes active in different contexts. The dominance model (with possible contribution of faster X-effects) apparently fully accounts for unviability aspects, and likely contributes to sterility aspects; but faster-male effects (and possibly also MSCI disruption) are also required to fully account for the latter (Coyne 1985; Wu & Davis 1993; Orr 1993, 1997; Schilthuizen et al. 2011).

### 6.5.2 Darwin's corollary

Darwin's corollary (Turelli & Moyle 2007) refers to asymmetric fitness in hybrid crosses: the fitness of crosses between two species might be more or less affected depending on which parental species the mother and father come from. Asymmetries are also best accounted for by the dominance model of DM incompatibilities. DM interactions between autosomal loci should affect both reciprocal crosses identically, but DM interactions between autosomal loci and uniparentally inherited factors (including cytoplasmic elements, maternal transcripts, and sex chromosomes in heterogametic hybrids), are specific to the direction of hybridization, and may thus induce asymmetric reproductive isolation. As already noted by Darwin (1859), asymmetries seem particularly common and widespread in angiosperms, where they may stem from nucleo-cytoplasmic interactions, gametophyte–sporophyte interactions, and conflicts between the haploid male and diploid female contributions to the triploid endosperm (Turelli & Moyle 2007; Phillips & Edmands 2012).

### 6.5.3 Large X-effects

Large X-effects (also known as Coyne's rule) refer to the disproportionately large effect of the X chromosome versus autosomes in driving hybrid sterility and unviability (Orr & Coyne 1989; Coyne 1992; Turelli & Moyle 2007; Fig. 6.6). QTL mapping in many animal taxa have shown that genetic factors with the largest effects on hybrid sterility tend to be X-linked (Coyne et al. 1991; Coyne & Orr 2004). Evidence also comes from backcross genetic analyses; substituting one species' X chromosome for that of another has a disproportionately large effect on hybrid fitness, compared to autosomal substitutions; genome-wide introgression of *Drosophila mauritiana* chromosome segments in a *D. sechellia* genetic background shows that the X chromosome harbours roughly four times as many hybrid male sterility factors as an average autosome of comparable size (Masly & Presgraves 2007). Causes for this accumulation are not well established (Masly & Presgraves 2007; Presgraves 2008); it may relate to faster X evolution, stemming from hemizygous exposure in males (Chapter 5), sex-ratio meiotic drive (Tao et al. 2001; Tao & Hartl 2003; Orr & Irving 2005; Orr et al. 2007), disruption of dosage compensation or MSCI mechanisms, or sex-specific movement of genes off and onto the X chromosome, possibly in combination with gene silencing (Moyle et al. 2010). Like Haldane's rule, the large X-effect seems thus a composite phenomenon with multiple causes that may differ between species.

### 6.5.4 Pre-mating isolation

Hybrid incompatibilities are expected to select for **pre-mating isolation** via female mate choice: premating **reinforcement** of the speciation process should prevent females from investing their reproductive effort into unfit offspring. However, gene flow between hybridizing populations might pose constraints on the evolution of female choice by disrupting linkage between signalling and recognition traits. Theory predicts that the level of reinforcement depends on the mode of inheritance of male trait and female preference; assortative mating might be facilitated by sex linkage of sexually selected traits. As shown by Hall & Kirkpatrick (2006), the dynamics of pre-mating isolation depend on whether reinforcement acts mostly on the male traits or on the female preferences; when reinforcement is driven mainly by selection in the male trait, and intrinsic hybrid incompatibilities are weak, sex-linked preferences and autosomal male traits are the most conducive to reinforcement, whereas autosomal preferences and X-linked traits are the least.

Linkages between assortative-mating and genetic-incompatibility loci also promote reinforcement, by generating a positive feedback loop between

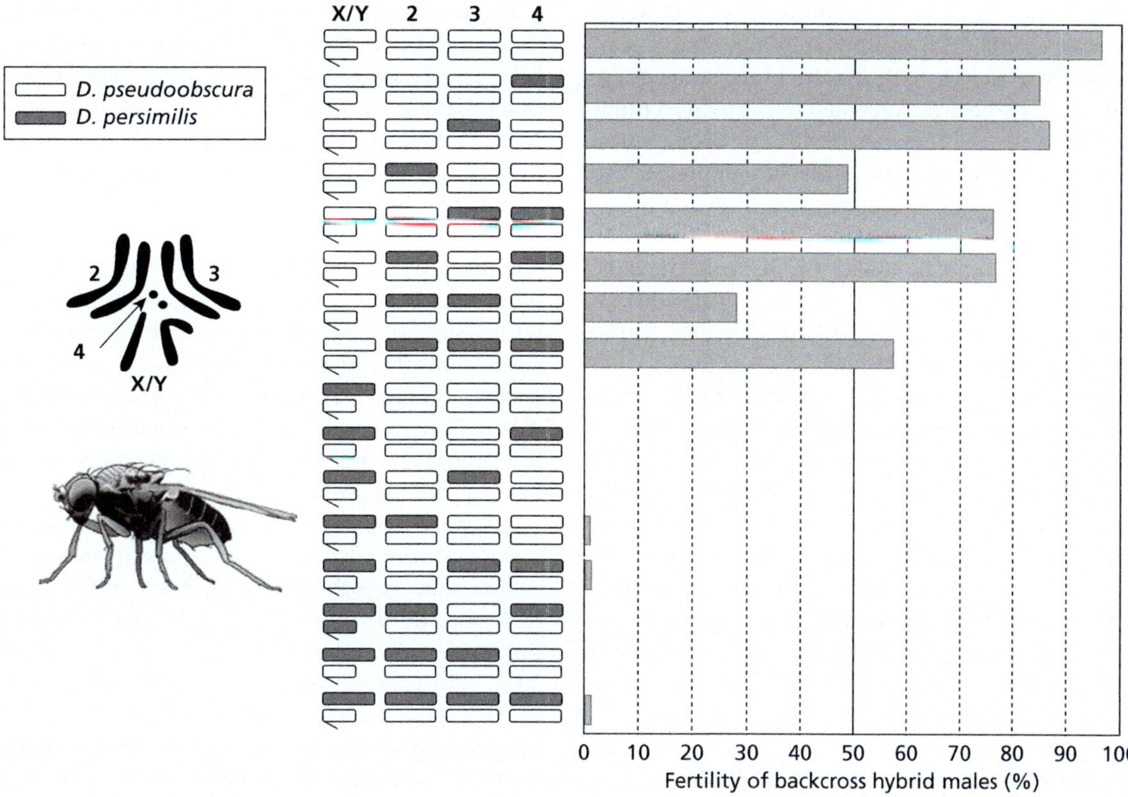


**Figure 6.6** The large X-effect. The X chromosome has a disproportionately large effect on male hybrid sterility, as shown from backcrosses between *Drosophila pseudoobscura* and *D. persimilis*. Bars indicate proportional fertility. Data from Orr (1987), figure adapted from Presgraves (2008). Reproduced with permission from Elsevier.

traits involved with pre- and **post-zygotic isolation** (Felsenstein 1981; Kirkpatrick 1982; Trickett & Butlin 1994; Servedio 2000; Noor et al. 2001). Sex linkage further strengthens this effect: arrested recombination in the heterogametic sex offers additional opportunities to strengthen the link between sexual selection and incompatibility genes (Servedio & Saetre 2003; Lemmon & Kirkpatrick 2006). There are, thus, ample arguments for an effect of the mode of sex determination on speciation reinforcement via assortative mating.

Empirical support, however, is still limited. In *Ficedula* flycatchers, the Z sex chromosome harbours genes causing hybrid incompatibility and other genes influencing the expression of diverged male plumage traits (Saetre et al. 2003). In addition, the Z chromosome contains genes controlling female mate choice, as shown by cross-fostering experiments (Saether et al. 2007). Similar cross-fostering experiments with Gouldian finches (*Erythrura*) show that mating preferences of $F_1$ and $F_2$ hybrids are consistent with inheritance on the Z chromosome, where the genes controlling colour expression and others causing low hybrid fitness are located (Pryke 2010).

### 6.5.5 Sex chromosome introgression in hybrid zones

Sex chromosomes are thus expected to contribute significantly to the early stages of speciation, by establishing both pre- and post-mating barriers to gene flow. Theoretical models of the dynamics of sex-biased hybrid incompatibilities predict strong effects of sex chromosome-associated incompatibilities on gene flow in hybrid zones (e.g. Wang &

Zhao 2008): sex-linked loci should pass less freely than autosomal loci, and possibly in a more asymmetric way, across hybrid zone boundaries. Such patterns have been documented across many natural hybrid zones, including both XY (*Drosophila*, mice; Box 6.6) and ZW systems (birds, Lepidoptera). The predictions appear to be borne out in several mammalian (Tucker et al. 1992; Dod et al. 1993, Payseur et al. 2004; Macholán et al. 2007; Teeter et al. 2008; Geraldes et al. 2008) and bird species pairs (Saetre et al. 2003; Storchova et al. 2010; Elgvin et al. 2011), although other processes such as sex differences in dispersal rates or effective population sizes may also affect the hybrid-zone dynamics of sex chromosomes (Petit & Excoffier 2009; Beysard et al. 2012).

The differential introgression of sex chromosomes has been well documented in the house mouse (*M. musculus*) across a European hybrid zone between *M. m. musculus* and *M. m. domesticus* (Box 6.6). Studies have consistently shown a narrower introgression of sex chromosomes (references in Macholán et al. 2011; Janoušek et al. 2012). The work on *Mus* hybridization also illustrates how speciation processes and secondary contacts may feedback on the evolution of sex chromosomes, for example, regarding the spread of a neo-Y sex chromosome (Veltsos et al. 2008).

### 6.5.6 What about systems lacking differentiated sex chromosomes?

Virtually all studies of sex-determination effects on speciation are restricted to a subset of organisms with a very specific mode of sex determination: gonochoric animals with dimorphic sex chromosomes. Studies of Haldane's and Coyle's rules have been largely neglected in organisms with alternative sex-determination systems (Schilthuizen et al. 2011; Phillips & Edmands 2012). The expectation that such taxa should display less rapid speciation has received some support from reptiles: diversification seems more rapid in squamates (lizards and snakes) than in turtles and crocodilians, which often have ESD; among lizards and turtles, families with GSD and differentiated sex chromosomes also seem to display higher diversification rates (references in Phillips & Edmands 2012). The expectation that male and female functions should accumulate unviability and sterility at comparable rates in taxa lacking differentiated sex chromosomes also received anecdotal support: male and female sterility factors accumulate at the same rate in hermaphroditic tomatoes (Moyle & Graham 2005; Moyle & Nakazato 2010) and *Tigriopus* copepods with polygenic sex determination (Ellison & Burton 2006; Willet 2008). Koevoets & Beukeboom (2009) and Schilthuizen et al. (2011) provide attempts to include haplodiploids and hermaphrodites in the study of **post-zygotic isolation**, and show that other factors such a nucleo-cytoplasmic and dosage effects may be important alternative causes of hybrid dysgenesis. Investigation on species lacking differentiated sex chromosomes are still too few to provide enough support for any generalization; we therefore wish to end this section with a plea to better use the diversity of sex-determination mechanisms to study the genetics and evolution of speciation processes.

CHAPTER 7

# Transitions among sex-determination systems

Because sexual reproduction is such a crucial life-history trait in most eukaryotes, one might, a priori, expect all processes involved in the sexual cycle to be evolutionarily conserved. Nearly 50 years ago, Ohno (1967) suggested that the sex chromosomes of mammals, birds, and snakes (despite presenting different patterns of heterogamety) had originated from the same ancestral pair of autosomes, based on their common involvement in sex determination. This notion has been now entirely dispelled by comparative gene mapping and sequence analyses (Matsubara et al. 2006; Bellott et al. 2010). Although conservation is certainly the rule for the core molecular mechanisms (meiotic genes and processes), those underlying the systems of sex determination are extraordinarily variable, and, in many groups, undergo surprisingly frequent transitions. This is already evident from the diversity of systems found among closely related species or even conspecific populations in many lineages. What kinds of transitions occur commonly and what might be the driving forces? Why do transitions occur more often in some groups than in others, and why are some sex-determination systems more frequent than others? Are there evolutionary traps for sex-determination systems? Such are some of the questions we are developing in this Chapter.

## 7.1 A diversity of transitions

Transitions between sex-determination systems are manifold, as already described throughout this book. They may occur between co-sexuality (hermaphroditism) and unisexuality (gonochorism), exemplified, for instance, by angiosperms (Box 3.10); these may (or not) parallel transitions between epigenetic and genetic initial triggers. In gonochorists, transitions have repeatedly occurred between environmental sex determination (ESD) and genotypic sex determination (GSD) (Chapter 4), as exemplified by the distribution of temperature sex determination (TSD) among reptiles: either temperature or genes provide the initial trigger in closely related species (Ewert & Nelson 1991; Pokorna & Kratochvil 2009; Ezaz et al. 2009) or even conspecific populations (Pen et al. 2010). Similar situations are found in fishes (e.g. Conover & Heins 1987a). In GSD species, transitions may happen between polygenic and monogenic systems, as documented in molluscs, annelids, and insects such as the housefly *Musca domestica* (Box 3.20). In monogenic systems, sex determination may switch between male- (XY) and female- (ZW) heterogamety. Eight of 26 families of teleost fishes display both XY and ZW systems (Mank et al. 2006), including Poeciliidae (Volff & Schartl 2001), Gasterosteidae (Ross et al. 2009), and Cichlidae: in Tilapiinae, male heterogamety on **linkage group** (LG)1 coexists with female heterogamety on LG3, sometimes within the same species (Lee et al. 2004; Cnaani et al. 2008), while in Haplochrominae, male heterogamety on LG7 coexists with female heterogamety on LG5, also sometimes within the same species (Ser et al. 2010). XY and ZW systems also co-occur in closely related species from the medaka lineage (Takehana et al. 2008). In the frog *Rana rugosa*, the same sex chromosome pair determines either an XY or a ZW system depending on populations (Miura 2007; Ogata et al. 2008).

*The Evolution of Sex Determination.* Leo W. Beukeboom and Nicolas Perrin.

Transitions may also occur without affecting the patterns of heterogamety: in the genus *Rana*, several pairs of chromosomes have been co-opted for sex determination depending on species or populations, but transitions have maintained the ancestral pattern of male heterogamety, with only the abovementioned exception of *R. rugosa* (Miura 2007). In the genus *Oryzias*, similarly, one single transition, out of the six documented, has diverged from the ancestral male heterogamety (Myosho et al. 2012). Other transitions include evolution from XY to XO males to haplodiploidy, observed in many ecdysozoans (Box 7.2). Less drastic transitions are represented by sex-chromosome fusions with autosomes, which happened many times, for example, in the phylogeny of birds and therians (Section 5.1.3).

A distinction has to be made between transitions in systems (e.g. towards a different pair of sex chromosomes) versus transitions in mechanisms. Such a distinction is nicely illustrated with the case of Salmonidae: all species from this family share a male heterogametic system (XY; Davidson et al. 2009). Although the chromosomal location of the sex locus varies largely (Woram et al. 2003; Li et al. 2011), the sex-determining locus itself (*SdY*), and likely the whole downstream cascade, is shared by most species of the family (Yano et al. 2012, 2013). Chromosomal transitions are driven by transposition of *SdY*, mediated by linkage of this gene with moveable elements (Yano et al. 2013). Jumping sex-determination factors have also been documented repeatedly in dipterans, including house flies (Box 3.20), Phoridae (Mainx 1964; Traut & Willhoeft 1990), and Chironomidae (Beermann 1955; Kraemer & Schmidt 1993), where a single gene (M) determines maleness, but varies in genomic position and copy number.

Transitions in molecular mechanisms also occur frequently; as further detailed in Section 3.1.3, comparative studies show that the main effectors at the bottom of cascades (e.g. DM-domain genes in animals) are much more conserved than initial triggers at the top (e.g. *Sxl*, is present in many insects but only takes a leading sex-determination role within the Drosophilidae; Serna et al. 2004). This suggests that cascades are often built bottom up during evolution, by sequential recruitment of novel regulators higher up the pathway (Wilkins 1995). Pomiankowski et al. (2004) proposed a possible scenario for the evolution of the sex-determination cascade in *Drosophila*, where additional steps are progressively added to the bottom effectors, in order to limit ambiguities in the sex-determination signal (see also MacCarthy et al. 2010). Additions may also result from conflicts over sex ratios: long cascades, such as found in *Caenorhabditis elegans* (Box 3.15), with genes alternatively on or off depending on sex, are suggestive of tug-of-war dynamics where male and female elements alternatively take the master role in sex determination (Section 7.2.3c). An alternative interpretation, however, relies on the fact that novel modifiers of sex determination are usually epistatically dominant over the ancestral sex-determining system. If a dominant male modifier successfully invades in a species with male heterogamety as the ancestral system, then it will replace the ancestral Y chromosome; in this case, the new sex-determination gene will take the position of the old sex-determination gene in the cascade without increasing the length of the pathway. If, however, a dominant feminizing mutation invades successfully, the ancestral Y chromosome will go to fixation, and the new feminizing gene will be added at the top of the existing cascade. Accordingly, the length of the cascade only increases after a heterogametic transition. A sequence of repeated transitions (with some of them changing the heterogametic sex) thus naturally leads to an alternation of masculinizing and feminizing genes in the cascade.

## 7.2 Evolutionary causes

Several evolutionary forces have been proposed as potential ultimate causes for transitions in sex-determination systems or mechanisms. Here we classify them into three main categories, namely (1) neutral processes, (2) fitness differences between sex phenotypes, and (3) sex-ratio selection. The main theories that have been developed to formalize the dynamics of these processes are discussed in more detail below.

### 7.2.1 Neutral processes

Bull & Charnov (1977) investigated transitions between different heterogametic systems, and between polygenic and monogenic systems. Computer simulations were used to predict the fate of masculinizing or feminizing mutations occurring either at an established sex locus or at an unlinked autosomal locus. Any such mutation, whether dominant or recessive, generates a new recurrent pair (i.e. a pair of genotypes which, when crossed, give rise to zygotes of these genotypes only). A strongly feminizing mutation *W* occurring at an ancestral sex locus with *X* and *Y* alleles, for instance, opens opportunity for a switch to the female heterogametic system WY/YY; the two recurrent pairs (XX/XY and WY/YY) are connected by a continuous path of neutral equilibria (illustrated on a de Finetti diagram in Fig. 4.2), consisting of all mixes of genotypes (WX, WY, XX, XY, and YY) that produce 1:1 sex ratios at the population level (though not necessarily at the family level). In the absence of fitness differences between sex genotypes, genetic drift may drive transitions between recurrent pairs along such paths, without any departure from Fisherian sex ratio at the population level (Scudo 1964, 1967; Bull & Charnov 1977; Bull 1983). Under neutral dynamics, however, multifactorial sex determination is expected to be only transient (at least when based on a few loci with major effects): depending on effective population size, drift should fix one or the other recurrent pair sooner or later. Neutrality is clearly a null model that has to be considered (and possibly falsified) in any kind of transition mentioned in Section 7.1.

### 7.2.2 Fitness differences between sex phenotypes

Bull & Charnov (1977) also simulated situations where one of the sex genotypes was assigned higher or lower fitness values. Such fitness effects might be intrinsic to the new mutation (via pleiotropic effects), or arise from linkage with other genes under selection; linkage with DDT-resistance genes, for instance, has been invoked in the spread of autosomal male-determining factors in the housefly (Kerr 1970; Franco et al. 1982). Later extensions of this model explicitly considered linkage with SA genes, where some alleles are beneficial to one sex-genotype but detrimental to another (Van Doorn & Kirkpatrick 2007, 2010; Section 7.2.2a). From Bull & Charnov's simulations, enhancing the fitness of a genotype belonging to one recurrent pair drives this pair to fixation; decreasing it causes its elimination. Outcomes differ when the affected genotype is not part of a recurrent pair (e.g. WX): a lowered fitness induces a disruptive selection against transition genotypes, so that any departure from a recurrent pair is counter-selected, preventing turnovers. Increased fitness, on the other hand, provides a head start to the mutation (either *W* in an XY/XX context, or *X* in a WY/YY context) allowing its initial spread; but transitions do not go to completion (due to lower fitness of recurrent pairs), inducing a stable polymorphism. This situation might account for the polymorphism found, for example, in platyfish (*Xiphophorus maculatus*), where a colouration gene is tightly linked to the sex locus, with a bright colouration allele fixed on the Y chromosome, a mild allele on the X, and a dull on the W (Kallman 1965, 1970). In a theoretical analysis, Orzack et al. (1980) provided necessary and sufficient conditions for a balanced polymorphism to be maintained, assuming complete linkage between the SA and sex-determination genes. However, as shown by Blaser et al. (2011), these conditions also induce female biases in primary sex ratios, so that sex-ratio selection makes the system collapse toward male or female heterogamety as soon as recombinant genotypes appear; polymorphism can nevertheless be maintained under scenarios comprising strong sexual selection against dull males, mild natural selection against bright females, and low recombination rates between SA and sex-determination genes.

In Section 7.2.2a, b, we detail two special cases of sex-chromosome transitions, driven respectively by sexually-antagonistic mutations appearing on autosomes (SA-driven transitions), and by the deleterious load of mutation accumulating on non-recombining sex chromosome. However, it is worth recalling that other kinds of transitions may also be driven by fitness benefits gained by specific sex phenotypes (e.g. transitions from co-sexuality towards unisexuality driven by sex specialization).

#### 7.2.2a SA-driven transitions

In male heterogametic systems, a male-benefiting mutation occurring on an autosome automatically induces a selective pressure favouring masculinizing mutations in its vicinity; Van Doorn & Kirkpatrick (2007) derived the conditions for such an autosomal sex-determination locus to spread. Fixation of the proto-sex chromosome is expected when the sex-determination and SA loci are loosely linked (in such a case, the conditions that favour the proto-Y chromosome, when rare, are the same that favour it when frequent). Tight linkage might generate bi-stable equilibria or protected polymorphisms, occurring when the strengths of SA selection are similar on the ancestral and proto-Y chromosomes. Rice (1986) pointed out that polygenic sex determination is unlikely to persist for long when only one sex-determination locus is linked to a SA gene. The polymorphism may be evolutionarily stable if the multiple sex-determination loci are each linked to a SA gene, but this requires a rather fine balance between the strengths of selection acting on the different SA loci; the system will collapse when one of these selection pressures change only slightly.

Van Doorn & Kirkpatrick (2010) further showed that SA selection can also induce heterogametic transitions (either homologous or non-homologous); however, as this process fixes the ancestral Y chromosome as an autosome (and not the X as in homogametic transitions), the load of recessive deleterious mutations accumulated in the absence of recombination may prevent the spread of the ZW system (or, depending on conditions, maintain a protected polymorphism with the ancestral XY). Heterogametic transitions are well illustrated by situations described in Cichlidae; in the genus *Oreochromis* (tilapias), both XY and ZW systems coexist on different chromosomes, sometimes within the same species or individuals (Cnaani et al. 2008). In Lake Malawi cichlids, a new ZW system recently invaded an ancestral XY system, triggered by a mutation on the proto-W chromosome conferring a blotched pattern of colouration. This mutation is beneficial to females because it confers a cryptic phenotype, but costly to males because it disrupts mating colouration (Roberts et al. 2009; Ser et al. 2010).

More tests of the predicted association between sexually antagonistic and sex-determination genes are, however, needed from a wider array of taxa (see also Section 5.1.1). Comparisons with the homologous autosomal regions in closely related species might allow disentangling causes from effects (i.e. whether sexual antagonism drives sex-determination transitions, or, vice versa, that sex-determining regions capture SA genes). Particularly promising systems are those in which new sex chromosomes have recently arisen and are known to carry sexually selected genes, such as in medaka fish (*Oryzias*), three-spined stickleback (*Gasterosteus aculeatus*), and poeciliid fish. Other possibilities include the housefly that harbours multiple proto-Y chromosomes, and the phorid fly *Megaselia scalaris* that has a jumping sex determiner. For these flies, annotated genomes in combination with information about SA genes from *Drosophila melanogaster* (Innocenti & Morrow 2010) might be the way forward.

#### 7.2.2b Mutation-load-driven transitions

Deleterious mutations accumulating on non-recombining chromosomes also specifically affect the fitness of one sex genotype (XY in male heterogametic systems). Depending on lineages, this pressure may have selected for some non-zero rate of X-Y recombination (very low rates are enough to counteract accumulation; Section 5.2.2d) or for the evolution of complex mechanisms of Y silencing and dosage compensation (Section 5.2.2e). As an alternative, the decaying Y chromosome might simply be trashed and replaced by a proto-Y chromosome. This may occur by takeover of a new mutation affecting another gene from the sex-determining cascade (as occurs in Ranidae), or by transposition of the ancestral sex-determination gene (as occurs in Salmonidae). Individual-based simulations by Blaser et al. (2013) showed that turnovers are expected as soon as benefits (alleviated mutation load) outweigh costs (loss of SA genes fixed on the ancestral Y chromosome). Note that, once established, the proto-Y chromosome is expected to progressively attract new SA mutations and stop recombining, thereby re-initiating a process of degeneration. As argued by Blaser et al. (2014), this 'hot-potato' dynamics might fuel a potentially infinite cycling among a limited set of chromosomes; one pair that has already been involved in the past is, a priori, more likely to take over this role again, because it harbours genes prone to mutate towards sex-determining and SA alleles. Another expectation is that transitions induced by this mechanism should keep the ancestral patterns of heterogamety (for obvious reasons, a decaying Y chromosome cannot become fixed as an autosome).

Both predictions nicely fit the patterns observed in anurans, where a limited set of chromosomes seem recurrently co-opted for sex, and transitions most often conserve the ancestral pattern of heterogamety (Miura 2007; Brelsford et al. 2013).

### 7.2.3 Sex-ratio selection

An important category of selective causes specifically involves sex-ratio selection, arguably the most important pressure influencing the evolution of sex-determining mechanisms (Section 1.2.3). Transitions may occur in order to track shifting optimal sex ratios when they differ from Fisherian values; alternatively, transitions may allow restoration of Fisherian sex ratios following biases induced either by changes in ecological settings under ESD, or by genetic conflicts stemming from nucleo-cytoplasmic interactions, meiotic-drive sex chromosomes, or other selfish elements (Section 6.3.2).

#### 7.2.3a Shifting optimal sex ratios

Optimal sex ratios may differ from Fisherian values, due, for example, to differences in the production costs of males relative to females, interdemic selection, or local mate competition (Section 1.2.3). Kozielska et al. (2006) investigated theoretically the effect of sex-ratio selection induced by cost differences in the production of sons and daughters on the dynamics of a three-locus system (inspired by the housefly), consisting of a male heterogametic sex locus (XY), an autosomal masculinizing gene (*M*), and an epistatic female determinant *F* at a third locus. Under such conditions, sex ratios could only evolve towards slight female biases, and only when the cost of producing males greatly exceeded that of females. In addition, transitions among monogenic systems were not complete, with more than one locus remaining polymorphic at equilibrium.

Under **metapopulation** settings, interdemic selection may in some conditions override individual selection; demes overproducing females might be at selective advantage if the females exported allow outcompeting of other demes (Colwell 1981; Wilson & Colwell 1981). Vuilleumier et al. (2007) investigated conditions favouring the invasion of a dominant female determiner (W) in an initially male heterogametic population (XX/XY). Female heterogamety (WY/YY) could be established more likely than at random, mostly at intermediate degrees of subdivision (i.e. fairly high rates of dispersal among demes). **Interdemic selection** for female-biased sex ratios has been invoked to account for the spread of feminizing mutations in rodents (e.g. wood lemming, *Myopus schisticolor*, Fredga et al. 1977; Fredga 1988; *Mus minutoides*, Veyrunes et al. 2010), where an X mutation (X*) interferes with sex-determination pathways downstream of *Sry*, such that X*Y individuals develop as females (and demic sex ratios are strongly female biased; Section S2.2.9f). Freedberg & Taylor (2007) also suggested that ESD might outcompete GSD strategies in the colonization of new environments, due to overproduction of females.

Optimal allocation is condition-dependent whenever the returns from investment in males or females depend on the environment (Charnov & Bull 1977; Bull 1981; Van Dooren & Leimar 2003). Relevant environmental variables include temperature (Valenzuela & Lance 2004), population density (Aparici et al. 1998), breeding opportunities (Charnov 1975; Drent & Daan 1980), local mate competition (Hamilton 1967; Reinhold 1996; Werren & Taylor 1984; Pen et al. 2010), inbreeding (Werren & Hatcher 2000; Werren et al. 2002), maternal condition (Trivers & Willard 1973), or mate attractiveness (Burley 1981). Chromosomal sex determination in principle offers little flexibility to adapt sex ratios to local conditions (even though adaptive plasticity is found in some birds and mammals; Section 6.3.1). The need to track shifting optima might have driven transitions to plastic sex-determination systems, possibly under maternal control. Epigenetic sex determination, widespread in metazoans, has been shown to allow quick local adaptation in several species of fishes and reptiles (Section 4.10). Sex-ratio adjustment is also facilitated by haplodiploidy, where egg fertilization by ovipositing females controls the proportion of males. Some haplodiploid hymenopterans show variable and often strongly female-biased offspring sex ratios consistent with theory (West 2009). It is not clear, however, whether the benefits of flexible sex allocation selected for the evolution of a permissive sex determining mechanism, or simply resulted as a by-product of pre-existing mechanisms (see also Box 7.2).

#### 7.2.3b Environmentally driven transitions

Although ESD may, in principle, allow tracking of shifts in optimal sex ratios, it also constitutes a major drawback when the fine-tuning of sex ratios is impeded. Changes in pivotal temperatures or oviposition site selection might accommodate long-term climatic trends (Chapter 4), but transitions to GSD may also result whenever genes and environment both affect the liability factor underlying sex determination (Chapter 4). Experimental exposure of Atlantic silversides (*Menidia menidia*) to sex-ratio biasing temperature may lead to the rapid elimination of ESD, allowing restoration of equal sex ratios (Conover & Van Voorhees 1990; Conover et al. 1992; Chapter 4). In addition, large sex-ratio fluctuations stemming from variable and unpredictable environment should be highly detrimental to short-lived organisms, thereby selecting for a transition to GSD (Bull 1981). Several TSD lineages of fishes and lizards have been shown to evolve GSD in northern or highland populations (e.g. Atlantic silverside; Lagomarsino & Conover 1993; *Niveoscincus*; Pen et al. 2010). Depending on the shape of reaction norms, a change in temperature (e.g. stemming from climatic change or range expansion) may also induce transitions between male heterogamety, female heterogamety, and TSD (through the loss of the XY or ZW genotype; Fig. 4.3a; Grossen et al. 2011). Transition clines between male- and female heterogamety in houseflies correlate worldwide with climatic conditions (Feldmeyer et al. 2008).

#### 7.2.3c Genetic conflicts

Genomes consist of multiple separate entities that, in sexually reproducing species, have different inheritance routes and are only temporarily associated. As a consequence, they may be subject to selection in opposite directions: any given element is selected to increase its own transmission, even if this is detrimental to the organism or to other elements of the genome. Conflicts are said to be intra-genomic when different genes in the same individual have divergent evolutionary interests (Fig. 7.1); they are called inter-genomic when occurring between individuals, for example, when the same locus is subject to divergent selective pressures depending on the context in which it is expressed (males versus females). Genetic conflicts can spark evolutionary arm races between different genomic components over basic biological processes and also contribute to the maintenance of heritable variation (Rice 1998; Partridge & Hurst 1998; Burt & Trivers 2009; Harris et al. 2008; Werren 2011). They are particularly important forces in the evolution of sex-determination mechanisms, as already illustrated by the role of SA selection (inter-genomic, intra-locus conflict; Section 7.2.2a). The long gene cascades underlying sex determination such as found in *C. elegans* (Box 3.15), with genes alternatively on and off, are highly suggestive of tug-of-war evolutionary games, where male and female elements alternatively seized the leading role in sex determination (but see Section 7.1). Table 7.1 provides an overview of the main theoretical studies that have investigated the dynamics of genetic conflict in the context of sex-determination evolution.

**Intragenomic conflicts** over sex ratios arise when different genetic elements that make up individual genomes have divergent interests regarding sex ratios (Jaenike 2001; Section 6.3.2). The autosomes gain equal transmission through both sexes, and thus favour Fisherian sex ratios; the X and Z chromosomes are normally more often inherited through the homogametic sex, and thus favour its production; the Y and W chromosomes are only transmitted through the heterogametic sex, and therefore selected to overproduce it; cytoplasmic elements (mitochondria, chloroplasts) are only transmitted maternally, and thus favour female-biased sex ratios. Vertically transmitted microbes constitute another important source of conflict, being often exclusively transmitted to one of the two sexes (e.g. through females via the egg cytoplasm).

Evolutionary arm races over the control of sex ratios may lead to transitions in sex-determination systems (Cosmides & Tooby 1981; Hurst et al. 1996; Partridge & Hurst 1998; Werren & Beukeboom 1998). **Meiotic-drive sex chromosomes** have been invoked to account for the weird sex ratios and dynamics of sex-determination systems in a variety of mammals, including the wood lemming (Fredga et al. 1976), creeping vole (*Microtus oregoni*; Charlesworth & Dempsey 2001), grass mouse (*Akodon azarae*; Hoekstra & Hoekstra 2001), pygmy mouse (*M. minutoides*; Veyrunes et al. 2010), and moles (*Talpa occidentalis*, *T. europaea*; McVean & Hurst 1996). They

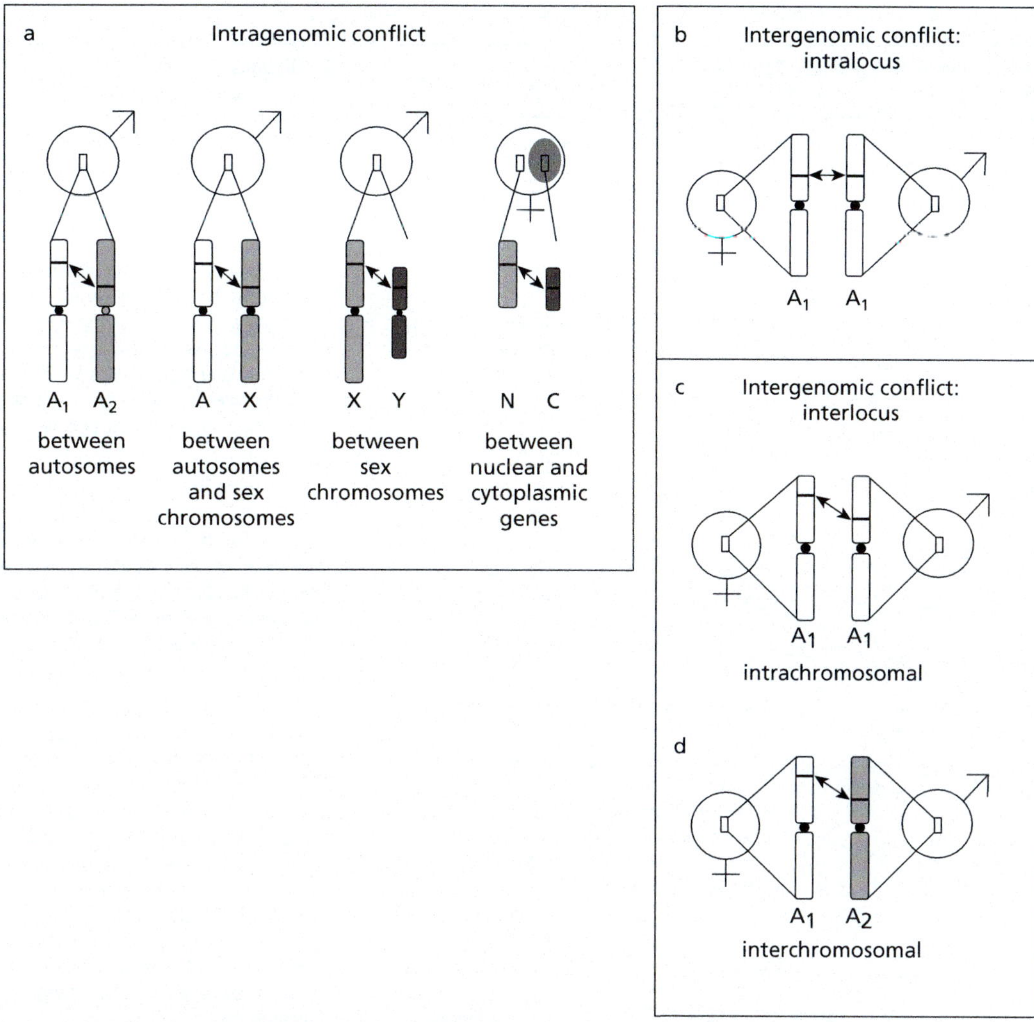


**Figure 7.1** Types of genetic conflicts. Intragenomic conflicts (a) refer to interactions between non-homologous loci within an individual. Intergenomic conflicts occur between homologous (b) or non-homologous loci from different individuals, either on homologous (c) or on non-homologous (d) chromosomes.

might also be at work in the heterogametic transitions documented in houseflies *M. domestica* (Clark 1999). Kozielska et al. (2010) developed recurrence equations to investigate the evolutionary dynamics of transitions mediated by meiotic drive; when distortion biases sex ratios towards females (driving X chromosomes), an autosomal masculinizing gene can invade, generating a new male heterogametic system (homogametic transition). Depending on the cost of the drive, the driving X ($X^d$) may eliminate the original X, or ultimately disappear after the transition. Conversely, a distortion biasing sex ratios towards males (driving Y chromosomes) may cause a switch to female heterogamety via an autosomal feminizing mutation (heterogametic transition). Here again, depending on the cost of the drive, the driving Y ($Y^d$) may eliminate the original Y or eventually disappear. An autosomal masculinizing drive, finally, should induce male-biased sex ratios, and thereby eliminate the original Y chromosome and favour the spread of a feminizing autosomal mutation, leading to a female heterogametic

**Table 7.1** Overview of models that investigated a role of genetic conflict in sex-determination evolution.

| Type of conflict | Conflicting parties | Reference | Topic | Predictions |
|---|---|---|---|---|
| Intragenomic | Autosomes–sex chromosomes | Kozielska et al. 2008 | Segregation distortion (meiotic-drive sex chromosomes) | Meiotic-drive sex chromosomes can select for evolution of novel **masculinizers** and **feminizers**, and switches in heterogamety |
| | Autosomes–cytoplasmic genes | Werren 1987 | Cytoplasmic sex-ratio distorters | Cytoplasmic sex-ratio distorters with incomplete transmission results in a population of monogenic individuals with all-male producing autosomes and all-female producing cytoplasmic genes |
| | | Caubet et al. 2000 | Cytoplasmic sex-ratio distorters | Interplay between cytoplasmic sex-ratio distorters and host masculinizers can cause heterogamety switching depending on type and cost of repressors and population structure |
| Intergenomic, intralocus | Male–female | Van Doorn & Kirkpatrick 2007 | Sexual antagonism and sex chromosomes | New sex-determining loci can invade as a result of linkage with a sexually antagonistic locus and replace the ancestral Y chromosome |
| | | Van Doorn & Kirkpatrick 2010 | Sexual antagonism and sex chromosomes | New sex-determining loci that are linked with a sexually antagonistic locus can result in transitions between XY and ZW systems depending on the degree of degeneration of the ancestral sex chromosome |
| Intergenomic interlocus | Maternal–offspring | Eshel & Sansone 1994 | Maternal and offspring control over sex allocation | Conflict between genes for maternal manipulation of sex ratio and zygotic responses when costs of sons and daughters differ selects for female control over sex ratio |
| | | Werren & Hatcher 2000 | Inbreeding and population subdivision with ESS approach | Maternal-zygotic conflict occurs under inbreeding and population subdivision and can causes changes in sex determination |
| | | Beukeboom et al. 2001 | Inbreeding and population subdivision with inclusive fitness approach | Maternal-zygotic conflict occurs under inbreeding and population subdivision and can causes changes in sex determination |
| | | Werren et al. 2002 | Progeny sex ratio and maternal or among progeny fitness | Fitness differences between sons and daughters leads to evolution of dominant zygotic and heterogametic sex-determination systems |
| | | Uller & Pen 2011 | Mating system and dispersal | Mating system and dispersal regime favour maternal over offspring control over sex ratio and restrict invasion of feminizers and masculinizers |

system (heterogametic transition). Depending on the cost of the drive, the original system will be either eliminated or maintained as a segregating polymorphism (coexistence of male and female-heterogamety at different loci, e.g. as observed in some Cichlidae). Interestingly (and worrying for the experimentalist), the distorter that induced the change is often lost from the population after the transition. Due to this predicted transience of distorters, the role of meiotic-drive sex chromosomes on the dynamics of transitions in sex-determination systems might have been underestimated.

Meiotic-drive sex chromosomes have also been invoked to account for transitions from XX/XO to paternal genome elimination and haplodiploidy in scale insects and sciarid flies (Haig 1993a,b); following Bull (1979), Haig (1993a) argued that a meiotic-drive X chromosome would select for chromosomal hitchhiking during male meiosis (i.e. autosomal genes causing their own chromosome to co-segregate with the X). The whole maternal set of autosomes would thereby become effectively X-linked and preferentially transmitted. Simultaneously, the overproduction of XX females would provide a selective advantage to females converting some of their XX offspring into males by eliminating the paternal genome. The ending point would be pure haplodiploidy (see Box 7.2 for an alternative scenario to the evolution of haplodiploidy).

Werren (1987) pioneered theoretical investigations on the interactions between autosomal and **cytoplasmic sex determiners**. Taylor (1990) simulated the selection of autosomal resistance alleles against cytoplasmic sex-ratio distorters. Caubet et al. (2000) showed evolutionary outcomes to depend on the type of repressor and associated costs. Cytoplasmic factors that kill males (rather than shifting the primary sex ratio) are not expected to cause transitions; a switch from female to male heterogamety is only expected if the dominant repressor acts on the feminizing effect of the distorter, and if the latter's transmission is very efficient. In most other cases, a multi-factor system for sex determination is expected. When distorter transmission is not complete, synergistic interactions between autosomal and cytoplasmic sex-ratio factors may also result in a polymorphic population made of monogenic individuals that produce either only sons (autosomally) or only daughters (cytoplasmically). A role for cytoplasmic sterility factors has similarly been proposed for the evolution of dioecy in plants (Maurice et al. 1994). Following Bull (1983), Juchault & Rigaud (1995) and Rigaud (1997) suggested that genomic conflict between cytoplasmic sex-ratio distorters and their hosts drive the dynamics of sex-determination systems in isopods. The woodlice *Armadillidium vulgare* nicely illustrates the potential for feminizing *Wolbachia* to induce turnovers (Box 7.1).

Female heterogamety might be more prone towards microbial manipulation than male heterogamety, because the time window for the microbes to accomplish their action is larger in females. In ZW systems, the Z-bearing (i.e. male-determining) eggs may already be targeted during female oogenesis, whereas in XY systems, offspring sex is signalled only after fertilization, which reduces the time for manipulation. Male heterogamety may thus provide partial immunity against cytoplasmic sex-ratio distorters (but see Box 7.2); in addition, female-biased sex ratios are likely to elicit the spread of dominant male determiners. The risk of microbial manipulation and response to it may thus partially account for the preponderance of male- over female heterogamety in several groups (though no formal models have investigated this hypothesis). Hamilton (1993) and Normark (2004) also suggested a role for endosymbionts in the evolution of haplodiploidy: male-killer endosymbionts operating through the destruction of the paternal set of chromosomes should select for host genes that rescue haploid embryos as viable males (Box 7.2). Similarly, Stouthamer et al. (2010) suggested that parthenogenesis-inducing endosymbionts have led to the irreversible loss of sex in some insect lineages.

**Intergenomic conflicts** occur between genetic elements that are present in different individuals, either at the same or at different loci (intra- versus inter-locus conflicts). Paternal–maternal genome interactions within offspring have features of both intra- and intergenomic conflict. As already pointed out (Section 7.2.2a), sex-antagonistic genes (intra locus intergenomic conflicts) might play an important role in generating turnovers. Other intergenomic conflicts may specifically act through sex-ratio selection. Trivers (1974) first noticed the potential for conflicts between parents and offspring over parental resource allocation. Except for clonal reproduction, individuals are always more related to themselves than to any partners, so that the trait value that maximizes their inclusive fitness may be suboptimal for relatives. Parent–offspring conflicts may arise when optimal sex ratios differ from parity, due, for example, to differences in production costs between sexes (Kozielska et al. 2006), local mate competition, or interdemic selection (West et al. 2000; reviewed in Charnov 1982, Karlin & Lessard 1986; Hardy 2002; West 2009). The zygotic optimum is normally less biased than the maternal one, which generates a conflict. Whenever parents invest more in one sex than in the other, members of the rare sex will enjoy a higher reproductive success; genes acting in offspring are therefore selected to increase the odds of residing

**Box 7.1 Turnover of sex determination in woodlice**

In female heterogametic populations of the isopod *Armadillidium vulgare*, maleness results from the development of an **androgenic gland**, under control of a masculinizing gene (*M*); in females, this gene is silenced by a dominant feminizing gene (*F*) on the W chromosome. ZW females produce highly female-biased progenies when infected by *Wolbachia*, which feminize ZZ offspring by preventing androgenic gland differentiation in the embryo. The resulting ZZ females will mate with ZZ males, and thus only produce ZZ offspring. As transmission rate is not complete (~90%), a minority of ZZ progeny develops as males. Under these settings, sex-ratio selection will quickly eliminate the W chromosome, because it only produces individuals of the more abundant (i.e. less wanted) sex; sex determination thus becomes purely cytoplasmic. In addition, sex-ratio selection generates strong pressure for autosomal suppressors of the *Wolbachia* feminizing effect. In

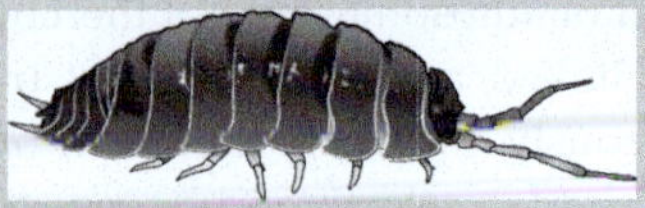

some strains, this has resulted in a new male heterogametic system via autosomal masculinizing genes. In other strains, the *Wolbachia* feminizing element (*f*) has been stably integrated in the host genome, generating a new W-like chromosome that re-establishes chromosomal sex determination with female heterogamety (Fig. 7.2). A single sex-ratio distorter can thus generate transitions to cytoplasmic, male heterogametic or female heterogametic sex determination (Juchault & Rigaud 1995; Rigaud 1997; Rigaud et al. 1997; Bouchon et al. 1998, 2008; Cordaux et al. 2004, 2011).

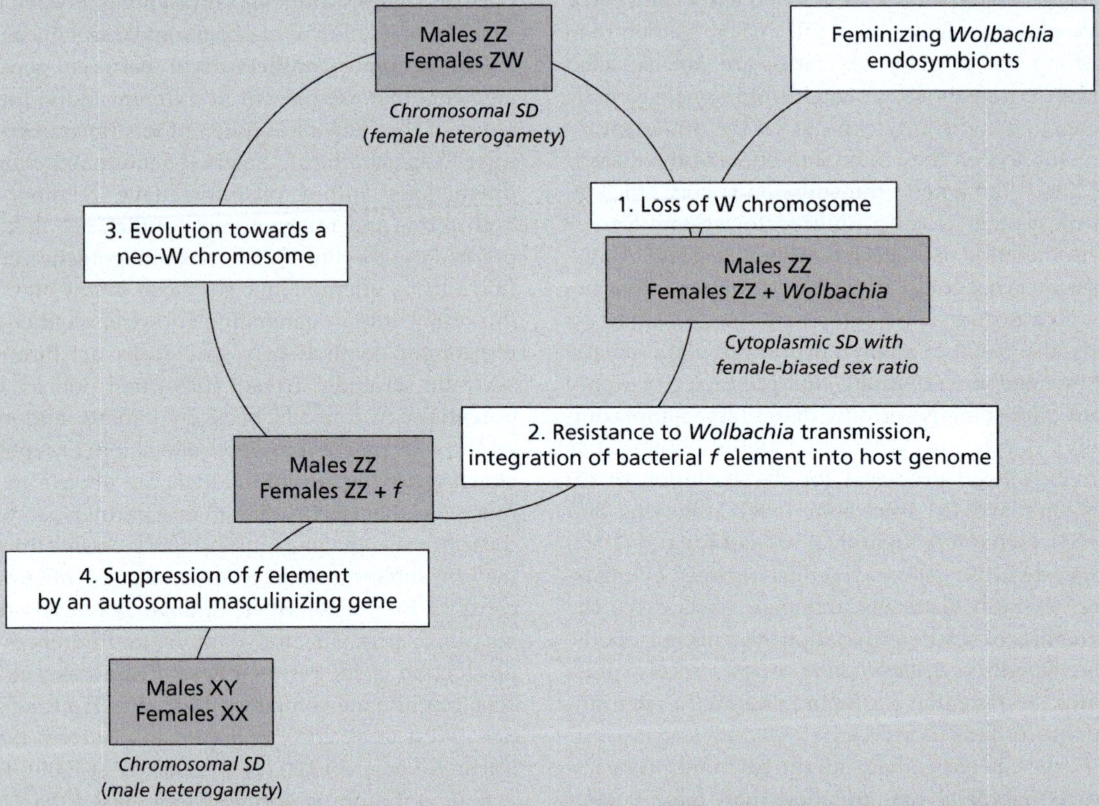


**Figure 7.2** Turnover of sex-determination systems in *Armadillidium vulgare*. 1: An initial population with female heterogamety becomes infected with *Wolbachia*, driving a shift from chromosomal to cytoplasmic sex determination with female-biased sex ratio. 2: Woodlice evolve resistance against *Wolbachia* and stably integrate its feminizing *f* element into a Z chromosome. 3. This integration generates a neo-W chromosome and restores chromosomal sex determination with female heterogamety (neo-W chromosome). 4. Alternatively, a masculinizing nuclear suppressor mutation generates male heterogamety (neo-Y chromosome). Modified after Cordaux et al. (2011).

**Box 7.2 How and why did haplodiploidy evolve?**

Haplodiploidy, including both arrhenotoky (in which males develop from unfertilized eggs) and paternal genome loss (in which the paternally inherited genome is eliminated from the fertilized egg) has evolved independently at least 20 times, 10 of which in insects (Otto & Jarne 2001; Normark 2003, 2004). It is mostly found in ecdysozoans (oxyuridian nematodes, mites, ticks, hymenopterans, bark beetles, micromalthid beetles, sciarid flies, cecidomyids, thrips, and coccoids) but also occurs in platyzoans (rotifers). Why haplodiploidy evolved in some groups and not in others, and what drives its evolution, remain controversial. Genetic hypotheses (maternal-gene transmission advantages or purge of deleterious mutations in hemizygous males) are not readily tied to any ecological correlate and thus fail to account for its specific distribution. Ecological hypotheses (sex-ratio control, and feminizing endosymbionts), by contrast, emerge from considerations on ecological correlates; lineages that evolved haplodiploidy tend to share two important features: maternally transmitted endosymbiotic bacteria (often with a mutualistic role in processing food) and gregarious broods. Hamilton (1993) first suggested that in small, highly structured, and inbred populations, male killing by endosymbionts would benefit both the bacteria (which are only transmitted by females) and its host (because of local competition for mates and resources). By eliminating the paternal genome upon fertilization when it carries a male-determining element (e.g. a Y chromosome), bacteria would render male zygotes haploid, and thus mostly unviable. Some males would however survive this haploidization and eventually evolve towards normal phenotypes. Normark (2004) and Engelstädter & Hurst (2006; see also Ubeda & Normark 2006) further formalized this idea, showing that male killing by haploidization may directly evolve towards paternal genome elimination, a primitive form of haplodiploidy. As further shown by Kuijper & Pen (2010), direct mutualistic benefits provided by endosymbionts are required for the long-term maintenance of this system. These models predict haplodiploidy to evolve from male-, rather than female, heterogamety and more specifically from sex-determination systems based on a sex-chromosome counting mechanism (dominant male determiners might restrict the evolution of haploid males, because their absence would result in female development).

in the rare sex, even if this reduces their relatives' inclusive fitness. Optimal sex ratios can also differ between the two parents and between father and offspring (Fisher 1930; Trivers 1974; Uller & Pen 2011), but theoretical models formalizing these scenarios are lacking.

Eshel & Sansone (1991, 1994) developed a three-locus diploid model to analyse the evolution of the primary sex ratio among offspring resulting from parental manipulation, and offspring response to parental manipulation, when the costs of rearing a male or female offspring differ; as they show, selection may favour either offspring or maternal control, depending on the information that mothers have regarding offspring response. Werren & Hatcher (2000) studied a model of a patch-structured population with local mate competition (offspring mate at random in their natal patch, after which mated females disperse to produce the next generation). This situation favours females producing more daughters, but optimal sex ratios are less biased for offspring. A maternal gene biasing sex ratios towards the maternal optimum may spread when at the offspring optimum, while dominant offspring genes increasing the production of the underrepresented sex will be favoured when at the maternal optimum (Werren & Hatcher 2000; Werren et al. 2002). Such conflict may drive the evolution of male or female heterogamety, when invading offspring genes have fully masculinizing or feminizing effects, respectively (Werren et al. 2002); the resulting population sex ratio will then eventually be 1:1 even though this is optimal neither for the mother nor for the offspring. As an alternative, the conflict can lead to the evolution of monogeny when a maternal gene has a fully feminizing effect (Werren et al. 2002).

There is ample scope for mother–offspring conflict over sex determination whenever sex determination potentially depends on both maternally and zygotically expressed genes (Werren & Beukeboom 1998). Examples include *C. elegans* (Ahringer et al. 1992), *D. melanogaster* (Steinmann-Zwicky et al. 1990; Cline 1993; Schütt & Nöthiger 2000), *M. domestica* (Schmidt et al. 1997; Dübendorfer & Hediger 1998), and *Nasonia vitripennis* (Verhulst et al. 2010a). In many coccids, sex of the progeny is determined by the maternal genotype, likely due to maternal-effect products placed into the egg (Nur 1989). The splicing regulator TRA often mediates mother–offspring conflicts over sex determination in insects, with amounts regulated by both maternal transmission to the egg and own zygotic production (Section 3.3.2c). Mutations have been found in maternally expressed genes, resulting in laboratory strains in which sex depends on the female genotype, as in *C. elegans* (Hodgkin 2002) and *M. domestica* (Este & Rovati 1982; Inoue & Hiroyoshi 1986). Although rare, monogeny has been reported from several insect species, including the dipterans *Chrysomya rufifacies* (Ullerich 1984) and *Sciara coprophila* (Brown & Chandra 1977; Sanchez & Perondini 1999; Sanchez 2008). As pure maternal control of sex is much scarcer than zygotic control, the maternal-offspring battle appears predominantly decided in favour of offspring. This may in part be due to the fact that offspring genes act later in the sex-determination process than maternal genes, which provides ample opportunity to overrule maternal decisions.

Arrhenotoky offers a large scope for maternal control over sex ratio via fertilization; the same may apply to ESD via nest site selection. In such systems, offspring control may be limited (Bulmer & Bull 1982), despite potentially strong conflicts. Many parasitic wasps produce highly female-biased sex ratios under local mate competition; any diploid offspring able to develop into a male would gain large benefits in reproductive output (because it could now mate with multiple sisters), a move likely prevented by the specific sex-determination mechanism (Beukeboom et al. 2001). However, workers in many insect societies (which usually prefer female-biased sex ratios due to the asymmetric relatedness values generated by haplodiploidy) often succeed in manipulating secondary sex ratios by eliminating their brothers.

Heterogamety might also offer a battlefield for control over sex ratios between paternal and maternal interests. This could potentially lead to adaptive over-production of one type of gametes in the heterogametic sex (Bull & Bulmer 1981), a phenomenon that has not been formally documented (but see Bulmer 1988; Komdeur & Pen 2002; Pryke & Rollins 2012). Sex determination in plants, which often show large variation in seed sex ratios and differential maternal and paternal control over sex ratio (e.g. *Silene latifolia*, Taylor et al. 1999), may be another promising group for study of parent–offspring conflict over sex determination.

To briefly sum up this long section: genetic conflicts, which occur widely both within and among genomes, have the potential to drive a whole range of transitions towards a large diversity of systems, via either sexually antagonistic or sex-ratio selection. This includes changes in heterogamety, evolution towards haplodiploidy or dioecy, transition to cytoplasmic, zygotic, or maternal control, or active maintenance of **polyfactorial sex determination**. Such conflicts also have the potential to account for some of the surprising aspects of sex-determination pathways, including long molecular cascades in which male and female determinants are alternatively switched on and off. Due to their ubiquity and the strong selective forces involved, genetic conflicts actually have the potential to account for much of the dynamic nature of sex-determination systems. Although the theoretical basis for a role of genetic conflicts in sex-determination evolution is now well established, the empirical support is still meagre and restricted to a few well-studied systems, such as *Armadillidium*. More systems need to be studied before the relative importance of different types of genetic conflict in sex-determination evolution can be judged.

## 7.3 Open questions on sex-determination transitions

This bewildering diversity of sex-determination systems and surprisingly high rates of turnover, which ongoing genomic studies will further uncover, open a whole range of new questions that will have to be addressed in the coming years. We identify a few that raise particular interest, and mention potential avenues for research in this promising area (Section 7.3.1. to 7.3.3.).

### 7.3.1 Why do turnover rates differ between lineages?

The dynamics of changes in sex-determination systems differ drastically among taxonomic groups. In vertebrates, for instance, birds and therians have maintained ancestral mechanisms of GSD for circa 120 and 165 My respectively, leaving ample time for their W or Y chromosomes to degenerate; in striking contrast, most cold-blooded vertebrates (fish, reptiles, and amphibians) display an astonishing diversity and lability in sex-determination systems and mechanisms. Even when determination is purely genetic (as is generally the case for amphibians), sex chromosomes show little or no decay, partly due to high turnover (and partly to occasional X-Y recombination; Section 5.2.2d). Transitions seem particularly common in teleost fishes. Interestingly, at lower levels also the dynamics differ strikingly between orders or families; Salmoniformes, for instance, strictly stick to male heterogamety, while Cypriniformes show a diversity of transitions between XY, ZW, ESD, and hermaphroditism (Mank et al. 2006; Mank & Avise 2009). In reptiles similarly, some groups show a predominant TSD (e.g. alligators, several families of turtles), others a predominant GSD (e.g. all snakes, often with differentiated W chromosomes), while still others transition widely among systems. Within Squamata, the Lacertidae, and Varanidae consistently display female heterogamety, other families (Iguanidae, Scincidae, Polychrotidae, Phrynosomatidae) are male heterogametic, while still others (Agamidae, Gekkonidae) switch between TSD, XY, and ZW systems (Ezaz et al. 2009; Pokorna & Kratochvil 2009). Similar contrasts are found among ecdysozoans (e.g. insects and crustaceans): all Lepidoptera and Trichoptera display a common pattern of female heterogamety, all Hymenoptera are haplodiploid, but Diptera and Hemiptera apparently cover the whole range of possible sex-determination systems, including TSD, polyfactorial, XY, ZW, maternal control, and haplodiploidy (Fig. 2.7, SOM Table 2.1).

This seemingly random taxonomic distribution poses a real challenge to the identification of possible common factors underlying the lability/stability of sex-determination systems, or the co-option of specific systems by specific groups. Causes may often be idiosyncratic. Furthermore, as developed below, consistency within groups strongly suggests non-independence; some sex-determination systems might be difficult to opt out of once chosen. Specific adaptations in ecology, life history, physiology, or genomic architecture might predispose some lineages to a limited set of sex-determination systems or transitions, and prevent others. Birds and mammals, for instance, display strong homeostatic abilities that protect developing embryos from environmental effects on sex determination; the efficient homeothermy of therians certainly offers little scope for TSD and, by the same token, for any environmentally-induced transition in sex-determination systems (Grossen et al. 2011; Section 7.2.3b). This might have contributed to the stable GSD systems found in endothermic vertebrates. The size and structure of populations might also matter: large and open populations offer little scope for genetic drift, deviations from Fisherian sex ratios, local mate competition, or interdemic selection, all features with a potential to induce transitions (Sections 7.2.1, 7.2.3a). Genomic features might render some groups more likely to evolve dosage compensation as an alternative to mutation-load-driven transitions (Section 7.2.2b), or more susceptible to meiotic-drive elements and/or cytoplasmic factors. *Wolbachia* bacteria, for instance, are only found in some groups of ecdysozoans, including isopods, where they are known to trigger transitions (Box 7.1). Certain genomic organizations may also facilitate the resolution of sexually antagonistic conflicts via differential regulation of autosomal genes, rather than via sex linkage, another source of transitions (Section 7.2.2a).

### 7.3.2 Are some sex-determination systems more labile or some transitions more likely?

Sex-determination systems may intrinsically differ in stability, being more or less vulnerable to potential causes of turnovers. Polyfactorial sex-determination systems are highly susceptible to drift and likely to collapse soon or later into a **monofactorial system**, at least when based on a few loci with major effects; balanced polymorphism can then only be maintained under stringent conditions (Rice 1986). If based on very many loci with small effects, however, polygenic sex determination might be maintained just by mutation pressure. Other systems are more likely to act as absorbing states or even as evolutionary traps. Such might be the case of dioecy in plants: dioecious lineages tend to be species poor, due to lower speciation rates and/or higher extinction rates (Heilbuth

2000; Vamosi & Vamosi 2005). Dioecy may lower the potential to colonize new habitats, and thereby limit adaptive radiations. Reversal to hermaphroditism is likely impeded (though not prevented), because it requires back mutations to restore male- or female fertility (complex functions are more readily lost than gained). Opting out of haplodiploidy also seems a difficult task: there have been at least 20 independent evolutions from male heterogamety to haplodiploidy (Box 7.2), but only two reversions are known, both in scale insects, and both from an incipient state (germline paternal genome elimination; Normark 2003); reversal to **diplodiploidy** from arrhenotoky likely poses insuperable difficulties.

Such might also be the case of GSD in amniotes: transitions from TSD to GSD seem plentiful in reptiles, while no case of reverse transition seems to be documented with certainty (Pokorna & Kratochvil 2009; but see Janzen & Krenz 2004). There are additional reasons to believe that sex chromosomes might constitute evolutionary traps. Ancient Y chromosomes, such as found in therians or *Drosophila*, show an accumulation of genes essential for male fertility (Section 5.2.2) and cannot be easily discarded (though the Y chromosome has been lost from a few rodent and *Drosophila* lineages). Feminization of the X chromosome might create a similar situation. The differentiation of sex chromosomes is furthermore often accompanied by sophisticated methods of dosage compensation (Section 5.2.2e); in such a case, the production of XX males through a turnover in sex-determination systems might generate drastic gene imbalances. Such fitness costs may be overcome, however: the dot chromosome of *Drosophila* provides a nice example of a highly differentiated sex chromosome that has reverted back to an autosome (Vicoso & Bachtrog 2013).

Heterogametic transitions (XY $\rightarrow$ ZW) are even less likely in lineages with differentiated sex chromosomes, because they require the fixation of Y (or W) chromosomes as autosomes, with dramatic consequences on viability due to the accumulated deleterious load. Heterogametic transitions are thus possibly limited to homomorphic sex chromosomes, and certainly not mutation-load-driven (Section 7.2.2b); SA-driven mechanisms (7.2.2a) or cytoplasmic sex-ratio distorters (Box 7.1) are more plausible candidate causes. Note, however, that the majority of transitions among undifferentiated sex chromosomes may not affect the original patterns of heterogamety, and are thus less likely to be detected. In *Rana*, a minimum of 12 transitions is required to account for the distribution of sex chromosomes documented so far in the genus, from which a single one has led to a change from XY to ZW (Miura 2007). In *Oryzias*, out of six documented transitions, a single one has changed from XY to ZW (Takehana et al. 2008). In Salmonidae, all species are male heterogametic, despite the many transitions documented (Phillips et al. 2001; Woram et al. 2003, 2004). In squamates, as already mentioned, many families appear to stick to one single pattern of heterogamety (e.g. ZW for Lacertidae and Varanidae, XY for Iguanidae and Scincidae; Ezaz et al. 2009; Pokorna & Kratochvil 2009). Such a consistency is expected from the mutation-load-driven transition mechanism (Section 7.2.2b), but others causes might be at work: the gene networks orchestrating sex differentiation (downstream components of the sex-determination cascade) are highly integrated and may thus offer significant resistance to a change in heterogamety, which would impose deeper reorganization. In Salmonids, the whole cascade is likely conserved, including the initial trigger; transitions occur just by switching the chromosomal localization of the top gene (Yano et al. 2013).

Another intriguing fact is that male heterogamety appears more widespread and abundant than female heterogamety across both animal and plant lineages. There is no suggestion that speciation or extinction rates depend on the patterns of heterogamety. Are XY systems more likely to appear, for example, due to the nature of mutations involved? For reasons shown in Chapter 3 (Box 3.10), dioecy in plants mostly evolved via gynodioecy (rather than androdioecy). Female heterogamety is expected to result if the initial male-sterile mutation is dominant, and male heterogamety if recessive; the latter might occur preferentially, because loss-of-function mutations are usually recessive (Section 3.3.1). In animals, the generally stronger sexual selection on males might favour the transition towards XY over ZW sex determination: a dominant sex-determining mutation is likely to sweep to fixation when this gene (or one linked to it) is beneficial to the heterogametic sex (Rice 1986; Bachtrog et al. 2011). Dominant masculinizing mutations might also constitute the best response to female biases induced by cytoplasmic sex-ratio distorters (Box 7.1), and XY systems offer better protection against meiotic-drive sex chromosomes (true meiotic drive is only possible in ZW systems; Section 6.3.2).

### 7.3.3 Are some chromosome pairs particularly good at sex?

Two recent reviews have suggested that some chromosomes might be more likely than others to carry the master sex-determination gene (Graves & Peichel 2010; O'Meally et al. 2012). Supporting empirical evidence is still limited, however. The same teleost protokaryotype chromosome appears as an XY sex chromosome in the nine-spined stickleback (*Pungitius pungitius*) and the tiger pufferfish (*Takifugu rubripes*), and as a ZW sex chromosome in Lake Malawi cichlids and the medaka *Oryzias hubbsi*. Likewise, another teleost protokaryotype chromosome has been independently used as a sex chromosome in three-spined sticklebacks, Nile tilapia (*Oreochromis niloticus*), and Lake Malawi cichlids (Graves & Peichel 2010). Still another chromosomal pair became the ZW sex chromosomes in a soft-shelled turtle (*Pelodiscus sinensis*) and the XY sex chromosomes in a green anole (*Anolis carolinensis*) (O'Meally et al. 2012). The avian Z chromosome (harbouring *Dmrt1*) has been co-opted independently by a gecko (also ZW; Kawai et al. 2009), as well as several species of *Rana* and other frogs from deeply divergent lineages, all XY (Brelsford et al. 2013). However, neither the snake nor the therian sex chromosomes are known to be sex linked in any other amniote (O'Meally et al. 2012). In insects, no homology is evident between the sex chromosomes of Diptera, Lepidoptera, or Coleoptera (Pease & Hahn 2012). More extensive coverage is expected in the coming years; a systematic scan should properly establish the patterns and statistically test whether they differ from random.

Recurrent convergence towards a limited number of genomic regions might result from the reuse of a small set of genes able to capture the top position in the pathway; *Dmrt1* might indeed belong to this limited set of key genes that are intrinsically better at triggering the initial steps of sex determination (Section 3.3.2a); other genes from the network (e.g. *Sox* genes, *AR*, *Amh*/*Amhr2*, and *Cyp19a*; Chapter 3) are likely to play a similar role, attracting sex determination to other chromosomes. Note that, if frequent sex chromosome turnovers are indeed biased toward certain genomic regions, this bias could become a self-reinforcing evolutionary process: if a chromosome has been sex-linked in the past, it may have accumulated genes likely to be involved in sexually antagonistic effects, which could in turn make it more likely to recapture the role of sex chromosome in a turnover event (Van Doorn & Kirkpatrick 2007; Section 7.2.2a). Combined with the deleterious load accumulating on non-recombining Y chromosomes (Section 7.2.2b), this process might lead to cyclical sex-chromosome turnovers among a limited set of chromosomes with high potential for sexual antagonism (Blaser et al. 2014).

## 7.4 Perspectives on sex-determination transitions

Investigations on the dynamics of sex-determination transitions are still in their infancy. Properly addressing the above questions will require progresses along several lines. There is first a need to better formalize the diversity of models tackling the potential causes of transitions, in order to derive predictions that are amenable to test through comparative analyses and experimental approaches. What kinds of transitions are expected in which conditions? What empirical patterns are predicted to result from which evolutionary causes? Regarding empirical data, the number of properly documented non-independent transitions is still extremely limited, which prevents generalizations and proper statistical tests. With the ongoing popularization of genomic tools, however, we expect a flurry of relevant studies in the coming years that promise to drastically affect our vision of the field.

We would like to end this chapter by pleading for the development of experimental-evolutionary approaches to the dynamics of sex-determination transitions, which have the potential to be remarkably informative. The first step in this approach consists in identifying relevant variants in sex-determination systems. As illustrated throughout this book (e.g. Chapter 4), genetic variance in sex-determination systems occurs widely among- and within natural populations of many organisms. Suffice it to recall, for example, the coexistence of genotypes with different sensitivities to temperature in natural populations of fishes or reptiles (Chapter 4), the geographic clines in autosomal masculinizing factors and dominant feminizing mutations in *M. domestica* (Box 3.20), the abundant genetic variation in transcript levels occurring at all main genes from the sex-determination cascade in *D. melanogaster* (Tarone et al. 2005), or the widespread occurrence of meiotic-drive sex chromosomes in *D. simulans* (Box 6.1). Several mutations affecting sex-determination systems have been identified and isolated in the laboratory. Laboratory strains of *M. domestica* present either zygotic or maternal sex determination (Box 3.20), and laboratory strains of *Poeciliopsis lucida* either GSD or TSD (Sullivan & Schultz 1986; Chapter 3); through a series of laboratory crosses, Winge (1932, 1934) induced a sex-chromosome transition in guppies, generating a strain with autosomal inheritance of the ancestral sex chromosome (together with all linked SA colour genes). Laboratory strains of the common carp, similarly, could be selected for a recessive masculinizing autosomal mutation that replaced the ancestral XY by a ZW system (Komen et al. 1992a, b, c; Chapter 3). Otake et al. (2010) succeeded in generating artificial proto-sex chromosomes in medaka fish: transgenic strains were established by microinjecting freshly spawned fertilized eggs with *DmY*-containing BAC clones; some XX individuals incorporated this exogenous copy of *DmY* and developed as males. Sex-linkage analysis on grown fishes showed the exogenous *DmY* to be located on LG23 in one strain, and on LG5 in another strain (the original sex chromosome being on LG1). In several plant models, transitions from monoecy to gynomonoecy, gynoecy, or hermaphroditism have been documented and associated to specific mutations (Boxes 3.7 to 3.9). Similar mutations have been identified from *C. elegans*, leading to changes from XO males/XX hermaphrodites to a variety of gonochoric, dominant ZW, monogenic maternal sex determination, or TSD (Hodgkin 2002; Box 3.15); seven different autosomal genes were shown to be able to hijack the role of initial trigger, making each of the five autosomes evolve as a sex chromosome. Other strains had sex determined by fragmentary chromosomes, episomes, compound chromosomes, or environmental factors. Although the *C. elegans* sex-determination system may possess a number of features that make it particularly amenable to manipulation (see Hodgkin 2002 for details), it shows the promise of experimental evolution with genetic toolkits becoming available for an increasing number of species.

The second step in the approach consists of evolutionary experiments under controlled population settings and specific selective environments. Table 7.2 lists some of the few currently available studies (see Section 4.9 for further examples). Lyttle (1981) generated laboratory populations of *D. melanogaster* with an engineered meiotic-drive Y chromosome, which caused extreme sex-ratio distortions with the potential to ultimately lead to population extinction. As expected (Section 6.3.2), several of the populations surveyed tended to accumulate unlinked polygenic suppressors at a roughly linear rate over time. One population, however, was able to neutralize the meiotic-drive Y chromosome through an unexpected qualitative change in the chromosomal complement, namely the formation of sex-chromosome aneuploid XXY females and XYY males, revealing the potential of sex-chromosome non-disjunction as an adaptive response to sex-ratio distortion.

Conover et al. (1992) exposed Atlantic silversides from different localities to either low or high temperatures in the laboratory (Section 4.9). The ensuing highly skewed sex ratios led to increases in the frequency of the minority sex in all five replicate populations within ten generations. Depending on population and treatment, the evolutionary response resulted in either a change in the pivotal temperature, or loss of TSD, showing how rapidly sex-determination systems can respond to environmentally induced sex-ratio selection.

Chandler et al. (2012) made use of the temperature-sensitive strain CB5362 of *C. elegans* from Hodgkin (2002; Box 3.15), in which all individuals are XX but develop primarily as hermaphrodites at cool

**Table 7.2** Overview of experimental evolution studies of sex determination.

| Studied process | Species | Reference | Methodology | Outcome |
|---|---|---|---|---|
| Response to meiotic drive sex chromosomes | *Drosophila melanogaster* | Lyttle 1981 | Artificial translocation of the segregation distorter (sex determination) locus on the Y chromosome | Accumulation of sex chromosome aneuploids (XXY and XYY) with XX eggs yielding females |
| Evolution of ESD | Atlantic silverside (*Menidia menidia*) | Conover et al. 1992 | Sex-ratio selection for changes in ESD under extreme temperatures | High temperature led to a change of ESD to GSD, low temperature led to shifts in temperature thresholds |
| Evolution of proto-sex chromosomes | Medaka (*Oryzias latipes*) | Otake et al. 2010 | Creation of transgenics by microinjection | Generation of proto-sex chromosomes by incorporation of exogeneous DMY sequences |
| Evolution of compensatory adaptations in sex-determination pathway | *Caenorhabditis elegans* | Hodgkin 2002 | Series of generated mutation lines characterized for their sex determination | An array of different loci spread across the genome can acquire sex-determination function |
| | | Chandler et al. 2012 | Temperature-sensitive mutations of sex-determination genes causing intersexuality at intermediate temperatures | Disruption of sex-determination genes result in changes in sex-determination gene transcription and compensatory adaptations in the sex-determination pathway |
| Spread of neo-Y chromosomes (autosomal M) | Housefly, (*Musca domestica*) | Hamm et al. 2009 | Population cages with mixtures of ancestral and neo-Y chromosomes at high temperature | Neo-Y chromosome strains ($M^{III}$) had higher fitness and increased in frequency |
| | | Kozielska 2008 | Population cages with mixtures of ancestral and neo-Y chromosomes at low and high temperature | Neo-Y chromosome $M^{II}$ but not $M^{III}$ increased in frequency independent of temperature |

temperatures (<15°C) and as functional males at warm temperatures (>20°C). When reared for 50 generations at 16 or 18°C (temperatures that yield intersexed and low-fertility animals in the ancestral strain), lines evolved a shift in the pivotal temperature as well as a dampening (or in some cases, nearly complete loss) of temperature sensitivity, in favour of universally more hermaphrodite-like phenotypes. The shift was apparently not due to transcriptional alterations in the mutated genes, but rather in other genes of the sex-determination pathway and beyond, unveiling genetic variation for compensatory adaptations to mutations that can be revealed if the system is challenged.

Hamm et al. (2009) designed cage experiments to estimate the fitness of male houseflies with the *M*-factor either on the Y chromosome or on chromosome III (Box 3.20). On all accounts measured, $M^{III}$ males had higher fitness than $M^{Y}$ males. As isofemale lines were used, these differences cannot be attributed to genetic background, but rather to the genomic position of the sex-determination locus. Kozielska (2008) set up similar population cage experiments where different male genotypes ($M^{II}$, $M^{III}$, and $M^{Y}$) competed at either 20 or 25°C, motivated by the latitudinal distribution documented from natural populations (Box 3.20). $M^{II}$ but not $M^{III}$ males increased in frequency over eight generations, though independently of temperature.

In line with the above studies, it is to be expected that future experimental manipulations, using engineered sex chromosomes or sex-determination systems, possibly via endosymbiotic bacteria, will uncover further flexibility in sex-determination pathways, provide tests of the conditions for transitions under varying environmental conditions, and thereby contribute substantially to our understanding of sex-determination evolution.

# Glossary

## A

**ABCE model:** a model of flower development describing the interactions between homeotic MADS-box genes in determining organ identity within the floral meristems (Box 3.6).

**Achiasmate:** characterized by the absence of chiasmata and crossovers during meiosis.

**Achiasmy:** absence of recombination in one sex, either males (as in *Drosophila*) or females (as in *Bombyx*).

**Additive genetic variance:** component of the genetic variance that is attributable to additive effects of alleles.

**Agamont:** asexual form in some unicellular eukaryotes.

**Allochronic isolation:** isolation resulting from individuals being reproductively active during different times of the year (having non-overlapping breeding seasons).

**Allocycly:** differences in the dynamics of chromosome coiling behaviour during meiosis.

**Alternative splicing:** processing of messenger RNA by which particular exons of a gene may be included within, or excluded from, the final messenger RNA, resulting in proteins with different function (Section 3.1.2b).

**Androgenic gland:** in crustaceans, gland secreting the androgenic gland hormone (AGH), which controls spermatogenesis and development of male-specific morphology and behaviour.

**Amphigenic:** producing offspring of different genders (males and females).

**Androecium:** collective term for all stamens in a flower.

**Androdioecy:** condition of having males and hermaphrodites coexisting in the same population (Fig. 1.2).

**Andromonoecy:** condition of having male and hermaphrodite flowers on the same plant (Fig. 1.2).

**Aneuploid:** possessing too many or too few copies of one or more chromosomes in a cell or organism.

**Anisogamonty:** sexual dimorphism of gamonts, with macrogamonts and microgamonts.

**Anisogamy:** condition of having gametes of different size (Section 2.1.2).

**Anisomotility:** condition of having isogametes of different motility (Section 2.1.2).

**Antheridium:** male sex organ (gametangium) of fungi, algae and non-seed plants.

**Antheridiogen:** sex pheromone of ferns that induces other spores in the neighbourhood to develop into male gametophytes.

**Anti-Müllerian hormone (AMH):** non-steroid hormone affecting the development of reproductive organs in vertebrates.

**Antisense RNA:** an RNA strand having a sequence complementary to a transcribed RNA strand.

**Apomeiosis:** imperfect or suppressed meiosis, resulting in unreduced eggs.

**Apomixis:** asexual reproduction, lacking fertilization and meiosis.

**Archegonium:** female sex organ on the gametophyte of bryophytes and pteridophytes.

**Aromatase:** enzyme encoded by the gene *Cyp19*, converting androgens into oestrogens.

**Arrhenotoky:** sex determination system under haplodiploid reproduction in which males arise from unfertilized eggs and females from fertilized eggs.

**Ascogonium:** female reproductive organ of ascomycetes.

**Ascospore:** meiospore of ascomycetes.

**Autogamy:** 1) self-fertilization 2) in ciliates, fusion of two identical haploid nuclei derived from the mitotic duplication of a meiotic product.

**Automixis:** fusion of nuclei derived both from the same meiotic event (e.g. fusion of egg cell with one polar body).

**Autosome:** a chromosome that is not a sex chromosome.

## B

**B chromosome:** supernumerary chromosome lacking a homologue, usually genetically inert and not essential for an organism.

**Background selection:** selection against deleterious mutations; in non-recombining genomic regions, contributes to Hill–Robertson interferences (Section 5.2).

**Basidiospore:** meiospore of basidiomycetes.

**Bipolar:** characterized by a single mating-type locus, so that meiosis produces spores of two mating types.

**Bisexual:** condition of having two sexes (syn. co-sexual or hermaphrodite).

**bZIP-domain:** DNA-binding domain characterizing a large family of transcription factors in eukaryotes, often with a role in sexual development or sex determination.

## C

**Canalization:** holding of a developmental process within narrow bounds despite both genetic and environmental disturbing factors.

**Carpel:** female reproductive organ of flowering plants, composed of ovule, style, and stigma.

**Carpogonium:** in red algae, female sex organ on the gametophyte.

**Carpophore:** in basidiomycetes, fruiting body in which karyogamy and meiosis take place.

**Carposporangium:** in red algae, reproductive organ of a carposporophyte, mitotically producing diploid carpospores.

**Carpospores:** in red algae, diploid spores mitotically produced by the carposporophyte, and developing into a tetrasporophyte.

**Carposporophyte:** in red algae, diploid stage developing on the female gametophyte after fertilization, and mitotically producing diploid dispersing carpospores in a carposporangium (Fig. 2.3).

**Caryonids:** in ciliates, cells resulting from conjugation and meiosis, containing genetically identical micronuclei but independently derived macronuclei; adjective caryonidal.

**Centromere:** specialized region of DNA on each eukaryotic chromosome that acts as a site for the binding of the kinetochore proteins during cell division.

**Chiasma:** cross-shaped structure occurring between non-sister chromatids in meiosis, the site of a crossover.

**Cis-regulatory element:** molecule affecting mRNA synthesis downstream on the same chromosome.

**Clade:** monophyletic group of organisms.

**Clone:** lineage of genetically identical individuals, reproducing asexually by mitotic division.

**CMS (= cytoplasmic male sterility):** abortion of pollen production in higher plants determined by mitochondrial genotype.

**Coenocyte:** cell with a multinucleate protoplasm formed by division of nuclei but not of cytoplasm (synonym, plasmodium).

**Co-evolution:** joint evolution of two or more interacting traits or species, each of which evolves in response to selection imposed by the other.

**Compensasome:** complex of proteins and RNAs that mediate dosage compensation in *Drosophila*.

**Conjugation:** direct contact between cells or thalli, allowing exchange of genetic material.

**Co-sexual:** with two sexes (syn. bisexual or hermaphroditic).

**Coyne's rule:** the disproportionately large effect of the X chromosome in causing hybrid sterility and unviability.

**Crossover (crossing-over):** exchange of genetic material between homologous chromosomes that results in recombinant chromosomes.

**CSD (= complementary sex determination):** in Hymenoptera, system of sex determination by complementation of alleles at a sex locus.

**Cyclical parthenogenesis:** form of reproduction in which sexual and parthenogenetic generations are alternated (e.g. in water fleas and aphids).

**Cytokinin:** phytohormone; growth substance that stimulates plant cell division.

**Cytoplasmic incompatibility:** inability of sperm and eggs to form viable offspring due to changes made to the DNA of gametes by intracellular parasites.

**Cytoplasmic inheritance:** inheritance of non-nuclear DNA, transmitted by organelles or endosymbionts.

## D

**D-GSD (= diplo-genotypic sex determination):** genotypic sex determination occurring at the diploid stage, implying some genomic differentiation between the interacting diploid individuals (e.g. sporophytes), and not between the gametes as in haplo-genotypic sex determination (Section 2.1.5).

**Di-haploidy:** condition of having two independent haploid genomes in a cell, as in the dikaryotic stage of some fungi.

**Dikaryotic:** refers to a cell with two nuclei (dikaryon) that divide synchronously; characteristic stage in Dikarya (basidiomycetes and ascomycetes).

**Dioecy:** condition of having separate male and female individuals in the diploid (sporophytic) phase. The term gonochory is used instead for animals, with the same meaning. Adjective dioecious.

**Dioicy:** condition of having separate male and female individuals in the haploid (gametophytic) phase (e.g. in bryophytes). Adjective dioicous.

**Diploid:** with two sets of chromosomes.

**Diplodiploidy:** as opposed to haplodiploidy, sexual cycle in which both females and males arise from fertilized eggs and are diploid.

**Diplontic cycle:** life cycle with no haploid mitoses; the haploid phase is limited to gametes (as in diatoms, Fucales, or animals).

**Directional selection:** selection for a value of a character that differs from its current mean value.

**Disruptive selection:** selection against intermediate phenotypes, leading to distinct phenotypes and increased variation.

**Distyly:** a form of heterostyly with two morphological types of flowers differing in style length (Fig. 3.11).

**DM domain:** protein domain involved in DNA sequence recognition, characterizing the DM family of transcription factors. DM-domain genes, including *Dsx*, *Mab3*, and *Dmrt1*, are involved in sex differentiation or determination throughout the Metazoa (Section 3.3.2a).

**dN/dS ratios:** ratio of the number of non-synonymous to synonymous substitutions; indicator of selective pressure acting on a protein-coding gene.

**Dobzhansky-Muller (DM) incompatibilities:** negative epistatic interactions between genes with different evolutionary histories leading to a decrease in fitness between individuals from two ancestrally identical populations.

**Double-strand break:** DNA break cleaving the sugar-phosphate backbones of both strands of the double helix.

## E

**Effective population size:** the average number of individuals in a population that effectively contributes genes to succeeding generations.

**Endosymbiont:** parasitic, commensal, or mutualistic organism that lives within the body of its host.

**ESD (= environmental sex determination):** determination of sex by environmental cues, experienced over a discrete sensitive period early in development.

**Epigenetic:** refers to variations in gene expression or organism phenotype that are not due to differences in DNA sequences; epigenetic effects may stem from DNA methylation or modifications of the chromatin structure, for example, via histone acetylation or methylation (Section 2.1.3).

**Epistasis:** interaction between alleles from two or more loci on phenotype expression.

**Evolutionary strata:** distinct regions of divergence between sex chromosomes as a result of differently timed events of recombination suppression.

**Exon:** part of a gene that is translated into a polypeptide (as opposed to introns).

## F

**Female heterogamety:** sex-determination system in which the female carries distinct sex chromosomes (Z and W).

**Feminizer:** factor that induces femaleness.

**Fisherian sex ratio:** sex ratio expected under Fisher's principle, according to which the same investment is made into the male and the female function.

**Fission:** division of a cell (or body) into two or more parts and the regeneration of those parts into separate cells (or bodies).

**FMF pattern of sex determination:** temperature-dependent sex determination in which males develop at intermediate temperatures and females at both low and high temperatures.

**Frequency-dependent selection:** mode of natural selection in which the fitness of each phenotype varies as a function of its frequency in the population; negative frequency dependence results in balancing selection.

## G

**Gametangium:** gamete-producing organ in haploid gametophytes.

**Gamete:** haploid cell specialized in mating. The fusion of two gametes produces a diploid zygote.

**Gametocyte:** cell that undergoes meiosis to produce gametes. Male gametocytes are called spermatocytes, female gametocytes are called oocytes.

**Gametogenesis:** formation of gametes.

**Gametologues:** homologous genes on non-recombining sex chromosomes.

**Gametophyte:** gamete-producing haploid phase in haplontic and haplo-diplontic cycles.

**Gamont:** gamete-producing sexual form in unicellular eukaryotes.

**Gamontogamy:** mating between gamonts.

**Gender:** 1) sexual identity of a multicellular organism, describing its relative allocation to the male and female function; syn. sexes (Section 1.2) 2) social construction of sexes in humans.

**Gene conversion:** meiotic process of directed change in which one allele directs the conversion of a partner allele into its own sequence and form.

**Genetic conflict:** antagonistic fitness relationships between alleles at homologous or non-homologous loci in a genome.

**Genetic drift:** random changes in the frequencies of two or more alleles or genotypes within a population, due to stochasticity in the patterns of survival and reproduction; the strength of drift is inversely proportional to the effective population size.

**Genetic hitchhiking:** change in the frequency of an allele due to linkage with a selected allele at another locus; contributes to Hill–Robertson interferences in non-recombining genomic regions.

**Genic balance theory:** model according to which sex is determined by the ratio of X chromosomes to autosomes; originally proposed by Bridges (1925) based on experiments in *Drosophila*.

**Genotype × environment (G × E) interaction:** phenotypic variance arising from the differential effect of environment on the expression of different genotypes.

**Genomic imprinting:** parental marking of genomic regions (e.g. gene silencing via DNA methylation) so that their expression in offspring depends on parental origin.

**Germ line:** as opposed to soma, differentiated cell line in animals that forms the gametes.

**Gonad:** sexual organ in animals, producing either sperm or ova.

**Gonochorism:** condition of having separate male and female individuals; applies to animals (the term dioecy is used instead for plants, with the same meaning).

**Good-genes hypothesis:** model of sexual selection according to which females choose mates on their ability to pass on genes that will increase offspring survival or fecundity (Box 6.3).

**Granulosa cell:** somatic cell closely associated with the developing female gamete in the ovary.

**GSD (= genotypic sex determination):** determination of sex by a genetic difference between individuals.

**GSI (= gametophytic self-incompatibility):** a type of incompatibility between the pollen and pistil determined by the haploid genotype of the pollen and the genotype of the stigma (Section 3.4.1a).

**Gynodioecy:** condition of having females and hermaphrodites in the same population.

**Gynoecium:** collective term for all carpels in a flower.

**Gynoecy:** condition of producing only female flowers.

**Gynomonoecy:** condition of producing female and hermaphroditic flowers, on the same plant.

## H

**Haldane's rule:** empirical rule first formulated by Haldane (1922) according to which the heterogametic sex is more susceptible to hybrid incompatibility than is the homogametic sex.

**Handicap selection:** model of sexual selection according to which evolution leads to honest or reliable signalling between individuals, despite conflicts (Box 6.3).

**Haplobiontic cycle:** life cycle in which mitoses are either haploid or diploid, but not both.

**Haplodiploidy:** life cycle in which males are haploid and females diploid; male haploidy arises from either non-fertilization of eggs (arrhenotoky) or paternal genome loss (Box 7.2).

**Haplo-diplontic cycle:** life cycle with both haploid and diploid mitoses, widespread among algae (Section 2.1.1).

**Haploid:** with one set of chromosomes.

**Haplontic cycle:** life cycle with no diploid mitoses; the diploid phase is limited to the zygote (Section 2.1.1).

**Haplotype:** a haploid genotype, a set of linked genes that are inherited together.

**HD (= homeodomain):** DNA-binding protein domain encoded by a homeobox.

**Hemizygosity:** condition of having a single copy of a gene expressed in an otherwise diploid context.

**Heritability:** proportion of the variance in a phenotypic trait that is due to genetic effects (broad-sense heritability) or genetic-additive effects (narrow-sense heritability) (Section 4.7).

**Hermaphroditism:** condition of having both male and female organs in the same individual (syn. bisexual or co-sexual)

**Heterochiasmy:** condition of having different recombination rates between sexes.

**Heterochromatin:** densely staining condensed chromosomal regions that are for the most part genetically inert.

**Heterodimer:** protein composed of two different polypeptide chains.

**Heterogametic sex:** under diplo-genotypic sex determination, characterizes the sex with distinct sex chromosomes (XY males, ZW females).

**Heterogamety:** condition of producing gametes of different sex-chromosome constitution (e.g. X and Y spermatozoa produced by XY males).

**Heterokaryotic:** refers to a cell that contains multiple, genetically different nuclei (e.g. in the mycelium of some fungi).

**Heteromorphic chromosomes:** morphologically dissimilar homologous chromosomes.

**Heteromorphic self-incompatibility:** a form of self-incompatibility based on morphological differences (Section 3.4.1a).

**Heteroplasmy:** condition of having a mixture of more than one type of an organellar genome (mitochondrial DNA or plastid DNA) within a cell.

**Heterosis:** hybrid vigour, increased fitness of the offspring resulting from a cross between two different inbred lines or congeneric species.

**Heterospory:** condition of having two kinds of spores, microspores and macrospores, giving rise to male and female gametophytes respectively.

**Heterostyly:** condition in which the length of the style differs in flowers of different plants of the same species.

**Heterothallism:** refers to a mating incompatibility between genetically identical individuals (Section 2.1.4).

**H-GSD (= haplo-genotypic sex determination):** genetic sex determination occurring at the haploid stage, implying some genomic differentiation between the interacting gametes (Section 2.1.4).

**Hill–Robertson interferences:** interactions between linked genes, resulting in a reduction in the efficiency of selection; come in different flavours such as Muller's ratchet, background selection, or selective sweeps (Section 5.2).

**Hitchhiking:** change in the frequency of an allele due to linkage with a selected allele at another locus; contributes to Hill–Robertson interferences in non-recombining genomic regions.

**HMG (= high-mobility group) genes:** genes coding for a family of transcription factors, with several members involved in sex-determination pathways, such as *Sry* or *Sox9* (Section 3.1.2a).

**Holliday junction:** a mobile junction between four strands of DNA during homologous recombination.

**Homeobox:** DNA sequence encoding a DNA binding domain (homeodomain), characterizing a family of transcription factors that organize mating in many fungi (Section 3.1.2).

**Homodimer:** protein composed of two identical polypeptide chains.

**Homogametic sex:** under diplo-genotypic sex determination, characterizes the sex with a single type of sex chromosomes (XX females, ZZ males).

**Homologous recombination:** genetic exchange between homologous chromosomes during meiosis.

**Homologue (or homologous):** refers to DNA sequences that are similar because of shared ancestry.

**Homomorphic chromosomes:** morphologically similar homologous chromosomes.

**Homospory:** condition in which male and female gametophytes stem from morphologically similar spores.

**Homothallism:** refers to a mating compatibility between genetically identical individuals (Section 2.1.6).

**Homozygosity:** condition of having two identical alleles at the same locus in a diploid organism.

**Hybrid dysgenesis:** reduced fitness of hybrids due to genetic abnormalities, including sterility, mutation, and chromosomal rearrangements.

**Hybrid zone:** a geographical region in which genetically distinct populations come into contact and produce offspring of mixed ancestry.

**I**

**Idiomorphs:** non-homologous alleles at the same locus.

**Imprinting:** see genomic imprinting.

**Inbreeding:** mating between relatives.

**Inbreeding depression:** decreased vigour in terms of growth, survival, or fecundity as a consequence of mating between relatives.

**Induction types:** secondary mating types found in oomycetes, encoding complementary pheromones and receptors, and allowing gametogenesis and meiosis to be induced by recognition of non-self.

**Interdemic selection:** natural selection that results from sub-populations (demes) that die out or give rise to new sub-populations at different rates.

**Interlocus sexual conflict:** conflict mediated by antagonistic alleles at different loci in males and females.

**Intersex:** individual with sexual characteristics that are intermediate between the two sexes due to failure in sex determination or development.

**Intragenomic conflict:** conflict between genes from the same genome that have different interests over transmission.

**Intralocus sexual conflict:** conflict mediated by antagonistic alleles at the same locus when sexes have different phenotypic optima.

**Intron:** part of a gene that is transcribed but not translated into a polypeptide (as opposed to exons).

**Inversion:** chromosomal rearrangement in which a segment is inverted; inversions prevent or limit recombination between homologous chromosomes.

**Isoform:** any of several different forms of the same protein.

**Isogamy:** condition of having gametes that do not differ morphologically.

**K**

**Karyogamy:** fusion of two haploid nuclei during syngamy (Fig.1.1).

**Karyotype:** set of chromosomes of a cell or organism.

**L**

**Leydig cells:** group of cells in vertebrate testis responsible for steroid production.

**Liability factor:** continuous quantitative factor underlying a threshold trait; different discrete phenotypes are expressed depending on whether the liability factor does or does not exceed a given threshold (Section 4.1).

**LINE (= long interspersed element):** a type of transposable element that encodes a reverse transcriptase and occurs in large numbers interspersed in the genome.

**Linkage:** association of genes on the same chromosome; genes in non-recombining regions are entirely linked.

**Linkage group:** a group of genes linked on a chromosome.

**Linkage disequilibrium:** association of alleles at two or more loci with frequencies that differ from Hardy-Weinberg expectations.

**LMC (= local mate competition):** competition between relatives for access to mates. Highly structured populations where brothers compete to mate with their sisters select for a female-biased sex allocation.

**LRS (= lifetime reproductive success):** total number of offspring produced during a lifetime.

**M**

**Macrogamete:** under anisogamy, larger-sized female gamete.

**Macronucleus:** in ciliates, large polyploid nucleus that divides by fission and organizes phenotypic expression (as opposed to micronucleus).

**Macrospore:** large meiospore developing into a female gametophyte (syn. megaspore).

**Macrosporangium:** conceptacle in which the large female spores are formed.

**MADS box:** conserved DNA sequence motif in the MADs-box gene family, encoding the DNA-binding MADS-domain of transcription factors that organize floral meristems in angiosperms (Sections 3.1.2a; 3.3.1).

**Male heterogamety:** sex determination system in which the male carries distinct sex chromosomes (X and Y).

**MAPK:** Mitogen-activated protein kinases, also known as MAP kinases, direct a variety of cellular processes in eukaryotes.

**Masculinizer:** factor that induces maleness.

**MAT locus:** locus that determines the mating type of haploid isogametes.

**Mating types:** in isogamous lineages, alternative genotypes that determine compatibility in the sexual cycle.

**Mating-type switching:** mechanism allowing the expression of alternative mating types by genetically identical cells (Section 3.2.3a).

**Megaspore, see macrospore.**

**Meiospore:** spore produced by meiosis.

**Meiotic drive:** unequal transmission of alleles during meiosis.

**Mendelian segregation:** ratio of progeny phenotypes expected according to Mendel's laws.

**Meristem:** plant tissue with undifferentiated cells, in regions of active growth and mitotic cell division.

**Metapopulation:** population fragmented into several sub-populations (or demes) connected by gene flow and undergoing extinctions and recolonizations.

**Methylation:** addition of a methyl group to a substrate.

**Microconidia:** in fungi, small spores produced on conidiophores, which can function either as asexual dispersing/reproductive structures or as spermatia (male gametes).

**Microgamete:** in anisogamous lineages, smaller-sized male gamete.

**Micronucleus:** in ciliates, small diploid nucleus that divides mitotically and organizes Mendelian transmission during meiosis (as opposed to macronucleus).

**MicroRNA:** a class of functional RNA that regulates gene expression.

**Microsporangium:** conceptacle containing microspores.

**Microspore:** small meiospore developing into a male gametophyte.

**Mitospore:** spore produced by mitosis.

**Monoecy:** in plants, condition of a sporophyte that produces male and female gametes on different flowers or conceptacles.

**Monofactorial sex determination:** sex determination mechanism in which sex is determined by a single factor, usually one locus with two alleles.

**Monogeny:** condition of producing offspring of only one sexual type: males, females, or hermaphrodites.

**Monoicy:** condition of a gametophyte that produces both male and female gametes; adjective monoicous.

**MSCI (= Meiotic sex chromosome inactivation):** inactivation of the X or Z chromosome during late meiosis in the heterogametic sex.

**Muller's ratchet:** accumulation by drift of deleterious mutations in non-recombining genomic regions, due to the random loss of the least-loaded class of haplotypes (Section 5.2; Box 5.2); contributes to Hill–Robertson interferences.

**Multifactorial sex determination:** sex determination based on several factors, often including several genes with small effects each.

**Multigenic sex determination:** sex determination based on several genes with small effects each.

**Multiple sex chromosomes:** condition in which more than one pair of chromosomes segregates with sex, each potentially carrying sex determination information.

**Multivalent:** association of three or more chromosomes, joined by chiasmata, during meiosis.

## N

**Neo functionalization:** functional divergence of a gene into a new function following a gene duplication event.

**Neo-sex chromosome:** sex chromosome newly formed by fusion of an autosome with an existing sex chromosome.

**Neutral evolution:** non-adaptive evolutionary changes due to non-selective causes (genetic drift, migration, recombination).

**Non-homologous recombination:** genetic exchange between non-homologous chromosomes.

**Non-synonymous mutation:** base pair substitution in DNA that induces an amino acid change in the resulting protein.

**Norm of reaction:** pattern of phenotypic expression of a single genotype across a range of environments.

**Nucleo–cytoplasmic conflict:** conflict arising from differences in the inheritance patterns of genes in the nucleus and in the cytoplasm.

**Nucleosomal array:** Building block of chromosomal structure consisting of core histone octamer-DNA complexes spaced at approximately 200 bp intervals along a DNA molecule.

## O

**Oocyte:** cell undergoing meiosis during oogenesis.

**Oogamy:** condition of having large non-motile female gametes fertilized by small motile male gametes.

**Oogonium:** female sexual organ containing one or several eggs (e.g. in algae and fungi).

**Operational sex ratio:** proportion of each sex that is effectively available for mating.

**Orthologues:** homologous genes in different species that originated by descent from the same ancestral sequence.

**Outcrossing:** mating between unrelated individuals.

**Ovipary:** laying of eggs in which the embryos have developed little or not at all.

**Ovovivipary:** laying of eggs in which the embryos have already developed within the female parent and are ready to hatch.

**Ovule:** in seed plants, structure containing an egg cell and developing as a seed after fertilization.

## P

**Palindromic sequence:** DNA sequence that is identical whether read 5' to 3' on one strand or 5' to 3' on the complementary strand. Palindromic sequences on the Y chromosome allow self-repair in absence of a homologous copy.

**PAR (= pseudo-autosomal region):** segment of sex chromosomes that still recombines with its homologue in the heterogametic sex.

**Paralogue:** copy of a gene that originated from a duplication event.

**Paraphyletic:** refers to a clade including some but not all descendants of a common ancestor.

**Parental-effect sex determiner:** gene expressed in the parent but acting in the zygote to affect its sex.

**Parthenogenesis:** reproduction through development of an egg without fertilization.

**Paternal genome loss:** mode of haplodiploid reproduction in which haploid males derive from fertilized eggs from which the paternal genome is expelled during early zygotic divisions.

**Perianth:** in flowering plants, structure consisting of the calyx and corolla of the flower.

**Pericentric inversion:** chromosomal inversion that includes the centromere.

**Pheromone:** signal molecule, released by some individuals to affect phenotypic expression in other individuals (Section 3.1.2c).

**Plasmodium:** 1) genus of apicomplexan parasites, agent of malaria, 2) amoeboid multinucleate unicell (coenocyte) resulting from multiple nuclear division without cell division; feeding stage of plasmodial slime moulds.

**Plasmogamy:** fusion of cytoplasms during syngamy. In ascomycetes and basidiomycetes, plasmogamy is not followed immediately by karyogamy, resulting in a dikaryotic (or di-haploid) stage.

**Pleiotropy:** refers to the multiple effects of one gene on a number of different phenotypic traits.

**Ploidy:** number of chromosome complements.

**Polyfactorial sex determination:** sex determination resulting from the combined effects of a number of genetic or environmental factors.

**Polygenic sex determination:** sex determination resulting from the combined effects of a number of genes at more than one locus.

**Polyploidy:** condition of having more than two sets (n) of homologous chromosomes, e.g. triploidy (3n) or tetraploidy (4n).

**Post-mating isolation:** barrier to hybridization that acts after mating has occurred, e.g. due to hybrid unviability or infertility.

**Post-zygotic isolation:** barrier to hybridization that acts after egg-sperm fusion has occurred, e.g. due to hybrid unviability or infertility.

**Potential reproductive success:** maximal fitness gain that can be expected through reproduction; usually larger in males than in females.

**Pre-mating (or pre-zygotic) isolation:** barrier to hybridization that acts before mating has occurred, e.g. due to mate choice by females.

**Primary mating types:** mating categories under haplogenotypic sex determination with isogamy (Section 1.2).

**Primary sex ratio:** ratio of males and females at fertilization.

**Primordial germ cells:** precursors of germ cells that proliferate on their way to the developing gonads, where they differentiate into gametes.

**Promoter:** regulatory region of DNA usually located upstream of a gene.

**Protandry:** in sequential hermaphrodites, condition of having male organs developing before female organs.

**Protogyny:** in sequential hermaphrodites, condition of having female organs developing before male organs.

**Proto-sex chromosome:** incipient sex chromosome, close to its ancestral autosomal condition.

**Pseudoarrhenotoky:** mode of haplodiploid reproduction in which males develop from fertilized eggs but the paternal genome is heterochromatized or lost in the somatic cells and not passed to their offspring (see paternal genome loss).

**Pseudogamy (pseudogamous parthenogenesis):** a form of reproduction where fertilization of an egg is required but the paternal chromosomes are subsequently expelled and do not contribute to the offspring.

**Pseudohomothallism:** refers to a mating compatibility between gametophytes stemming from the same spore, or gametes from the same gametophyte, despite the presence of mating types (Fig. 2.3).

**Purifying selection:** elimination by natural selection of deleterious alleles from a population

## Q

**Quantitative trait locus (QTL):** DNA sequence containing or linked to genes that affect phenotypic variation at a quantitative trait.

## R

**Recombination:** genetic exchange between chromosomes during meiosis or mitosis. Recombination can be homologous or non-homologous, and can result in crossovers or not (Section 1.1.1).

**Recombination load:** drop in fitness due to the break up, by recombination, of beneficial associations between interacting alleles.

**Red Queen hypothesis:** evolutionary principle holding that organisms must constantly evolve to adapt to ever-changing environment, as a result of arms race between conflicting parties such as hosts and parasites. The Red Queen has been invoked to account for the evolution of sex (Section 1.1.2b).

**Reinforcement:** evolution of enhanced reproductive isolation between populations due to natural selection against the formation of hybrids.

**Repetitive DNA:** redundant DNA sequences that do not code for proteins and are present in many copies in the genome.

**RNase:** enzyme that hydrolyses RNA by breaking its phosphodiester bonds.

**Robertsonian translocation:** chromosomal rearrangement involving fusion of two chromosome arms.

S

**Same-clone mating:** mating between genetically identical individuals stemming from the same clone.

**Schizont:** asexual life cycle stage in some unicellular eukaryotes

**Secondary mating types:** mating categories under diploid sex determination, genetically determined by self-incompatibility systems or induction types that evolved as a way to prevent or limit inbreeding (Sections 1.2; 2.1.7; 3.4).

**Secondary sex ratio:** sex ratio at birth.

**Segregation distorter:** genetic element that segregates at a higher than regular Mendelian ratio at the cost of its homologue.

**Selection coefficient:** measure of the relative difference in fitness between phenotypes in a population.

**Selective sweep:** reduction or elimination of DNA sequence variation in the vicinity of a mutation that has been recently fixed by natural selection; contributes to Hill–Robertson interferences in non-recombining genomic regions (Section 5.2).

**Self-incompatibility (SI) systems:** secondary mating types (e.g. in angiosperms and ascidians) evolved to prevent selfing (Sections 2.17; 3.4).

**Selfing:** self-fertilization.

**Selfish genetic element:** DNA sequence enhancing its own transmission to the detriment of the rest of the genome.

**Sequential hermaphroditism:** condition of having both male and female organs in the same individual at different stages during the lifetime (Section 2.1.6b).

**Sertoli cells:** nurse cells of the testes, secrete the anti-Müllerian hormone.

**Sex allocation:** the allocation of resources towards the male or female function.

**Sex determination:** developmental process that directs the fate of an individual or cellular lineage towards a male or female condition (Section 1.4).

**Sex linkage:** condition for a gene of being located on the sex chromosome.

**Sex ratio:** relative number of males and females in a progeny or population, typically expressed as the proportion of males.

**Sex-ratio gene:** gene affecting the relative proportion of sons and daughters in the progeny.

**Sex-ratio selection:** natural selection on the relative number of males to females in a population.

**Sex reversal:** in species with predominantly genotypic sex determination, mismatch between the genotypic and phenotypic sex.

**Sexes:** 1) at the gametic level, refers to male *versus* female gametes. 2) at the multicellular level (gametophytes or sporophytes), refers to the relative allocation to the male *versus* female function; the term gender is also used with this latter meaning (Section 1.2).

**Sexual antagonism (= sexual conflict):** difference between sexes in optimal fitness strategies, usually over reproduction.

**Sexual differentiation:** the process of development of the differences between males and females from an undifferentiated zygote.

**Sexual dimorphism:** phenotypic differences between males and females.

**Sexual selection:** differential reproduction as a result of variation in the ability to obtain mates.

**SI (= self-incompatibility):** genetic mechanism that prevents self-fertilization (Section 2.1.7).

**Simultaneous hermaphroditism:** condition of having both male and female organs in the same individual at any one time (Section 2.1.6a).

**SNP (= single nucleotide polymorphism):** a nucleotide-pair difference at a given location in the genomes of two individuals.

**Social sex determination:** condition in which the sex is determined by social interactions (such as the proportion of males and females in the population).

**Soma:** as opposed to germ line, any cells of a multicellular organism other than those giving rise to gametes.

**Sonkiller:** genetic element, such as an endosymbiont, that kills male offspring of infected females.

**Spermatogonium:** undifferentiated male germ cell within testes giving rise to spermatocytes.

**Sporangium:** organ of a sporophyte where spores are formed.

**Spore:** A dormant, reproductive cell, resistant to survive under unfavorable conditions and giving rise to a new individual when conditions improve.

**Sporophyte:** spore-producing diploid multicellular stage.

**Sporulation:** formation of spores.

**SSI (= sporophytic self-incompatibility):** form of self-incompatibility between pollen and pistil determined by the diploid genotype of the anther and the genotype of the stigma (Section 3.4.1).

**Stabilizing selection:** selection against phenotypes that deviate in either direction from an optimal value of a character.

**Stamen:** pollen-producing reproductive organ of flowers.

**Sub-diplontic:** close to diplontic, with limited and non-independent haploid phase (as in angiosperms).

**Subfunctionalization:** functional divergence of a gene into a related function following a gene duplication event.

**Supergene:** group of loci that are tightly linked on the same chromosome and together affect a particular phenotype.

**Synaptonemal complex:** complex protein structure that unites homologues during the prophase of meiosis.

**Synclonal:** in ciliates, characterizes caryonids resulting from a conjugant pair that inherit the same mating type; synclonal mating-type determination occurs when macronucleus differentiation is entirely controlled by the newly formed micronucleus (Section 3.4.2a).

**Syncytium:** multinuclear cell resulting from multiple fusion of unicellular cells; adjective syncytial.

**Syngamy:** fusion of two gametes to produce a diploid zygote; initiated by plasmogamy (fusion of cytoplasms), followed by karyogamy (fusion of nuclei) (Fig. 1.1).

**Synonymous substitution (= silent substitution):** fixation of a base pair change that does not alter the amino acid sequence in the protein product of a gene

**Synteny:** physical co-localization of the same genes on the same chromosomes in different species.

**Syzygy:** in apicomplexa, union of gamonts.

## T

**Teliospore:** thick-walled resting spore formed by some basidiomycetes (rusts and smuts).

**Tetrapolar:** characterized by two mating-type loci, so that meiosis produces spores of four mating types.

**Tetrasporangium:** in red algae, reproductive organ of the tetrasporophyte, producing tetrapores.

**Tetraspore:** in red algae, tetrad of meiospores produced by the diploid tetrasporophyte.

**Tetrasporophyte:** in red algae, diploid thallus meiotically producing tetraspores.

**Thallus:** vegetative plant-like multicellular organism lacking differentiation into root, stem and leaf.

**Thelytoky:** mode of parthenogenetic reproduction in which females reproduce without fertilization, and males are rare or absent.

**Theria:** subclass of mammals that give birth to life young without using a shelled egg; comprises marsupials and placental mammals (eutheria).

**Threshold trait:** a trait whose expression depends on a critical intensity of an underlying liability factor (Section 4.1).

**Topoisomerase:** enzyme affecting the topology and winding of DNA.

**Transcription factor:** regulatory protein that recruits RNA polymerase to DNA promoter regions during transcription.

**Trans-regulatory element:** diffusible factors affecting mRNA synthesis of downstream genes on distant locations, e.g. on different chromosomes.

**Transformation:** the directed modification of a genome by the external application of DNA from a cell of different genotype.

**Transposon (= transposable element):** a DNA sequence that can replicate and insert at various sites in a genome

**Trichogyne:** in ascomycetes, part of the ascogonium that serves as fertilization tube for fusion of male and female hyphal strains.

**Trioecy:** condition of having male, female, and hermaphrodite flowers on different plants (adjective trioecious).

**Tristyly:** a form of heterostyly with three morphological types of flowers differing in style lengths (figure 3.11).

**TSD (= temperature-dependent sex determination):** a form of environmental sex determination in which the temperatures experienced during embryonic development determine the sex of the offspring.

## U

**U chromosome:** female-determining sex chromosome under haploid genotypic sex determination.

**Uniparental:** relative to a single parent; transmitted by a single parent.

**Unisexual:** 1) condition of having separate sexes; the terms dioecious and gonochoric are used for plants and animals respectively, with the same meaning. 2) parthenogenetic.

## V

**V chromosome:** male-determining sex chromosome under haploid genotypic sex determination.

**Vegetative growth:** asexual form of reproduction that does not involve recombination of genetic material (e.g. by detachment of a part of the body that develops into a new individual).

**Vivipary:** reproduction by which the embryos develop within the maternal body, leading to live birth.

## W

**W chromosome:** in female heterogametic systems (as in birds), sex chromosome that occurs singly in females only.

**Wild-type:** characterizes the allele, genotype, or phenotype that is most prevalent in wild populations.

## X

**X chromosome:** in male heterogametic systems, sex chromosome that occurs in double copy in females, and single one in males.

**XSE (= X-signal element):** group of genes on the X chromosome that communicate the X-chromosome dose.

## Y

**Y chromosome:** in male heterogametic systems (as in mammals), sex chromosome that occurs singly in males only.

## Z

**Z chromosome:** in female heterogametic systems, sex chromosome that occurs in double copy in males, and single one in females.

**Zinc-finger protein:** protein with a small structural DNA- or RNA-binding motif characterized by the co-ordination of one or more zinc ions that stabilize the fold. Zinc-finger proteins are active in the transcriptional and post-transcriptional regulation of gene expression (Section 3.1.2).

**Zygote:** unicellular diploid cell resulting from the fusion of two gametes.

**Zygotic sex-determination gene:** gene expressed in the zygote to affect its sex.

# References


Abucay JS, Mair GC, Skibinski DO, Beardmore JA (1999) Environmental sex determination: the effect of temperature and salinity on sex ratio in *Oreochromis niloticus* L. *Aquaculture* **173**: 219–234.

Acosta IF, Laparra H, Romero SP, Schmelz E et al. (2009) Tasselseed1 is a lipoxygenase affecting jasmonic acid signaling in sex determination of maize. *Science* **323**: 262–265.

Ahringer J, Rosenquist TA, Lawson DN, Kimble J (1992) The *Caenorhabditis elegans* sex determining gene fe m-3 is regulated post-transcriptionally. *EMBO J* **11**: 2303–2310.

Aida T (1921) On the inheritance of color in a fresh-water fish, *Aplocheilus latipes* Temmick and Schlegel, with special reference to sex-linked inheritance. *Genetics* **6**: 554–573.

Airoldi CA (2010) Determination of sexual organ development. *Sex Plant Reprod* **23**: 53–62.

Aitken RJ, Marshall Graves JA (2002) The future of sex. *Nature* **415**: 963.

Akada R (1985) Mating types and mating-inducing factors (gamones) in the ciliate *Euplotes patella* syngen. *Gen Res* 2. **46**: 125–132.

Akerib CC, Meyer BJ (1994) Identification of X chromosome regions in *Caenorhabditis elegans* that contain sex-determination signal elements. *Genetics* **138**: 1105–1125.

Albert AYK, Otto SP (2005) Sexual selection can resolve sex-linked sexual antagonism. *Science* **310**: 119–121.

Alekseyenko AA, Peng S, Larschan E, Gorchakov AA et al. (2008) A sequence motif within chromatin entry sites directs MSL establishment on the *Drosophila* X chromosome. *Cell* **134**: 599–609.

Allison JD, Roff DA, Cardé RT (2008) Genetic independence of female signal form and male receiver design in the almond moth, *Cadra cautella*. *J Evol Biol* **21**: 1666–1672.

Ammermann D (1982) Mating types in *Stylonychia mytilus* EHRBG. *Archiv für Protistenkunde* **126**: 373–381.

Amrein H, Gorman M, Nöthiger R (1988) The sex-determining gene tra-2 of *Drosophila* encodes a putative RNA binding protein. *Cell* **55**: 1025–1035.

Andersen SL, Sekelsky J (2010) Meiotic versus mitotic recombination: two different routes for double-strand break repair: the different functions of meiotic versus mitotic DSB repair are reflected in different pathway usage and different outcomes. *Bioessays* **32**: 1058–1066.

Aparici E, Carmona MJ, Serra M (1998) Sex allocation in haplodiploid cyclical parthenogens with density-dependent proportion of males. *Am Nat* **152**: 652–657.

Armbrust EV, Ferris PJ, Goodenough UW (1993) A mating type-linked gene cluster expressed in *Chlamydomonas* zygotes participates in the uniparental inheritance of the chloroplast genome. *Cell* **74**: 801–811.

Arnqvist G, Rowe L (2005) *Sexual Conflict*. Princeton University Press, Princeton.

Ar-Rushdi AH (1958) The polygenic basis of sex-ratio in *Tigriopus*. *Proceedings of the Xth International Congress on Genetics*, vol. 2, pp. 8–9.

Arunkumar KP, Mita K, Nagaraju J (2009) The silkworm Z chromosome is enriched in testis-specific genes. *Genetics* **182**: 493–501.

Atlan A, Capillon C, Derome N, Couvet D et al. (2003) The evolution of autosomal suppressors of sex-ratio drive in *Drosophila simulans*. *Genetica* **117**: 47–58.

Atlan A, Merçot H, Landre C, Montchamp-Moreau C (1997) The sex-ratio trait in *Drosophila simulans*: geographical distribution of distortion and resistance. *Evolution* **51**: 1886–1895.

Avner P, Heard E (2001) X-chromosome inactivation: counting, choice and initiation. *Nat Rev Genet* **2**: 59–67.

Baccetti B (1985) Evolution of the sperm cell. In Metz C, Monroy A eds. *Biology of fertilization*, Volume 2. Academic Press, New York. pp. 3–58.

Bacci G (1965) Sex determination and genic balance of *Ophryotrocha puerilis*, a hermaphrodite polychaete worm. *Nature* **207**: 1208–1209.

Bacci G, Bortesi O (1961) Pure males and females from hermaphroditic strains of *Ophryotrocha puerilis*. *Experientia* **17**: 229–230.

Bacci G (1978) Genetics of sex determination in Ophryotrocha (Annelida, Polychaeta). In: Battaglia B, Beardmore J. eds. *Marine organisms: Genetics, ecology, and evolution*. Plenum Press, New York. pp. 549–571.

Bachtrog D (2003) Adaptation shapes patterns of genome evolution on sexual and asexual chromosomes in *Drosophila*. *Nat Genet* **34**: 215–219.

Bachtrog D (2004) Evidence that positive selection drives Y-chromosome degeneration in *Drosophila miranda*. *Nat Genet* **36**: 518–522.

Bachtrog D (2006) A dynamic view of sex chromosome evolution. *Curr Opin Genet Dev* **16**: 578–585.

Bachtrog D (2008) The temporal dynamics of processes underlying Y chromosome degeneration. *Genetics* **179**: 1513–1525.

Bachtrog D (2013) Y-chromosome evolution: emerging insights into processes of Y-chromosome degeneration. *Nat Rev Genet* **14**: 113–124.

Bachtrog D, Charlesworth B (2000) Reduced levels of microsatellite variability on the neo-Y chromosome of *Drosophila miranda*. *Curr Biol* **10**: 1025–1031.

Bachtrog D, Charlesworth B (2002) Reduced adaptation of a non-recombining neo-Y chromosome. *Nature* **416**: 323–326.

Bachtrog D, Hom E, Wong KM, Maside X et al. (2008) Genomic degradation of a young Y chromosome in *Drosophila miranda*. *Genome Biol* **9**: R30.

Bachtrog D, Jensen JD, Zhang Z (2009) Accelerated adaptive evolution on a newly formed X chromosome. *PLoS Biol* **7**: e82.

Bachtrog D, Kirkpatrick M, Mank JE, McDaniel SF et al. (2011) Are all sex chromosomes created equal? *Trends Genet* **27**: 350–357.

Bagheri-Fam S, Sinclair AH, Koopman P, Harley VR (2010) Conserved regulatory modules in the Sox9 testis-specific enhancer predict roles for SOX, TCF/LEF, Forkhead, DMRT, and GATA proteins in vertebrate sex determination. *Int J Biochem Cell Biol* **42**: 472–477.

Bailey RI, Innocenti P, Morrow EH, Friberg U et al. (2011) Female *Drosophila melanogaster* gene expression and mate choice: the X chromosome harbours candidate genes underlying sexual isolation. *PLoS ONE* **6**: e17358.

Baker BS, Ridge KA (1980) Sex and the single cell. I. On the action of major loci affecting sex determination in *Drosophila melanogaster*. *Genetics* **94**: 383–423.

Bakker TCM, Pomiankowski A (1995) The genetic basis of female mate preferences. *J Evol Biol* **8**: 129–171.

Bakkeren G, Kronstad JW (1994) Linkage of mating-type loci distinguishes bipolar from tetrapolar mating in basidiomycetous smut fungi. *Proc Natl Acad Sci U S A* **91**: 7085–7089.

Banham AH, Asante-Owusu RN, Gottgens B, Thompson S et al. (1995) An N-terminal dimerization domain permits homeodomain proteins to choose compatible partners and initiate sexual development in the mushroom *Coprinus cinereus*. *Plant Cell* **7**: 773–783.

Banks JA (2008) MicroRNA, sex determination and floral meristem determinacy in maize. *Genome Biol* **9**: 204.

Banks JA, Hickok L, Webb MA (1993) The programming of sexual phenotype in the homosporous fern *Ceratopteris richardii*. *Int J Plant Sci* **154**: 522–534.

Banuett F (1995) Genetics of *Ustilago maydis*, a fungal pathogen that induces tumors in maize. *Annu Rev Genet* **29**: 179–208.

Barksdale AW (1967) The sexual hormones of the fungus *Achlya*. *Ann N Y Acad Sci* **144**: 313–319.

Baroiller JF, Chourrout D, Fostier A, Jalabert B (1995) Temperature and sex chromosomes govern sex ratios of the mouthbrooding cichlid fish *Oreochromis niloticus*. *J Exp Zool* **273**: 216–223.

Baroiller JF, D'Cotta H (2001) Environment and sex determination in farmed fish. *Comp Biochem Physiol C Toxicol Pharmacol* **130**: 399–409.

Baroiller JF, D'Cotta H, Saillant E (2009) Environmental effects on fish sex determination and differentiation. *Sex Dev* **3**: 118–135.

Barrett SCH (1992) Heterostylous genetic polymorphisms: model systems for evolutionary analysis. In: Barrett SCH ed. *Evolution and function of heterostyly*. Springer Verlag, Berlin, pp. 1–29.

Barrett L, Fletcher S, Wilton S (2012) Regulation of eukaryotic gene expression by the untranslated gene regions and other non-coding elements. *Cell Mol Life Sci* **69**: 3613–3634.

Barske LA, Capel B (2008) Blurring the edges in vertebrate sex determination. *Curr Opin Genet Dev* **18**: 499–505.

Barsoum E, Martinez P, Astrom SU (2010) Alpha3, a transposable element that promotes host sexual reproduction. *Genes Dev* **24**: 33–44.

Barton N (1995) A general model for the evolution of recombination. *Genet Res* **65**: 123–144.

Bastide H, Cazemajor M, Ogereau D, Derome N et al. (2011) Rapid rise and fall of selfish sex-ratio X chromosomes in *Drosophila simulans*: spatiotemporal analysis of phenotypic and molecular data. *Mol Biol Evol* **28**: 2461–2470.

Bateman AJ (1948) Intra-sexual selection in *Drosophila*. *Heredity* **2**: 349–368.

Bateson W (1909) Heredity and variation in modern lights. *Darwin and modern science*: 85–101.

Battaglia B (1958) Balanced polymorphism in *Tisbe reticulata*, a marine copepod. *Evolution* **12**: 358–364.

Battaglia B, Malesani L (1959) Ricerche sulla determinazione del sesso nel Copepode *Tisbe gracilis* (T. Scott). *Bol Zool* **26**: 423–433.

Baudat F, Buard J, Grey C, Fledel-Alon A et al. (2010) PRDM9 is a major determinant of meiotic recombination hotspots in humans and mice. *Science* **327**: 836–840.

Beck JB, Windham MD, Pryer KM (2011) Do asexual polyploid lineages lead short evolutionary lives? A case study from the fern genus *Astrolepis*. *Evolution* **65**: 3217–3229.

Beermann W (1955) Cytologische Analyse eines Camptochironomus-Artbastards. 1. Kreuzungsergebnisse und die Evolution des Karyotypus. *Chromosoma* **7**: 198–259.

Begun DJ, Whitley P (2000) Reduced X-linked nucleotide polymorphism in *Drosophila simulans*. *Proc Natl Acad Sci U S A* **97**: 5960–5965.

Beladjal L, Vandekerckhove TT, Muyssen B, Heyrman J et al. (2002) B-chromosomes and male-biased sex ratio with paternal inheritance in the fairy shrimp *Branchipus schaefferi* (Crustacea, Anostraca). *Heredity* **88**: 356–360.

Bell G (1978) The evolution of anisogamy. *J Theor Biol* **73**: 247–270.

Bell G (1982) *The masterpiece of nature: the evolution and genetics of sexuality*. CUP Archive.

Bell G (1985) The origin and early evolution of germ cells as illustrated by the Volvocales. In: Halvorson, H. and A. Monroy, eds. The origin and evolution of sex, 1984 Wood's *Hole Symposium*. *MBL Lectures in Biology*, volume 7, Liss, New York. pp. 221–256.

Bell G (1997) The evolution of the life cycle of brown seaweeds. *Biol J Linn Soc* **60**: 21–38.

Bellott DW, Skaletsky H, Pyntikova T, Mardis ER et al. (2010) Convergent evolution of chicken Z and human X chromosomes by expansion and gene acquisition. *Nature* **466**: 612–616.

Berec L, Schembri PJ, Boukal DS (2005) Sex determination in *Bonellia viridis* (Echiura: Bonelliidae): population dynamics and evolution. *Oikos* **108**: 473–484.

Bergero R, Charlesworth D (2011) Preservation of the Y transcriptome in a 10-million-year-old plant sex chromosome system. *Curr Biol* **21**: 1470–1474.

Bergero R, Charlesworth D, Filatov DA, Moore RC (2008) Defining regions and rearrangements of the *Silene latifolia* Y chromosome. *Genetics* **178**: 2045–2053.

Berlin S, Ellegren H (2004) Chicken W: a genetically uniform chromosome in a highly variable genome. *Proc Natl Acad Sci U S A* **101**: 15967–15969.

Bernstein H, Bernstein C (2010) Evolutionary origin of recombination during meiosis. *Bioscience* **60**: 498–505.

Bernstein H, Bernstein C, Michod RE (2011) Meiosis as an evolutionary adaptation for DNA repair. In Kruman I. ed; *DNA Repair*. InTech Publ. p. 357–382.

Berta P, Hawkins JR, Sinclair AH, Taylor A et al. (1990) Genetic evidence equating SRY and the testis-determining factor. *Nature* **348**: 448–450.

Bessho K, Iwasa Y (2010) Optimal seasonal schedules and the relative dominance of heteromorphic and isomorphic life cycles in macroalgae. *J Theor Biol* **267**: 201–212.

Betran E, Thornton K, Long M (2002) Retroposed new genes out of the X in *Drosophila*. *Genome Res* **12**: 1854–1859.

Beukeboom LW (2012) Microbial manipulation of host sex determination. Endosymbiotic bacteria can directly manipulate their host's sex determination towards the production of female offspring. *Bioessays* **34**: 484–488.

Beukeboom LW, de Jong TJ, Pen I (2001) Why girls want to be boys. *Bioessays* **23**: 477–480.

Beukeboom LW, Pijnacker LP (2000) Automictic parthenogenesis in the parasitoid *Venturia canescens* (Hymenoptera: Ichneumonidae) revisited. *Genome* **43**: 939–944.

Beukeboom LW, van de Zande L (2010) Genetics of sex determination in the haplodiploid wasp *Nasonia vitripennis* (Hymenoptera: Chalcidoidea). *J Genet* **89**: 333–339.

Beukeboom LW, Vrijenhoek RC (1998) Evolutionary genetics and ecology of sperm-dependent parthenogenesis. *J Evol Biol* **11**: 755–782.

Beye M, Hasselmann M, Fondrk MK, Page RE et al. (2003) The gene csd is the primary signal for sexual development in the honeybee and encodes an SR-type protein. *Cell* **114**: 419–429.

Beysard M, Perrin N, Jaarola M, Heckel G et al. (2012) Asymmetric and differential gene introgression at a contact zone between two highly divergent lineages of field voles (*Microtus agrestis*). *J Evol Biol* **25**: 400–408.

Bezault E (2005) *Etude du système de déterminisme du sexe au sein de populations naturelles de tilapia du Nil,* Oreochromis niloticus: *Importance des composantes génétiques et environnementales*. PhD thesis. Orsay, France: Université Paris XI.

Bezault E, Clota F, Derivaz M, Chevassus B et al. (2007) Sex determination and temperature-induced sex differentiation in three natural populations of Nile tilapia (*Oreochromis niloticus*) adapted to extreme temperature conditions. *Aquaculture* **272**: 3–16.

Billiard S, Lopez-Villavicencio M, Devier B, Hood ME et al. (2011) Having sex, yes, but with whom? Inferences from fungi on the evolution of anisogamy and mating types. *Biol Rev Camb Philos Soc* **86**: 421–442.

Blackmore MS, Charnov EL (1989) Adaptive variation in environmental sex determination in a nematode. *Am Nat* **134**: 817–823.

Blakeslee AF (1904) Zygospore formation a sexual process. *Science* **19**: 864–866.

Blaser O, Grossen C, Neuenschwander S, Perrin N (2013) Sex-chromosome turnovers induced by deleterious mutation load. *Evolution* **67**: 635–645.

Blaser O, Neuenschwander S, Perrin N (2011) On the maintenance of sex chromosome polymorphism by sex-antagonistic selection. *Am Nat* **178**: 515–524.

Blaser O, Neuenschwander S, Perrin N (2014) Sex-chromosome turnovers: the hot-potato model. *Am Nat* **183**: 140–146.

Bloomfield G (2011) Genetics of sex determination in the social amoebae. *Dev Growth Differ* **53**: 608–616.

Bloomfield G, Skelton J, Ivens A, Tanaka Y et al. (2010) Sex determination in the social amoeba *Dictyostelium discoideum*. *Science* **330**: 1533–1536.

Bonduriansky R, Chenoweth SF (2009) Intralocus sexual conflict. *Trends Ecol Evol* **24**: 280–288.

Bone JR, Kuroda MI (1996) Dosage compensation regulatory proteins and the evolution of sex chromosomes in *Drosophila*. *Genetics* **144**: 705–713.

Boualem A, Fergany M, Fernandez R, Troadec C et al. (2008) A conserved mutation in an ethylene biosynthesis enzyme leads to andromonoecy in melons. *Science* **321**: 836–838.

Bouchon D, Cordaux R, Greve P (2008) Feminizing *Wolbachia* and the evolution of sex determination in isopods. In: Bourtzis K, Miller T. eds. *Insect symbiosis (Vol. 3)*. Taylor and Francis, Boca Raton, FL. pp. 273–294.

Bouchon D, Rigaud T, Juchault P (1998) Evidence for widespread *Wolbachia* infection in isopod crustaceans: molecular identification and host feminization. *Proc Biol Sci* **265**: 1081–1090.

Boursot P, Auffray J-, Britton-Davidian J, Bonhomme F (1993) The evolution of house mice. *Annu Rev Ecol Syst* **24**: 119–152.

Bowden RM, Ewert MA, Nelson CE (2000) Environmental sex determination in a reptile varies seasonally and with yolk hormones. *Proc Biol Sci* **267**: 1745–1749.

Bowles J, Schepers G, Koopman P (2000) Phylogeny of the SOX family of developmental transcription factors based on sequence and structural indicators. *Dev Biol* **227**: 239–255.

Bradley KM, Breyer JP, Melville DB, Broman KW et al. (2011) An SNP-based linkage map for zebrafish reveals sex determination loci. *G3 (Bethesda)* **1**: 3–9.

Breeuwer JA, Werren JH (1990) Microorganisms associated with chromosome destruction and reproductive isolation between two insect species. *Nature* **346**: 558–560.

Brelsford A, Stöck M, Betto-Colliard C, Dubey S et al. (2013) Homologous sex chromosomes in three deeply divergent anuran species. *Evolution* **67**: 2434–2440.

Breton S, Beaupre HD, Stewart DT, Hoeh WR et al. (2007) The unusual system of doubly uniparental inheritance of mtDNA: isn't one enough? *Trends Genet* **23**: 465–474.

Bridges CB (1916) Non-disjunction as proof of the chromosome theory of heredity (concluded). *Genetics* **1**: 107.

Bridges CB (1925) Sex in relation to chromosomes and genes. *Am Nat* **59**: 127–137.

Bridges CB (1939) Cytological and genetical basis of sex. In: Allen A. ed. *Sex and internal secretion*. Balliere, Tindall and Cox, London. pp. 15–63.

Brooks LD (1988) The evolution of recombination rates. In: Michod RE, Levin BR. eds. *The evolution of sex: An examination of current ideas*. Sinauer, Sunderland, MA. pp. 87–105.

Brooks LD, Marks RW (1986) The organization of genetic variation for recombination in *Drosophila melanogaster*. *Genetics* **114**: 525–547.

Brothers AN, Delph LF (2010) Haldane's rule is extended to plants with sex chromosomes. *Evolution* **64**: 3643–3648.

Brown SW, Chandra HS (1977) Chromosome imprinting and the differential regulation of homologous chromosomes. In: Goldstein L, Prescott DM. eds. *Cell biology: A comparative treatise, volume 1; Genetic mechanisms of cells*. Academic Press, New York. pp. 108–189.

Bull JJ (1978) Sex chromosomes in haploid dioecy: a unique contrast to Muller's theory for diploid dioecy. *Am Nat* **112**: 245–250.

Bull JJ (1979) An advantage for the evolution of male haploidy and systems with similar genetic transmission. *Heredity* **43**: 361–381.

Bull JJ (1981) Sex ratio evolution when fitness varies. *Heredity* **46**: 9–26.

Bull JJ (1983) *Evolution of sex determining mechanisms*. The Benjamin/Cummings Publishing Company, Inc., Menlo Park, CA.

Bull JJ, Bulmer MG (1981) The evolution of XY females in mammals. *Heredity* **47**: 347–365.

Bull JJ, Bulmer MG (1989) Longevity enhances selection of environmental sex determination. *Heredity* **63** (Pt 3): 315–320.

Bull JJ, Charnov EL (1977) Changes in the heterogametic mechanism of sex determination. *Heredity* **39**: 1–14.

Bull JJ, Charnov E (1988) How fundamental are Fisherian sex ratios? *Oxford Surv Evol Biol* **5**: 96–135.

Bull JJ, Vogt R, Bulmer M (1982) Heritability of sex ratio in turtles with environmental sex determination. *Evolution* **36**: 333–341.

Bulmer MG (1988) Sex ratio evolution in lemmings. *Heredity* **61**: 231–233.

Bulmer MG (1994) *Theoretical evolutionary ecology*. Sinauer Associates, Sunderland, MA.

Bulmer MG, Bull JJ (1982) Models of polygenic sex determination and sex ratio control. *Evolution* **36**: 13–26.

Bulmer MG, Luttikhuizen P, Parker G (2002) Survival and anisogamy. *Trends Ecol Evol* **17**: 357–358.

Bulmer MG, Parker GA (2002) The evolution of anisogamy: a game-theoretic approach. *Proc Biol Sci* **269**: 2381–2388.

Burd M (2011) Are relationships between pollen-ovule ratio and pollen and seed size explained by sex allocation? *Evolution* **65**: 3002–3005.

Burley N (1981) Sex ratio manipulation and selection for attractiveness. *Science* **211**: 721–722.

Burt A (2000) Perspective: sex, recombination, and the efficacy of selection—was Weismann right? *Evolution* **54**: 337–351.

Burt A, Bell G (1987) Mammalian chiasma frequencies as a test of two theories of recombination. *Nature* **326**: 803–805.

Burt A, Bell G, Harvey PH (1991) Sex differences in recombination. *J Evol Biol* **4**: 259–277.

Burt A, Carter DA, Koenig GL, White TJ et al. (1996) Molecular markers reveal cryptic sex in the human pathogen *Coccidioides immitis*. *Proc Natl Acad Sci U S A* **93**: 770–773.

Burt A, Trivers R (2009) *Genes in conflict: the biology of selfish genetic elements*. Harvard University Press, Cambridge, MA.

Burtis KC, Baker BS (1989) *Drosophila* doublesex gene controls somatic sexual differentiation by producing alternatively spliced mRNAs encoding related sex-specific polypeptides. *Cell* **56**: 997–1010.

Busch JW, Schoen DJ (2008) The evolution of self-incompatibility when mates are limiting. *Trends Plant Sci* **13**: 128–136.

Butler B, Pirrotta V, Irminger-Finger I, Nothiger R (1986) The sex-determining gene *tra* of *Drosophila*: molecular cloning and transformation studies. *EMBO J* **5**: 3607–3613.

Caballero A (1995) On the effective size of populations with separate sexes, with particular reference to sex-linked genes. *Genetics* **139**: 1007–1011.

Çakir S, Kence A (1996) The distribution of males having XY and XX chromosomes in housefly populations (Diptera: Muscidae) of Turkey. *Genetica* **98**: 205–210.

Camacho JP, Schmid M, Cabrero J (2011) B chromosomes and sex in animals. *Sex Dev* **5**: 155–166.

Campbell P, Good JM, Dean MD, Tucker PK et al. (2012) The contribution of the Y chromosome to hybrid male sterility in house mice. *Genetics* **191**: 1271–1281.

Campbell P, Good JM, Nachman MW (2013) Meiotic sex chromosome inactivation is disrupted in sterile hybrid male house mice. *Genetics* **193**: 819–828.

Campos-Ramos R, Harvey S, Penman D (2009) Sex-specific differences in the synaptonemal complex in the genus *Oreochromis* (Cichlidae). *Genetica* **135**: 325–332.

Carmi I, Kopczynski JB, Meyer BJ (1998) The nuclear hormone receptor SEX-1 is an X-chromosome signal that determines nematode sex. *Nature* **396**: 168–173.

Carrel L, Willard HF (2005) X-inactivation profile reveals extensive variability in X-linked gene expression in females. *Nature* **434**: 400–404.

Carvalho AB (2002) Origin and evolution of the *Drosophila* Y chromosome. *Curr Opin Genet Dev* **12**: 664–668.

Carvalho AB, Clark AG (2005) Y chromosome of *D. pseudoobscura* is not homologous to the ancestral *Drosophila* Y. *Science* **307**: 108–110.

Carvalho AB, Dobo BA, Vibranovski MD, Clark AG (2001) Identification of five new genes on the Y chromosome of *Drosophila melanogaster*. *Proc Natl Acad Sci U S A* **98**: 13225–13230.

Carvalho AB, Koerich LB, Clark AG (2009) Origin and evolution of Y chromosomes: *Drosophila* tales. *Trends Genet* **25**: 270–277.

Carvalho AB, Lazzaro BP, Clark AG (2000) Y chromosomal fertility factors kl-2 and kl-3 of *Drosophila melanogaster* encode dynein heavy chain polypeptides. *Proc Natl Acad Sci U S A* **97**: 13239–13244.

Carvalho AB, Sampaio MC, Varandas FR, Klaczko LB (1998) An experimental demonstration of Fisher's principle: evolution of sexual proportion by natural selection. *Genetics* **148**: 719–731.

Casselton LA (2002) Mate recognition in fungi. *Heredity* **88**: 142–147.

Casselton LA, Olesnicky NS (1998) Molecular genetics of mating recognition in basidiomycete fungi. *Microbiol Mol Biol Rev* **62**: 55–70.

Castric V, Vekemans X (2004) Plant self-incompatibility in natural populations: a critical assessment of recent theoretical and empirical advances. *Mol Ecol* **13**: 2873–2889.

Caubet Y, Hatcher M, Mocquard J, Rigaud T (2000) Genetic conflict and changes in heterogametic mechanisms of sex determination. *J Evol Biol* **13**: 766–777.

Cazemajor M, Landre C, Montchamp-Moreau C (1997) The sex ratio trait in *Drosophila simulans*: genetic analysis of distortion and supression. *Genetics* **147**: 635–642.

Ceballos CP, Adams DC, Iverson JB, Valenzuela N (2013) Phylogenetic patterns of sexual size dimorphism in turtles and their implications for Rensch's rule. *Evol Biol* **40**: 194–208.

Chadwick DJ, Goode JA (2002) *The genetics and biology of sex determination*. John Wiley & Sons, Chichester.

Chandler J (2011) The hormonal regulation of flower development. *J Plant Growth Regul* **30**: 242–254.

Chandler CH, Chadderdon GE, Phillips PC, Dworkin I et al. (2012) Experimental evolution of the *Caenorhabditis elegans* sex determination pathway. *Evolution* **66**: 82–93.

Chandler CH, Phillips PC, Janzen FJ (2009) The evolution of sex-determining mechanisms: lessons from temperature-sensitive mutations in sex-determination genes. *J Evol Biol* **22**: 192–200.

Charlat S, Hornett EA, Fullard JH, Davies N et al. (2007) Extraordinary flux in sex ratio. *Science* **317**: 214–214.

Charlesworth B (1978a) The population genetics of anisogamy. *J Theor Biol* **73**: 347–357.

Charlesworth B (1978b) Model for evolution of Y chromosomes and dosage compensation. *Proc Natl Acad Sci U S A* **75**: 5618–5622.

Charlesworth B (1996) The evolution of chromosomal sex determination and dosage compensation. *Curr Biol* **6**: 149–162.

Charlesworth D (2013) Plant sex chromosome evolution. *J Exp Bot* **64**: 405–420.

Charlesworth B, Charlesworth D (1978) A model for the evolution of dioecy and gynodioecy. *Am Nat* **112**: 975–997.

Charlesworth D, Charlesworth B (1980) Sex differences in fitness and selection for centric fusions between sex-chromosomes and autosomes. *Genet Res* **35**: 205–214.

Charlesworth D, Charlesworth B (1981) Allocation of resources to male and female functions in hermaphrodites. *Biol J Linn Soc* **15**: 57–74.

Charlesworth B, Charlesworth D (1985) Genetic variation in recombination in *Drosophila*. I. Responses to selection and preliminary genetic analysis. *Heredity* **54**: 71–83.

Charlesworth B, Charlesworth D (2000) The degeneration of Y chromosomes. *Philos Trans R Soc Lond B Biol Sci* **355**: 1563–1572.

Charlesworth D, Charlesworth B (2010) Evolutionary biology: the origins of two sexes. *Curr Biol* **20**: R519–521.

Charlesworth D, Mank JE (2010) The birds and the bees and the flowers and the trees: lessons from genetic mapping of sex determination in plants and animals. *Genetics* **186**: 9–31.

Charlesworth D, Charlesworth B, Marais G (2005) Steps in the evolution of heteromorphic sex chromosomes. *Heredity* **95**: 118–128.

Charlesworth B, Coyne J, Barton N (1987) The relative rates of evolution of sex chromosomes and autosomes. *Am Nat* **130**: 113–146.

Charlesworth B, Dempsey ND (2001) A model of the evolution of the unusual sex chromosome system of *Microtus oregoni*. *Heredity* **86**: 387–394.

Charlesworth D, Morgan MT (1991) Allocation of resources to sex functions in flowering plants. *Philos Trans R Soc Lond B Biol Sci* **332**: 91–102.

Charlesworth B, Wall J (1999) Inbreeding, heterozygote advantage and the evolution of neo-X and neo-Y sex chromosomes. *Royal Soc. Proc. Biol. Sci.* **266**: 51–56.

Charnier M (1966) Action of temperature on the sex ratio in the *Agama agama* (Agamidae, Lacertilia) embryo. *C R Seances Soc Biol Fil* **160**: 620–622.

Charnov EL (1975) Sex ratio selection in an age-structured population. *Evolution* **29**: 366–368.

Charnov EL (1982) *The theory of sex allocation*. *Monogr Popul Biol* **18**: 1–355.

Charnov EL, Bull J (1977) When is sex environmentally determined? *Nature* **266**: 828–830.

Charnov EL, Los-den Hartogh RL, Jones WT, van den Assem J (1981) Sex ratio evolution in a variable environment. *Nature* **289**: 27–33.

Chase CD (2007) Cytoplasmic male sterility: a window to the world of plant mitochondrial-nuclear interactions. *Trends Genet* **23**: 81–90.

Chase CD, Gabay-Laughnan S (2004) Cytoplasmic male sterility and fertility restoration by nuclear genes. In: Daniell H, Chase CD. eds; *Molecular biology and biotechnology of plant organelles*. Springer, Dordrecht. pp. 593–621.

Chibalina MV, Filatov DA (2011) Plant Y chromosome degeneration is retarded by haploid purifying selection. *Curr Biol* **21**: 1475–1479.

Chinnici JP (1971) Modification of recombination frequency in *Drosophila*. I. Selection for increased and decreased crossing over. *Genetics* **69**: 71.

Chue J, Smith CA (2011) Sex determination and sexual differentiation in the avian model. *FEBS J* **278**: 1027–1034.

Civetta A, Singh RS (1995) High divergence of reproductive tract proteins and their association with postzygotic reproductive isolation in *Drosophila melanogaster* and *Drosophila virilis* group species. *J Mol Evol* **41**: 1085–1095.

Clancy DJ, Hime GR, Shirras AD (2011) Cytoplasmic male sterility in *Drosophila melanogaster* associated with a mitochondrial CYTB variant. *Heredity* **107**: 374–376.

Clark AB (1978) Sex ratio and local resource competition in a prosimian primate. *Science* **201**: 163–165.

Clark ME (1999) *The evolution of a neo-Y chromosome in the housefly,* Musca domestica. PhD thesis, University of Houston.

Clayton MN (1988) Evolution and life histories of brown-algae. *Bot Mar* **31**: 379–387.

Cline TW (1993) The *Drosophila* sex determination signal: how do flies count to two? *Trends Genet* **9**: 385–390.

Cline TW, Dorsett M, Sun S, Harrison MM et al. (2010) Evolution of the *Drosophila* feminizing switch gene Sex-lethal. *Genetics* **186**: 1321–1336.

Clutton-Brock T, Albon S, Guinness F (1984) Maternal dominance, breeding success and birth sex ratios in red deer. *Nature* **308**: 358–360.

Clutton-Brock T, Albon S, Guinness F (1986) Great expectations: dominance, breeding success and offspring sex ratios in red deer. *Anim Behav* **34**: 460–471.

Cnaani A, Lee BY, Zilberman N, Ozouf-Costaz C et al. (2008) Genetics of sex determination in tilapiine species. *Sex Dev* **2**: 43–54.

Cocquet J, Ellis PJ, Mahadevaiah SK, Affara NA et al. (2012) A genetic basis for a postmeiotic X versus Y chromosome intragenomic conflict in the mouse. *PLoS Genet* **8**: e1002900.

Coelho SM, Peters AF, Charrier B, Roze D *et al.* (2007) Complex life cycles of multicellular eukaryotes: new approaches based on the use of model organisms. *Gene* **406**: 152–170.

Coelho SM, Scornet D, Rousvoal S, Peters NT et al. (2012) *Ectocarpus*: A model organism for the brown algae. *Cold Spring Harb Protoc* 2012: 193–198.

Coenen A, Kevei F, Hoekstra RF (1997) Factors affecting the spread of double-stranded RNA viruses in *Aspergillus nidulans*. *Genet Res* **69**: 1–10.

Coimbra MRM, Kobayashi K, Koretsugu S, Hasegawa O et al. (2003) A genetic linkage map of the Japanese flounder, *Paralichthys olivaceus*. *Aquaculture* **220**: 203–218.

Colwell RK (1981) Group selection is implicated in the evolution of female-biased sex ratios. *Nature* **290**: 401–404.

Conover DO, Heins SW (1987a) Adaptive variation in environmental and genetic sex determination in a fish. *Nature* **326**: 496–498.

Conover DO, Heins SW (1987b) The environmental and genetic components of sex ratio in *Menidia menidia*. *Copeia* **1987**: 732–743.

Conover DO, Kynard BE (1981) Environmental sex determination: interaction of temperature and genotype in a fish. *Science* **213**: 577–579.

Conover DO, Van Voorhees DA (1990) Evolution of a balanced sex ratio by frequency-dependent selection in a fish. *Science* **250**: 1556–1558.

Conover DO, Van Voorhees DA, Ehtisham A (1992) Sex ratio selection and the evolution of environmental sex determination in laboratory populations of *Menidia menidia*. *Evolution* **46**: 1722–1730.

Conrad T, Akhtar A (2012) Dosage compensation in *Drosophila melanogaster*: epigenetic fine-tuning of chromosome-wide transcription. *Nat Rev Genet* **13**: 123–134.

Cook JM, Crozier RH (1995) Sex determination and population biology in the hymenoptera. *Trends Ecol Evol* **10**: 281–286.

Coppin E, Debuchy R, Arnaise S, Picard M (1997) Mating types and sexual development in filamentous ascomycetes. *Microbiol Mol Biol Rev* **61**: 411–428.

Cordaux R, Bouchon D, Greve P (2011) The impact of endosymbionts on the evolution of host sex-determination mechanisms. *Trends Genet* **27**: 332–341.

Cordaux R, Michel-Salzat A, Frelon-Raimond M, Rigaud T et al. (2004) Evidence for a new feminizing *Wolbachia* strain in the isopod *Armadillidium vulgare*: evolutionary implications. *Heredity* **93**: 78–84.

Cordaux R, Pichon S, Hatira HB, Doublet V *et al*. (2012) Widespread *Wolbachia* infection in terrestrial isopods and other crustaceans. *Zookeys* **176**: 123–131.

Corl A, Ellegren H (2012) The genomic signature of sexual selection in the genetic diversity of the sex chromosomes and autosomes. *Evolution* **66**: 2138–2149.

Cosmides LM, Tooby J (1981) Cytoplasmic inheritance and intragenomic conflict. *J Theor Biol* **89**: 83–129.

Counterman BA, Ortiz-Barrientos D, Noor MA (2004) Using comparative genomic data to test for fast-X evolution. *Evolution* **58**: 656–660.

Coyne JA (1985) The genetic basis of Haldane's rule. *Nature* **314**: 736–738.

Coyne JA (1992) Genetics of sexual isolation in females of the *Drosophila simulans* species complex. *Genet Res* **60**: 25–31.

Coyne JA, Orr HA (1993) Further evidence against meiotic-drive models of hybrid sterility. *Evolution* **47**: 685–687.

Coyne JA, Orr HA (2004) *Speciation*. Sinauer Associates, Sunderland MA.

Coyne JA, Rux J, David JR (1991) Genetics of morphological differences and hybrid sterility between *Drosophila sechellia* and its relatives. *Genet Res* **57**: 113–122.

Crouse HV (1960) The controlling element in sex chromosome behavior in *Sciara*. *Genetics* **45**: 1429–1443.

Crow JF (1993) How much do we know about spontaneous human mutation rates? *Environ Mol Mutagen* **21**: 122–129.

Crow JF, Kimura M (1965) Evolution in sexual and asexual populations. *Am Nat* **99**: 439–450.

Crow J, Simmons M (1983) The mutation load in *Drosophila*. In: Ashburner M, Carson HL, Thompson JN. eds. *The genetics and biology of* Drosophila, *Vol. 3c*. Academic Press, New York. pp. 1–35.

Czaja W, Miller KY, Miller BL (2011) Complex mechanisms regulate developmental expression of the matA (HMG) mating type gene in homothallic *Aspergillus nidulans*. *Genetics* **189**: 795–808.

Daan S, Dijkstra C, Weissing FJ (1996) An evolutionary explanation for seasonal trends in avian sex ratios. *Behav Ecol* **7**: 426–430.

Dacks J, Roger AJ (1999) The first sexual lineage and the relevance of facultative sex. *J Mol Evol* **48**: 779–783.

Darwin CR (1877) *The different forms of flowers on plants of the same species*. Murray, London. 352 pp.

Darwin CR (1859) *On the origins of species by means of natural selection, or the preservation of the favoured races in the struggle for life*. London: John Murray.

Davey AH, Jellyman D (2005) Sex determination in freshwater eels and management options for manipulation of sex. *Rev Fish Biol Fish* **15**: 37–52.

Davidson WS, Huang TK, Fujiki K, von Schalburg KR et al. (2009) The sex determining loci and sex chromosomes in the family Salmonidae. *Sex Dev* **3**: 78–87.

De Franceschi P, Dondini L, Sanzol J (2012) Molecular bases and evolutionary dynamics of self-incompatibility in the *Pyrinae* (Rosaceae). *J Exp Bot* **63**: 4015–4032.

De Vienne DM, Giraud T, Gouyon PH (2013) Lineage selection and the maintenance of Sex. *PLoS One* **8**: e66906.

De Villena FP, Sapienza C (2001a) Female meiosis drives karyotypic evolution in mammals. *Genetics* **159**: 1179–1189.

De Villena FP, Sapienza C (2001b) Nonrandom segregation during meiosis: the unfairness of females. *Mamm Genome* **12**: 331–339.

De Visser JA, Elena SF (2007) The evolution of sex: empirical insights into the roles of epistasis and drift. *Nat Rev Genet* **8**: 139–149.

Deakin JE, Hore TA, Koina E, Marshall Graves JA (2008) The status of dosage compensation in the multiple X chromosomes of the platypus. *PLoS Genet* **4**: e1000140.

Dean R, Perry JC, Pizzari T, Mank JE et al. (2012) Experimental evolution of a novel sexually antagonistic allele. *PLoS Genet* **8**: e1002917.

Decarpentrie F, Vernet N, Mahadevaiah SK, Longepied G et al. (2012) Human and mouse ZFY genes produce a conserved testis-specific transcript encoding a zinc finger protein with a short acidic domain and modified transactivation potential. *Hum Mol Genet* **21**: 2631–2645.

Denholm I, Franco M, Rubini P, Vecchi M (1986) Geographical variation in house-fly (*Musca domestica* L.) sex determinants within the British Isles. *Genet Res* **47**: 19–27.

Derelle E, Ferraz C, Rombauts S, Rouze P *et al.* (2006) Genome analysis of the smallest free-living eukaryote *Ostreococcus tauri* unveils many unique features. *Proc Natl Acad Sci U S A* **103**: 11647–11652.

Destombe C, Godin J, Lefebvre C, Dehorter O et al. (1992) Differences in dispersal abilities of haploid and diploid spores of *Gracilaria verrucosa* (Gracilariales, Rhodophyta). *Bot Mar* **35**: 93–98.

Destombe C, Godin J, Nocher M, Richerd S et al. (1993) Differences in response between haploid and diploid isomorphic phases of *Gracilaria verrucosa* (Rhodophyta: Gigartinales) exposed to artificial environmental conditions. *Hydrobiologia* **260/261**: 131–137.

Devlin RH, Nagahama Y (2002) Sex determination and sex differentiation in fish: an overview of genetic, physiological, and environmental influences. *Aquaculture* **208**: 191–364.

Diggle PK, Di Stilio VS, Gschwend AR, Golenberg EM et al. (2011) Multiple developmental processes underlie sex differentiation in angiosperms. *Trends Genet* **27**: 368–376.

Dijkstra C, Daan S, Buker JB (1990) Adaptive seasonal variation in the sex ratio of Kestrel broods. *Funct Ecol* **4**: 143–147.

Dixson BJ, Haywood J, Lester PJ, Ormsby DK (2011) Whatever the weather: ambient temperature does not influence the proportion of males born in New Zealand. *PLoS One* **6**: e25064.

Dobzhansky T (1937) *Genetics and the origin of species*. Columbia University Press, New York.

Dobzhansky T, Dobzhansky TG (1970) *Genetics of the evolutionary process*. Columbia University Press, New York.

Dod B, Jermiin LS, Boursot P, Chapman VH et al. (1993) Counterselection on sex chromosomes in the *Mus musculus* European hybrid zone. *J Evol Biol* **6**: 529–546.

Dodd KL, Murdoch C, Wibbels T (2006) Interclutch variation in sex ratios produced at pivotal temperature in the red-eared slider, a turtle with temperature-dependent sex determination. *J Herpet* **40**: 544–549.

Döpp W (1950) Eine die Antheridienbildung bei farnen fordernde Substanz in den Prothallien von *Pteridium aquilinum* L. *Ber Dtsch Bot Ges* **63**: 139–147.

Drent R, Daan S (1980) The prudent parent: energetic adjustments in avian breeding. *Ardea* **68**: 225–252.

Droop MR (1962) *Heteramoeba clara* n.gen., n.sp., a sexual biphasic Amoeba. *Arch Mikrobiol* **42**: 254–266.

Dübendorfer A, Hediger M (1998) The female-determining gene F of the housefly, *Musca domestica*, acts maternally to regulate its own zygotic activity. *Genetics* **150**: 221–226.

Dübendorfer A, Hediger M, Burghardt G, Bopp D (2002) *Musca domestica*, a window on the evolution of sex-determining mechanisms in insects. *Int J Dev Biol* **46**: 75–79.

Duharcourt S, Butler A, Meyer E (1995) Epigenetic self-regulation of developmental excision of an internal eliminated sequence on *Paramecium tetraurelia*. *Genes Dev* **9**: 2065–2077.

Dunn A, Hatcher M, Terry R, Tofts C (1995) Evolutionary ecology of vertically transmitted parasites: transovarial transmission of a microsporidian sex ratio distorter in *Gammarus duebeni*. *Parasitology* **111**: 91–109.

Durand B, Durand R (1991) Sex determination and reproductive organ differentiation in *Mercurialis*. *Plant Sci* **80**: 49–65.

Duron O, Bouchon D, Boutin S, Bellamy L et al. (2008) The diversity of reproductive parasites among arthropods: *Wolbachia* do not walk alone. *BMC Biol* **6**: 27–7007–6–27.

Dyer PS, Paoletti M, Archer DB (2003) Genomics reveals sexual secrets of *Aspergillus*. *Microbiology* **149**: 2301–2303.

Ebersold WT (1967) *Chlamydomonas reinhardi*: heterozygous diploid strains. *Science* **157**: 447–449.

Eggert C (2004) Sex determination: the amphibian models. *Reprod Nutr Dev* **44**: 539–549.

Elgvin TO, Hermansen JS, Fijarczyk A, Bonnet T et al. (2011) Hybrid speciation in sparrows II: a role for sex chromosomes? *Mol Ecol* **20**: 3823–3837.

Ellegren H (2009) The different levels of genetic diversity in sex chromosomes and autosomes. *Trends Genet* **25**: 278–284.

Ellegren H, Parsch J (2007) The evolution of sex-biased genes and sex-biased gene expression. *Nat Rev Genet* **8**: 689–698.

Ellison CK, Burton RS (2006) Disruption of mitochondrial function in interpopulation hybrids of *Tigriopus californicus*. *Evolution* **60**: 1382–1391.

Emerson JJ, Kaessmann H, Betran E, Long M (2004) Extensive gene traffic on the mammalian X chromosome. *Science* **303**: 537–540.

Endler JA (1980) Natural selection on color patterns in *Poecilia reticulata*. *Evolution* **34**: 76–91.

Engelstädter J (2008) Constraints on the evolution of asexual reproduction. *Bioessays* **30**: 1138–1150.

Engelstädter J, Haig D (2008) Sexual antagonism and the evolution of X chromosome inactivation. *Evolution* **62**: 2097–2104.

Engelstädter J, Hurst G (2006) Can maternally transmitted endosymbionts facilitate the evolution of haplodiploidy? *J Evol Biol* **19**: 194–202.

Enjalbert J, Duan X, Giraud T, Vautrin D et al. (2002) Isolation of twelve microsatellite loci, using an enrichment protocol, in the phytopathogenic fungus *Puccinia striiformis* f. sp. *tritici*. *Mol Ecol* **2**: 563–565.

Eppley SM, Jesson LK (2008) Moving to mate: the evolution of separate and combined sexes in multicellular organisms. *J Evol Biol* **21**: 727–736.

Erickson JW, Quintero JJ (2007) Indirect effects of ploidy suggest X chromosome dose, not the X: A ratio, signals sex in *Drosophila*. *PLoS Biol* **5**: e332.

Eshel I, Sansone E (1991) Parent-offspring conflict over the sex ratio in a diploid population with different investment in male and in female offspring. *Am Nat* **138**: 954–972.

Eshel I, Sansone E (1994) Parent-offspring conflict over sex ratio. II. Offspring response to parental manipulation. *Am Nat* **143**: 987–1006.

Este SV, Rovati C (1982) Inheritance of the arrhenogenic factor *Ag* of *Musca domestica* L. *Ital J Zool* **49**: 269–278.

Ewert MA, Lang JW, Nelson CE (2005) Geographic variation in the pattern of temperature-dependent sex determination in the American snapping turtle (*Chelydra serpentina*). *J Zool* **265**: 81.

Ewert MA, Nelson CE (1991) Sex determination in turtles—diverse patterns and some possible adaptive values. *Copeia* **1991**: 50–69.

Ezaz T, Quinn AE, Miura I, Sarre SD et al. (2005) The dragon lizard *Pogona vitticeps* has ZZ/ZW micro-sex chromosomes. *Chromosome Res* **13**: 763–776.

Ezaz T, Sarre S, O'Meally D, JA MG et al. (2009) Sex chromosome evolution in lizards: independent origins and rapid transitions. *Cytogenet Genome Res* **127**: 249–260.

Ezov TK, Chang S, Frenkel Z, Segre AV et al. (2010) Heterothallism in *Saccharomyces cerevisiae* isolates from nature: effect of HO locus on the mode of reproduction. *Mol Ecol* **19**: 121–131.

Fabritius AL, Judelson HS (1997) Mating-type loci segregate aberrantly in *Phytophthora infestans* but normally in *Phytophthora parasitica*: implications for models of mating-type determination. *Curr Genet* **32**: 60–65.

Fairbairn DJ, Roff DA (2006) The quantitative genetics of sexual dimorphism: assessing the importance of sex-linkage. *Heredity* **97**: 319–328.

Feder JL, Roethele JB, Filchak K, Niedbalski J et al. (2003) Evidence for inversion polymorphism related to sympatric host race formation in the apple maggot fly, *Rhagoletis pomonella*. *Genetics* **163**: 939–953.

Fedler M, Luh KS, Stelter K, Nieto-Jacobo F et al. (2009) The a2 mating-type locus genes lga2 and rga2 direct uniparental mitochondrial DNA (mtDNA) inheritance and constrain mtDNA recombination during sexual development of *Ustilago maydis*. *Genetics* **181**: 847–860.

Feldmeyer B, Kozielska M, Kuijper B, Weissing FJ et al. (2008) Climatic variation and the geographical distribution of sex-determining mechanisms in the housefly. *Evol Ecol Res* **10**: 797–809.

Felsenstein J (1974) The evolutionary advantage of recombination. *Genetics* **78**: 737–756.

Felsenstein J (1981) Skepticism towards Santa Rosalia, or why are there so few kinds of animals? *Evolution* **35**: 124–138.

Fernandez AA, Morris MR (2008) Mate choice for more melanin as a mechanism to maintain a functional oncogene. *Proc Natl Acad Sci U S A* **105**: 13503–13507.

Fernandino JI, Hattori RS, Shinoda T, Kimura H et al. (2008) Dimorphic expression of dmrt1 and cyp19a1 (ovarian aromatase) during early gonadal development in pejerrey, *Odontesthes bonariensis*. *Sex Dev* **2**: 316–324.

Ferris PJ, Armbrust EV, Goodenough UW (2002) Genetic structure of the mating-type locus of *Chlamydomonas reinhardtii*. *Genetics* **160**: 181–200.

Ferris PJ, Goodenough UW (1994) The mating-type locus of *Chlamydomonas reinhardtii* contains highly rearranged DNA sequences. *Cell* **76**: 1135–1145.

Ferris P, Olson BJ, De Hoff PL, Douglass S et al. (2010) Evolution of an expanded sex-determining locus in *Volvox*. *Science* **328**: 351–354.

Fierst J, TerHorst C, Kübler JE, Dudgeon S (2005) Fertilization success can drive patterns of phase dominance in complex life histories. *J Phycol* **41**: 238–249.

Filatov DA, Laporte V, Vitte C, Charlesworth D (2001) DNA diversity in sex-linked and autosomal genes of the plant species *Silene latifolia* and *Silene dioica*. *Mol Biol Evol* **18**: 1442–1454.

Filatov DA, Moneger F, Negrutiu I, Charlesworth D (2000) Low variability in a Y-linked plant gene and its implications for Y-chromosome evolution. *Nature* **404**: 388–390.

Fishelson L (1970) Protogynous sex reversal in fish *Anthias squamipinnis* (Teleostei, Anthiidae) regulated by presence or absence of a male fish. *Nature* **227**: 90–&.

Fisher RA (1930) *The genetical theory of natural selection*. Clarendon Press, Oxford.

Fisher RA (1931) The evolution of dominance. *Biol Rev* **6**: 345–368.

Fishman L, Willis JH (2006) A cytonuclear incompatibility causes anther sterility in *Mimulus* hybrids. *Evolution* **60**: 1372–1381.

Fitzpatrick MJ (2004) Pleiotropy and the genomic location of sexually selected genes. *Am Nat* **163**: 800–808.

Flexon PB, Rodell CF (1982) Genetic recombination and directional selection for DDT resistance in *Drosophila melanogaster*. *Nature* **298**: 672–674.

Fouvry L, Ogereau D, Berger A, Gavory F et al. (2011) Sequence analysis of the segmental duplication responsible for Paris sex-ratio drive in *Drosophila simulans*. *G3 (Bethesda)* **1**: 401–410.

Francis RC, Barlow GW (1993) Social control of primary sex differentiation in the Midas cichlid. *Proc Natl Acad Sci U S A* **90**: 10673–10675.

Franco MG, Rubini PG, Vecchi M (1982) Sex-determinants and their distribution in various populations of *Musca domestica* L. of Western Europe. *Genet Res* **40**: 279–293.

Frank SA (1986) Hierarchical selection theory and sex ratios I. General solutions for structured populations. *Theor Popul Biol* **29**: 312–342.

Frank SA (1991) Divergence of meiotic drive-suppression systems as an explanation for sex-biased hybrid sterility and inviability. *Evolution* **45**: 262–267.

Franke H-D (1986) Sex ratio and sex change in wild and laboratory populations of *Typosyllis prolifera* (Polychaeta). *Mar Biol* **90**: 197–208.

Franklin-Tong, VE, Franklin, FCH (2003) Gametophytic self-incompatibility inhibits pollen tube growth using different mechanisms. *Trends Plant Sci* **8**: 598–605.

Fraser JA, Diezmann S, Subaran RL, Allen A et al. (2004) Convergent evolution of chromosomal sex-determining regions in the animal and fungal kingdoms. *PLoS Biol* **2**: e384.

Fraser JA, Heitman J (2004) Evolution of fungal sex chromosomes. *Mol Microbiol* **51**: 299–306.

Fraser JA, Heitman J (2005) Chromosomal sex-determining regions in animals, plants and fungi. *Curr Opin Genet Dev* **15**: 645–651.

Fredga K (1988) Aberrant chromosomal sex-determining mechanisms in mammals, with special reference to species with XY females. *Philos Trans R Soc Lond B Biol Sci* **322**: 83–95.

Fredga K, Gropp A, Winking H, Frank F (1976) Fertile XX- and XY-type females in the wood lemming *Myopus schisticolor*. *Nature* **261**: 225–227.

Fredga K, Gropp A, Winking H, Frank F (1977) A hypothesis explaining the exceptional sex ratio in the wood lemming (*Myopus schisticolor*). *Hereditas* **85**: 101–104.

Freedberg S, Taylor DR (2007) Sex ratio variance and the maintenance of environmental sex determination. *J Evol Biol* **20**: 213–220.

Friberg U, Stewart AD, Rice WR (2012) X- and Y-chromosome linked paternal effects on a life-history trait. *Biol Lett* **8**: 71–73.

Fricke HW (1979) Mating system, resource defence and sex change in the anemonefish *Amphiprion akallopisos*. *Ethology* **50**: 313–326.

Fulton C (1993) *Naegleria*: a research partner for cell and developmental biology. *J Eukaryot Microbiol* **40**: 520–532.

Galagan JE, Selker EU (2004) RIP: the evolutionary cost of genome defense. *Trends Genet* **20**: 417–423.

Galgoczy DJ, Cassidy-Stone A, Llinas M, O'Rourke SM et al. (2004) Genomic dissection of the cell-type-specification circuit in *Saccharomyces cerevisiae*. *Proc Natl Acad Sci U S A* **101**: 18069–18074.

Galimov Y, Walser B, Haag CR (2011) Frequency and inheritance of non-male producing clones in *Daphnia magna*: evolution towards sex specialization in a cyclical parthenogen? *J Evol Biol* **24**: 1572–1583.

Galitski T, Saldanha AJ, Styles CA, Lander ES et al. (1999) Ploidy regulation of gene expression. *Science* **285**: 251–254.

Galloway RE, Goodenough UW (1985) Genetic analysis of mating locus linked mutations in *Chlamydomonas reinhardii*. *Genetics* **111**: 447–461.

Galton F (1889). *Natural inheritance*. MacMillan, London.

Gamble T (2010) A review of sex determining mechanisms in geckos (Gekkota: Squamata). *Sex Dev* **4**: 88–103.

Gellatly C (2009) Trends in population sex ratios may be explained by changes in the frequencies of polymorphic alleles of a sex ratio gene. *Evol Biol* **36**: 190–200.

Gempe T, Beye M (2011) Function and evolution of sex determination mechanisms, genes and pathways in insects. *Bioessays* **33**: 52–60.

Gempe T, Hasselmann M, Schiøtt M, Hause G et al. (2009) Sex determination in honeybees: two separate mechanisms induce and maintain the female pathway. *PLoS Biol* **7**: e1000222.

Gendrel AV, Heard E (2011) Fifty years of X-inactivation research. *Development* **138**: 5049–5055.

Geraldes A, Basset P, Gibson B, Smith KL et al. (2008) Inferring the history of speciation in house mice from autosomal, X-linked, Y-linked and mitochondrial genes. *Mol Ecol* **17**: 5349–5363.

Gerstein AC, Cleathero L, Mandegar M, Otto S (2011) Haploids adapt faster than diploids across a range of environments. *J Evol Biol* **24**: 531—540.

Gherna RL, Werren JH, Weisburg W, Cote R et al. (1991) *Arsenophonus nasoniae* gen. nov., sp. nov., the causative agent of the son-killer trait in the parasitic wasp *Nasonia vitripennis*. *Int J Syst Bacteriol* **41**: 563–565.

Ghiselin MT (1969) The evolution of hermaphroditism among animals. *Q Rev Biol* **44**: 189–208.

Gibson W (2001) Sex and evolution in trypanosomes. *Int J Parasitol* **31**: 643–647.

Gibson JR, Chippindale AK, Rice WR (2002) The X chromosome is a hot spot for sexually antagonistic fitness variation. *Proc Biol Sci* **269**: 499–505.

Gilles R, Jaenicke L (1984) Extracellular phosphoproteins are involved in sexual differentiation of *Volvox carteri*. *H-S Z Physiol Chem* **365**: 990–990.

Gillissen B, Bergemann J, Sandmann C, Schroeer B et al. (1992) A two-component regulatory system for self/non-self recognition in *Ustilago maydis*. *Cell* **68**: 647–657.

Giorgini M, Monti M, Caprio E, Stouthamer R *et al.* (2009) Feminization and the collapse of haplodiploidy in an asexual parasitoid wasp harboring the bacterial symbiont *Cardinium*. *Heredity* **102**: 365–371.

Giraud T, Yockteng R, López-Villavicencio M, Refrégier G et al. (2008) Mating system of the anther smut fungus *Microbotryum violaceum*: selfing under heterothallism. *Eukaryotic Cell* **7**: 765–775.

Gladyshev E, Meselson M (2008) Extreme resistance of bdelloid rotifers to ionizing radiation. *Proc Natl Acad Sci U S A* **105**: 5139–5144.

Glawe GA, de Jong TJ (2009) Complex sex determination in the stinging nettle *Urtica dioica*. *Evol Ecol* **23**: 635–649.

Gleason J, Nuzhdin S, Ritchie M (2002) Quantitative trait loci affecting a courtship signal in *Drosophila melanogaster*. *Heredity* **89**: 1–6.

Goddard MR, Godfray HC, Burt A (2005) Sex increases the efficacy of natural selection in experimental yeast populations. *Nature* **434**: 636–640.

Godwin J, Luckenbach JA, Borski RJ (2003) Ecology meets endocrinology: environmental sex determination in fishes. *Evol Dev* **5**: 40–49.

Goldberg EE, Kohn JR, Lande R, Robertson KA et al. (2010) Species selection maintains self-incompatibility. *Science* **330**: 493–495.

Goldman DA, Willson MF (1986) Sex allocation in functionally hermaphroditic plants: A review and critique. *Bot Rev* **52**: 157–194.

Goldstein DB (1992) Heterozygote advantage and the evolution of a dominant diploid phase. *Genetics* **132**: 1195–1198.

Good JM, Handel MA, Nachman MW (2008) Asymmetry and polymorphism of hybrid male sterility during the early stages of speciation in house mice. *Evolution* **62**: 50–65.

Good JM, Vanderpool D, Smith KL, Nachman MW (2011) Extraordinary sequence divergence at *Tsga8*, an X-linked gene involved in mouse spermiogenesis. *Mol Biol Evol* **28**: 1675–1686.

Goodenough U, Lin H, Lee JH (2007) Sex determination in *Chlamydomonas*. *Semin Cell Dev Biol* **18**: 350–361.

Gordo I, Charlesworth B (2000) The degeneration of asexual haploid populations and the speed of Muller's ratchet. *Genetics* **154**: 1379–1387.

Gordo I, Charlesworth B (2001) The speed of Muller's ratchet with background selection, and the degeneration of Y chromosomes. *Genet Res* **78**: 149–161.

Gorelick R, Carpinone J (2009) Origin and maintenance of sex: the evolutionary joys of self sex. *Biol J Linn Soc* **98**: 707–728.

Gorodetskii VP, Zhuchenko AA, Korol' AB (1990) Efficiency of feedback selection for recombination in *Drosophila*. *Genetika* **26**: 1942–1952.

Gouyon P, Vichot F, Van Damme J (1991) Nuclear-cytoplasmic male sterility: single-point equilibria versus limit cycles. *Am Nat*: 498–514.

Gramzow L, Ritz MS, Theissen G (2010) On the origin of MADS-domain transcription factors. *Trends Genet* **26**: 149–153.

Graves JAM (1995) The evolution of mammalian sex chromosomes and the origin of sex determining genes. *Philos Trans R Soc Lond B Biol Sci* **350**: 305–311.

Graves JAM (2005) Recycling the Y chromosome. *Science* **307**: 50–51.

Graves JAM (2006) Sex chromosome specialization and degeneration in mammals. *Cell* **124**: 901–914.

Graves JAM, Foster JW (1994) Evolution of mammalian sex chromosomes and sex-determining genes. *Int Rev Cytol* **154**: 191–259.

Graves JAM, Peichel CL (2010) Are homologies in vertebrate sex determination due to shared ancestry or to limited options. *Genome Biol* **11**: 205–216.

Green DM, Zeyl CW, Sharbel TF (1993) The evolution of hypervariable sex and supernumerary (B) chromosomes in the relict New Zealand frog, *Leiopelma hochstetteri*. *J Evol Biol* **6**: 417–441.

Greig D, Borts RH, Louis EJ (1998) The effect of sex on adaptation to high temperature in heterozygous and homozygous yeast. *Proc Biol Sci* **265**: 1017–1023.

Gribnau J, Grootegoed JA (2012) Origin and evolution of X chromosome inactivation. *Curr Opin Cell Biol* **24**: 397–404.

Grossen C, Neuenschwander S, Perrin N (2011) Temperature-dependent turnovers in sex-determination mechanisms: a quantitative model. *Evolution* **65**: 64–78.

Grossen C, Neuenschwander S, Perrin N (2012) The evolution of XY recombination: sexually antagonistic selection versus deleterious mutation load. *Evolution* **66**: 3155–3166.

Gruetzner F, Ashley T, Rowell DM, Marshall Graves JA (2006) How did the platypus get its sex chromosome chain? A comparison of meiotic multiples and sex chromosomes in plants and animals. *Chromosoma* **115**: 75–88.

Grula JW, Taylor OR, Jr. (1980) The effect of X-chromosome inheritance on mate-selection behavior in the sulfur butterflies, *Colias eurytheme* and *C. philodice*. *Evolution* **34**: 688–695.

Gschwend AR, Weingartner LA, Moore RC, Ming R (2012) The sex-specific region of sex chromosomes in animals and plants. *Chromosome Res* **20**: 57–69.

Guerrero RF, Kirkpatrick M, Perrin N (2012) Cryptic recombination in the ever-young sex chromosomes of Hylid frogs. *J Evol Biol* **25**: 1947–1954.

Guerrero-Estévez S, Moreno-Mendoza N (2010) Sexual determination and differentiation in teleost fish. *Rev Fish Biol Fisheries* **20**: 101–121.

Gupta V, Parisi M, Sturgill D, Nuttall R et al. (2006) Global analysis of X-chromosome dosage compensation. *J Biol* **5**: 3.

Gurbich TA, Bachtrog D (2008) Gene content evolution on the X chromosome. *Curr Opin Genet Dev* **18**. 493–498.

Haag ES (2007) Why two sexes? Sex determination in multicellular organisms and protistan mating types. *Semin Cell Dev Biol* **18**: 348–349.

Haag ES, Doty AV (2005) Sex determination across evolution: connecting the dots. *PLoS Biol* **3**: e21.

Haber JE (2012) Mating-type genes and MAT switching in *Saccharomyces cerevisiae*. *Genetics* **191**: 33–64.

Hackstein JH, Hochstenbach R (1995) The elusive fertility genes of *Drosophila*: the ultimate haven for selfish genetic elements. *Trends Genet* **11**: 195–200.

Hackstein JH, Hochstenbach R, Hauschteck-Jungen E, Beukeboom LW (1996) Is the Y chromosome of *Drosophila* an evolved supernumerary chromosome? *Bioessays* **18**: 317–323.

Hagimori T, Abe Y, Date S, Miura K (2006) The first finding of a *Rickettsia* bacterium associated with parthenogenesis induction among insects. *Curr Microbiol* **52**: 97–101.

Haig D (1993a) The evolution of unusual chromosomal systems in coccoids: extraordinary sex ratios revisited. *J Evol Biol* **6**: 69–77.

Haig D (1993b) The evolution of unusual chromosomal systems in sciarid flies: intragenomic conflict and the sex ratio. *J Evol Biol* **6**: 249–261.

Haig D (2006) Self-imposed silence: parental antagonism and the evolution of X-chromosome inactivation. *Evolution* **60**: 440–447.

Haig D, Grafen A (1991) Genetic scrambling as a defence against meiotic drive. *J Theor Biol* **153**: 531–558.

Haig D, Westoby M (1988) Inclusive fitness, seed resources, and maternal care. In: Doust JL, Doust LL. eds. *Plant reproductive ecology: Pattern and strategy*. Oxford University Press, Oxford. pp. 60–79.

Haig D, Wilczek A (2006) Sexual conflict and the alternation of haploid and diploid generations. *Philos Trans R Soc Lond B Biol Sci* **361**: 335–343.

Halary S, Malik SB, Lildhar L, Slamovits CH *et al.* (2011) Conserved meiotic machinery in *Glomus* spp., a putatively ancient asexual fungal lineage. *Genome Biol Evol* **3**: 950–958.

Haldane JBS (1922) Sex ratio and unisexual sterility in hybrid animals. *J Genet* **12**: 101–109.

Haldane JBS (1933) The part played by recurrent mutation in evolution. *Am Nat* **67**: 5–19.

Haley LE (1977) Sex determination in the American oyster. *J Heredity* **68**: 114–116.

Haley LE (1979) Genetics of sex determination in the American oyster. *Proc Natl Shellfish Assoc* **69**: 54–57.

Hall DW, Kirkpatrick M (2006) Reinforcement and sex linkage. *Evolution* **60**: 908–921.

Hallmann A (2008) VCRPs, small cysteine-rich proteins, might be involved in extracellular signaling in the green alga *Volvox*. *Plant Signal Behav* **3**: 124–127.

Hamaji T, Ferris PJ, Coleman AW, Waffenschmidt S et al. (2008) Identification of the minus-dominance gene ortholog in the mating-type locus of *Gonium pectorale*. *Genetics* **178**: 283–294.

Hamilton WD (1967) Extraordinary sex ratios. *Science* **156**: 477–488.

Hamilton WD (1993) Inbreeding in Egypt and in this book: a childish perspective. In: Thornhill NW. ed. *The natural history of inbreeding and outbreeding: theoretical and empirical perspectives*. University of Chicago Press, Chicago. pp. 429–450.

Hamm RL, Gao JR, Lin GG, Scott JG (2009) Selective advantage for IIIM males over YM males in cage competition, mating competition, and pupal emergence in *Musca domestica* L. (Diptera: Muscidae). *Environ Entomol* **38**: 499–504.

Hamm RL, Scott JG (2008) Changes in the frequency of YM versus IIIM in the housefly, *Musca domestica* L., under field and laboratory conditions. *Genet Res (Camb)* **90**: 493–498.

Hamm RL, Shono T, Scott JG (2005) A cline in frequency of autosomal males is not associated with insecticide resistance in house fly (Diptera: Muscidae). *J Econ Entomol* **98**: 171–176.

Handley LJ, Berset-Brändli L, Perrin N (2006) Disentangling reasons for low Y chromosome variation in the greater white-toothed shrew (*Crocidura russula*). *Genetics* **173**: 935–942.

Handley LJ, Ceplitis H, Ellegren H (2004) Evolutionary strata on the chicken Z chromosome: implications for sex chromosome evolution. *Genetics* **167**: 367–376.

Hannach G, Santelices B (1985) Ecological differences between the isomorphic reproductive phases of two species of *Iridaea* (Rhodophyta: Gigartinales). *Mar Ecol Prog Ser* **22**: 291–303.

Hanson MR, Bentolila S (2004) Interactions of mitochondrial and nuclear genes that affect male gametophyte development. *Plant Cell* **16 Suppl**: S154–169.

Harada Y, Sawada H (2008) Allorecognition mechanisms during ascidian fertilization. *Int J Dev Biol* **52**: 637–645.

Harada Y, Takagaki Y, Sunagawa M, Saito T et al. (2008) Mechanism of self-sterility in a hermaphroditic chordate. *Science* **320**: 548–550.

Hardy I (2002) *Sex Ratios: concepts and research methods*. Cambridge University Press, Cambridge.

Harris WE, McKane AJ, Wolf JB (2008) The maintenance of heritable variation through social competition. *Evolution* **62**: 337–347.

Haskins C, Haskins E, McLaughlin J, Hewitt R (1961) Polymorphism and population structure in *Lebistes reticulatus*, an ecological study. In: Blair WF. ed; *Vertebrate Speciation*. University of Texas Press, Austin, TX. pp. 320–395.

Hasselmann M, Gempe T, Schiott M, Nunes-Silva CG et al. (2008) Evidence for the evolutionary nascence of a novel sex determination pathway in honeybees. *Nature* **454**: 519–522.

Hastings IM (1994) Manifestations of sexual selection may depend on the genetic bases of sex determination. *Proc Biol Sci* **258**: 83–87.

Hattori A (1991) Socially controlled growth and size-dependent sex change in the anemonefish *Amphiprion frenatus* in Okinawa, Japan. *Jap J Ichthyol* **38**: 165–177.

Hattori RS, Gould RJ, Fujioka T, Saito T et al. (2007) Temperature-dependent sex determination in Hd-rR medaka *Oryzias latipes*: gender sensitivity, thermal threshold, critical period, and *DMRT1* expression profile. *Sex Dev* **1**: 138–146.

Hattori RS, Murai Y, Oura M, Masuda S et al. (2012) A Y-linked anti-Mullerian hormone duplication takes over a critical role in sex determination. *Proc Natl Acad Sci U S A* **109**: 2955–2959.

Haughn GW, Somerville CR (1988) Genetic control of morphogenesis in *Arabidopsis*. *Dev Genet* **9**: 73–89.

He CL, Du JL, Wu GC, Lee YH et al. (2003) Differential *Dmrt1* transcripts in gonads of the protandrous black porgy, *Acanthopagrus schlegeli*. *Cytogenet Genome Res* **101**: 309–313.

Heard E, Carrel L (2009) Foreword: Coping with sex chromosome imbalance. *Chromosome Res* **17**: 579–583.

Heard E, Disteche CM (2006) Dosage compensation in mammals: fine-tuning the expression of the X chromosome. *Genes Dev* **20**: 1848–1867.

Hediger M, Henggeler C, Meier N, Perez R et al. (2010) Molecular characterization of the key switch F provides a basis for understanding the rapid divergence of the sex-determining pathway in the housefly. *Genetics* **184**: 155–170.

Heilbuth JC (2000) Lower species richness in dioecious clades. *Am Nat* **156**: 221–241.

Heimel K, Scherer M, Vranes M, Wahl R et al. (2010) The transcription factor Rbf1 is the master regulator for b-mating type controlled pathogenic development in *Ustilago maydis*. *PLoS Pathog* **6**: e1001035.

Heimpel GE, de Boer JG (2008) Sex determination in the hymenoptera. *Annu Rev Entomol* **53**: 209–230.

Heisler IL (1984) Inheritance of female mating propensities for yellow locus genotypes in *Drosophila melanogaster*. *Genet Res* **44**: 133–149.

Hellborg L, Ellegren H (2004) Low levels of nucleotide diversity in mammalian Y chromosomes. *Mol Biol Evol* **21**: 158–163.

Henking H (1891) Untersuchungen uber die ersten Entwicklungsvorigoenge in den Eiern der Insektern II Uber spermatogenese und Beziehung zur Eientwicklung bei *Pyrrhocoris apterus*. *Z Wiss Zool* **51**: 685–736.

Hense W, Baines JF, Parsch J (2007) X chromosome inactivation during *Drosophila* spermatogenesis. *PLoS Biol* **5**: e273.

Herpin A, Schartl M (2011a) *Dmrt1* genes at the crossroads: a widespread and central class of sexual development factors in fish. *FEBS J* **278**: 1010–1019.

Herpin A, Schartl M (2011b) Sex determination: switch and suppress. *Curr. Biol.* **21**: R656–R659.

Herre EA, West SA, Cook JM, Compton SG et al. (1997) Fig wasp mating systems: pollinators and parasites, sex ratio adjustment and male polymorphism, population structure and its consequences. In: Choe J, Crespi B. eds. *Social competition and cooperation in insects and arachnids, Volume I: The evolution of mating systems*. Cambridge University Press, Cambridge. pp. 226–239.

Heslop-Harrison J (1961) The experimental control of sexuality and inflorescence structure in *Zea mays* L. *Proc Linn Soc London* **172**: 108–123.

Heslot H (1958) Contribution à l'étude cytogénétique et génétique des Sordariacées. *Rev Cytol Biol Veg* **19**: 1–235.

Hildreth PE (1965) Doublesex, recessive gene that transforms both males and females of *Drosophila* into intersexes. *Genetics* **51**: 659–678.

Hill WG, Robertson A (1966) The effect of linkage on limits to artificial selection. *Genet Res* **8**: 269–294.

Hobbs JP, Munday PL, Jones GP (2004) Social induction of maturation and sex determination in a coral reef fish. *Proc Biol Sci* **271**: 2109–2114.

Hodgkin J (2002) Exploring the envelope. Systematic alteration in the sex-determination system of the nematode *Caenorhabditis elegans*. *Genetics* **162**: 767–780.

Hoekstra RF (1982) On the asymmetry of sex: Evolution of mating types in isogamous populations. *J Theor Biol* **98**: 427–451.

Hoekstra HE, Hoekstra JM (2001) An unusual sex-determination system in South American field mice (genus *Akodon*): The role of mutation, selection, and meiotic drive in maintaining XY females. *Evolution* **55**: 190–197.

Hoffmann AA, Rieseberg LH (2008) Revisiting the impact of inversions in evolution: from population genetic markers to drivers of adaptive shifts and speciation? *Annu Rev Ecol Evol Syst* **39**: 21–42.

Holliday R (1990) DNA methylation and epigenetic inheritance. *Philos Trans R Soc Lond B Biol Sci* **326**: 329–338.

Holmes AM, Kaykov A, Arcangioli B (2005) Molecular and cellular dissection of mating-type switching steps in *Schizosaccharomyces pombe*. *Mol Cell Biol* **25**: 303–311.

Hood ME, Petit E, Giraud T (2013) Extensive divergence between mating-type chromosomes of the anther-smut fungus. *Genetics* **193**: 309–315.

Hori T, Asakawa S, Itoh Y, Shimizu N et al. (2000) *Wpkci*, encoding an altered form of *PKCI*, is conserved widely on the avian W chromosome and expressed in early female embryos: implication of its role in female sex determination. *Mol Biol Cell* **11**: 3645–3660.

Hornett EA, Charlat S, Duplouy AM, Davies N et al. (2006) Evolution of male-killer suppression in a natural population. *PLoS Biol* **4**: e283.

Hoshijima K, Inoue K, Higuchi I, Sakamoto H et al. (1991) Control of doublesex alternative splicing by transformer and transformer-2 in *Drosophila*. *Science* **252**: 833–836.

Howe K, Clark MD, Torroja CF, Torrance J et al. (2013) The zebrafish reference genome sequence and its relationship to the human genome. *Nature* **496**: 498–503.

Hughes JS, Otto SP (1999) Ecology and the evolution of biphasic life cycles. *Am Nat* **154**: 306–320.

Hughes JF, Skaletsky H, Brown LG, Pyntikova T et al. (2012) Strict evolutionary conservation followed rapid gene loss on human and rhesus Y chromosomes. *Nature* **483**: 82–86.

Hughes JF, Skaletsky H, Pyntikova T, Graves TA et al. (2010) Chimpanzee and human Y chromosomes are remarkably divergent in structure and gene content. *Nature* **463**: 536–539.

Huigens M, Stouthamer R (2003) Parthenogenesis associated with *Wolbachia*. In: Bourtzis K, Miller TA. eds; *Insect symbiosis*. CRC Press, Boca Raton, FL. pp. 247–266.

Hurst LD (1995) Selfish genetic elements and their role in evolution: the evolution of sex and some of what that entails. *Philos Trans R Soc Lond B Biol Sci* **349**: 321–332.

Hurst LD, Atlan A, Bengtsson BO (1996) Genetic conflicts. *Q Rev Biol* **71**: 317–364.

Hurst LD, Hamilton WD (1992) Cytoplasmic fusion and the nature of sexes. *Proc Biol Sci* **247**: 189–194.

Hurst LD, Pomiankowski A (1991) Causes of sex ratio bias may account for unisexual sterility in hybrids: a new explanation of Haldane's rule and related phenomena. *Genetics* **128**: 841–858.

Hurst G, Hurst L, Majerus M (1997) Cytoplasmic sex ratio distorters. In: O'Neill SL, Hoffmann AA, Werren JH. eds; *Influential passengers*. Oxford University Press, Oxford. pp. 125–154.

Husby A, Schielzeth H, Forstmeier W, Gustafsson L et al. (2013) Sex chromosome linked genetic variance and the evolution of sexual dimorphism of quantitative traits. *Evolution* **67**: 609–619.

Idnurm A, Walton FJ, Floyd A, Heitman J (2008) Identification of the sex genes in an early diverged fungus. *Nature* **451**: 193–196.

Igic B, Lande R, Kohn JR (2008) Loss of self-incompatibility and its evolutionary consequences. *Int J Plant Sci* **169**: 93–104.

Inderbitzin P, Harkness J, Turgeon BG, Berbee ML (2005) Lateral transfer of mating system in *Stemphylium*. *Proc Natl Acad Sci U S A* **102**: 11390–11395.

Inderbitzin P, Shoemaker RA, O'Neill NR, Turgeon BG et al. (2006) Systematics and mating systems of two fungal pathogens of opium poppy: the heterothallic *Crivellia papaveracea* with a *Brachycladium penicillatum* asexual state and a homothallic species with a *Brachycladium papaveris* asexual state. *Can J Bot* **84**: 1304–1326.

Innocenti P, Morrow EH (2010) The sexually antagonistic genes of *Drosophila melanogaster*. *PLoS Biol* **8**: e1000335.

Inoue H, Fukumori Y, Hiroyoshi T (1983) Mapping of autosomal male-determining factors of the housefly, *Musca domestica* L., by means of sex-reversal. *Jap J Genet* **58**: 451–461.

Inoue H, Hiroyoshi T (1986) A maternal-effect sex-transformation mutant of the housefly, *Musca domestica* L. *Genetics* **112**: 469–482.

Inoue K, Hoshijima K, Sakamoto H, Shimura Y (1990) Binding of the *Drosophila* sex-lethal gene product to the alternative splice site of transformer primary transcript. *Nature* **344**: 461–463.

Irish EE (1996) Regulation of sex determination in maize. *Bioessays* **18**: 363–369.

Irish EE, Nelson T (1989) Sex determination in monoecious and dioecious plants. *Plant Cell* **1**: 737–744.

Ironside JE (2010) No amicable divorce? Challenging the notion that sexual antagonism drives sex chromosome evolution. *Bioessays* **32**: 718–726.

Isberg SR, Johnston SM, Chen Y, Moran C (2006) First evidence of higher female recombination in a species with temperature dependent sex determination: the saltwater crocodile. *J. Hered* **97**: 599–602.

Ishizaki K, Shimizu-Ueda Y, Okada S, Yamamoto M et al. (2002) Multicopy genes uniquely amplified in the Y chromosome-specific repeats of the liverwort *Marchantia polymorpha*. *Nucleic Acids Res* **30**: 4675–4681.

Itoh Y, Melamed E, Yang X, Kampf K et al. (2007) Dosage compensation is less effective in birds than in mammals. *J Biol* **6**: 2.

Itoh Y, Replogle K, Kim YH, Wade J et al. (2010) Sex bias and dosage compensation in the zebra finch versus chicken genomes: general and specialized patterns among birds. *Genome Res* **20**: 512–518.

Iwase M, Satta Y, Hirai Y, Hirai H et al. (2003) The amelogenin loci span an ancient pseudoautosomal boundary in diverse mammalian species. *Proc Natl Acad Sci U S A* **100**: 5258–5263.

Iyengar VK, Reeve HK (2010) Z linkage of female promiscuity genes in the moth *Utetheisa ornatrix*: support for the sexy-sperm hypothesis? *Evolution* **64**: 1267–1272.

Iyengar VK, Reeve HK, Eisner T (2002) Paternal inheritance of a female moth's mating preference. *Nature* **419**: 830–832.

Jaenike J (1996) Sex-ratio meiotic drive in the *Drosophila quinaria* group. *Am Nat* **148**: 237–254.

Jaenike J (2001) Sex chromosome meiotic drive. *Annu Rev Ecol Syst* **32**: 25–49.

Jahnke M, Smith JE, Dubuffet A, Dunn AM (2013) Effects of feminizing microsporidia on the masculinizing function of the androgenic gland in *Gammarus duebeni*. *J Invertebr Pathol* **112**: 146–151.

Jakob S, Lovell-Badge R (2011) Sex determination and the control of *Sox9* expression in mammals. *FEBS J* **278**: 1002–1009.

James WH (2010) The sex ratios of monozygotic and dizygotic twins. *Twin Res Hum Genet* **13**: 381–382.

Janoušek V, Wang L, Luzynski K, Dufková P et al. (2012) Genome-wide architecture of reproductive isolation in a naturally occurring hybrid zone between *Mus musculus musculus* and *M. m. domesticus*. *Mol Ecol* **21**: 3032–3047.

Janzen FJ (1992) Heritable variation for sex ratio under environmental sex determination in the common snapping turtle (*Chelydra serpentina*). *Genetics* **131**: 155–161.

Janzen FJ (1994) Climate change and temperature-dependent sex determination in reptiles. *Proc Natl Acad Sci U S A* **91**: 7487–7490.

Janzen FJ, Krenz JG (2004) Phylogenetics: Which was first, TSD or GSD?. In: Valenzuela N, Lance VA. eds. *Temperature dependent sex determination in vertebrates*. Smithsonian Books, Washington. pp. 121–130.

Jaquiéry J, Stoeckel S, Rispe C, Mieuzet L et al. (2012) Accelerated evolution of sex chromosomes in aphids, an X0 system. *Mol Biol Evol* **29**: 837–847.

Javaux EJ, Knoll AH, Walter MR (2001) Morphological and ecological complexity in early eukaryotic ecosystems. *Nature* **412**: 66–69.

Johnson NA (2010) Hybrid incompatibility genes: remnants of a genomic battlefield? *Trends Genet* **26**: 317–325.

Johnson NA, Hollocher H, Noonburg E, Wu CI (1993) The effects of interspecific Y chromosome replacements on hybrid sterility within the *Drosophila simulans* clade. *Genetics* **135**: 443–453.

Johnson SG, Howard RS (2007) Contrasting patterns of synonymous and nonsynonymous sequence evolution in asexual and sexual freshwater snail lineages. *Evolution* **61**: 2728–2735.

Johnson NA, Perez DE, Cabot EL, Hollocher H et al. (1992) A test of reciprocal X-Y interactions as a cause of hybrid sterility in *Drosophila*. *Nature* **358**: 751–753.

Jokela J, Dybdahl MF, Lively CM (2009) The maintenance of sex, clonal dynamics, and host-parasite coevolution in a mixed population of sexual and asexual snails. *Am Nat* **174** Suppl 1: S43–53.

Jones WT (1982) Sex ratio and host size in a parasitoid wasp. *Behav Ecol Sociobiol* **10**: 207–210.

Joshi A, Moody ME (1998) The cost of sex revisited: effects of male gamete output of hermaphrodites that are asexual in their female capacity. *J Theor Biol* **195**: 533–542.

Juchault P, Rigaud T (1995) Evidence for female heterogamety in two terrestrial crustaceans and the problem of sex chromosome evolution in isopods. *Heredity* **75**: 466–471.

Judelson HS (2007) Sexual reproduction in plant pathogenic oomycetes: biology and impact on disease. In: Heitman, J, Kronstad, J, Casselton, L, et al. eds; Sex in fungi: Molecular determination and evolutionary implications. ASM Press, Washington. pp. 445–458.

Judelson HS (2009) Sexual reproduction in Oomycetes: biology, diversity and contributions to fitness. In: Lamour K, Kamoun S. eds. *Oomycete genetics and genomics: Diversity, interactions and research tools*. Wiley, Chichester. pp. 121–138.

Judelson HS, Spielman LJ, Shattock RC (1995) Genetic mapping and non-Mendelian segregation of mating type loci in the oomycete, *Phytophthora infestans*. *Genetics* **141**: 503–512.

Judson OP, Normark BB (1996) Ancient asexual scandals. *Trends Ecol Evol* **11**: 41–46.

Julien P, Brawand D, Soumillon M, Necsulea A et al. (2012) Mechanisms and evolutionary patterns of mammalian and avian dosage compensation. *PLoS Biol* **10**: e1001328.

Just W, Baumstark A, Hameister H, Schreiner B *et al.* (2002) The sex determination in *Ellobius lutescens* remains bizarre. *Cytogenet Genome Res* **96**: 146–153.

Just W, Rau W, Vogel W, Akhverdian M et al. (1995) Absence of *Sry* in species of the vole *Ellobius*. *Nat Genet* **11**: 117–118.

Kaiser VB, Charlesworth B (2009) The effects of deleterious mutations on evolution in non-recombining genomes. *Trends Genet* **25**: 9–12.

Kaiser VB, Charlesworth B (2010) Muller's ratchet and the degeneration of the *Drosophila miranda* neo-Y chromosome. *Genetics* **185**: 339–348.

Kaiser VB, Ellegren H (2006) Nonrandom distribution of genes with sex-biased expression in the chicken genome. *Evolution* **60**: 1945–1951.

Kaiser VB, Zhou Q, Bachtrog D (2011) Nonrandom gene loss from the *Drosophila miranda* neo-Y chromosome. *Genome Biol Evol* **3**: 1329–1337.

Kallman K (1965) Genetics and geography of sex determination in the poeciliid fish, *Xiphophorus maculatus*. *Zoologica* **50**: 151–190.

Kallman K (1970) Sex determination and the restriction of sex-linked pigment patterns to the X and Y chromosomes in populations of a poeciliid fish, *Xiphophorus maculatus*, from the Belize and Sibun rivers of British Honduras. *Zoologica* **55**: 1–16.

Kallman KD (1975) The platyfish, *Xiphophorus maculatus*. In: King RC. ed. *Handbook of Genetics*. Plenum Press, New York. pp. 81–132.

Kamiya T, Kai W, Tasumi S, Oka A et al. (2012) A trans-species missense SNP in Amhr2 is associated with sex determination in the tiger pufferfish, *Takifugu rubripes* (fugu). *PLoS Genet* **8**: e1002798.

Kämper J, Reichmann M, Romeis T, Bolker M et al. (1995) Multiallelic recognition: nonself-dependent dimerization of the bE and bW homeodomain proteins in *Ustilago maydis*. *Cell* **81**: 73–83.

Karlin S, Lessard S (1986) Theoretical studies on sex ratio evolution. *Monogr Popul Biol* **22**: 1–313.

Kato Y, Kobayashi K, Oda S, Tatarazako N et al. (2010) Sequence divergence and expression of a transformer gene in the branchiopod crustacean, *Daphnia magna*. *Genomics* **95**: 160–165.

Kato Y, Kobayashi K, Watanabe H, Iguchi T (2011) Environmental sex determination in the branchiopod crustacean *Daphnia magna*: deep conservation of a *Doublesex* gene in the sex-determining pathway. *PLoS Genet* **7**: e1001345.

Katsura Y, Satta Y (2012) No evidence for a second evolutionary stratum during the early evolution of mammalian sex chromosomes. *PLoS One* **7**: e45488.

Kawai A, Ishijima J, Nishida C, Kosaka A *et al.* (2009) The ZW sex chromosomes of *Gekko hokouensis* (Gekkonidae, Squamata) represent highly conserved homology with those of avian species. *Chromosoma* **118**: 43–51.

Kawano S, Anderson RW, Nanba T, Kuroiwa T (1987) Polymorphism and uniparental inheritance of mitochondrial DNA in *Physarum polycephalum*. *J Gen Microbiol* **133**: 3175–3182.

Kazancioğlu E, Alonzo SH (2010) Classic predictions about sex change do not hold under all types of size advantage. *J Evol Biol* **23**: 2432–2441.

Keeling PJ, Burger G, Durnford DG, Lang BF et al. (2005) The tree of eukaryotes. *Trends Ecol Evol* **20**: 670–676.

Kenigsbuch D, Cohen Y (1990) The inheritance of gynoecy in muskmelon. *Genome* **33**: 317–320.

Kennison JA (1981) The genetic and cytological organization of the Y chromosome of *Drosophila melanogaster*. *Genetics* **98**: 529–548.

Kerr R (1970) Inheritance of DDT resistance in a laboratory colony of the housefly, *Musca domestica*. *Aust J Biol Sci* **23**: 377–400.

Kettlewell JR, Raymond CS, Zarkower D (2000) Temperature-dependent expression of turtle *Dmrt1* prior to sexual differentiation. *Genesis* **26**: 174–178.

Khaitovich P, Hellmann I, Enard W, Nowick K et al. (2005) Parallel patterns of evolution in the genomes and transcriptomes of humans and chimpanzees. *Science* **309**: 1850–1854.

Khil PP, Smirnova NA, Romanienko PJ, Camerini-Otero RD (2004) The mouse X chromosome is enriched for sex-biased genes not subject to selection by meiotic sex chromosome inactivation. *Nat Genet* **36**: 642–646.

Kidwell MG (1972) Genetic change of recombination value in *Drosophila melanogaster*. I. Artificial selection for high and low recombination and some properties of recombination-modifying genes. *Genetics* **70**: 419–432.

Kim Y, Capel B (2006) Balancing the bipotential gonad between alternative organ fates: a new perspective on an old problem. *Dev Dyn* **235**: 2292–2300.

Kim Y, Kobayashi A, Sekido R, DiNapoli L et al. (2006) *Fgf9* and *Wnt4* act as antagonistic signals to regulate mammalian sex determination. *PLoS Biol* **4**: e187.

Kimmerer RW (1991) Reproductive ecology of *Tetraphis pellucida*. I. Population density and reproductive mode. *Bryologist* **94**: 255–260.

Kimura M, Maruyama T (1966) The mutational load with epistatic gene interactions in fitness. *Genetics* **54**: 1337–1351.

Kirkpatrick M (1982) Sexual selection and the evolution of female choice. *Evolution* **36**: 1–12.

Kirkpatrick M (2010) How and why chromosome inversions evolve. *PLoS Biol* **8**: e1000501.

Kirkpatrick M, Hall DW (2004) Sexual selection and sex linkage. *Evolution* **58**: 683–691.

Kitano J, Peichel C (2012) Turnover of sex chromosomes and speciation in fishes. *Environ Biol Fishes* **94**: 549–558.

Kitano J, Kawagishi Y, Mori S, Peichel CL *et al.* (2011) Divergence in sex steroid hormone signaling between sympatric species of Japanese threespine stickleback. *PLoS One* **6**: e29253.

Kitano J, Ross JA, Mori S, Kume M et al. (2009) A role for a neo-sex chromosome in stickleback speciation. *Nature* **461**: 1079–1083.

Kitano T, Takamune K, Nagahama Y, Abe SI (2000) Aromatase inhibitor and 17alpha-methyltestosterone cause sex-reversal from genetical females to phenotypic males and suppression of P450 aromatase gene expression in Japanese flounder (*Paralichthys olivaceus*). *Mol Reprod Dev* **56**: 1–5.

Knowlton N (1974) A note on the evolution of gamete dimorphism. *J Theor Biol* **46**: 283–285.

Ko W (1988) Hormonal heterothallism and homothallism in *Phytophthora*. *Annu Rev Phytopathol* **26**: 57–73.

Kobayashi A, Tanaka Y, Shimada M (2003) Genetic variation of sex allocation in the parasitoid wasp *Heterospilus prosopidis*. *Evolution* **57**: 2659–2664.

Kobayashi T, Matsuda M, Kajiura-Kobayashi H, Suzuki A et al. (2004) Two DM domain genes, DMY and DMRT1, involved in testicular differentiation and development in the medaka, *Oryzias latipes*. *Dev Dyn* **231**: 518–526.

Koerich LB, Wang X, Clark AG, Carvalho AB (2008) Low conservation of gene content in the *Drosophila* Y chromosome. *Nature* **456**: 949–951.

Koevoets T, Beukeboom LW (2009) Genetics of postzygotic isolation and Haldane's rule in haplodiploids. *Heredity* **102**: 16–23.

Komdeur J (1996) Facultative sex ratio bias in the offspring of Seychelles warblers. *Proc R Soc Lond B Biol Sci* **263**: 661–666.

Komdeur J, Magrath MJ, Krackow S (2002) Pre-ovulation control of hatchling sex ratio in the Seychelles warbler. *Proc R Soc Lond B Biol Sci* **269**: 1067–1072.

Komdeur J, Daan S, Tinbergen J, Mateman C (1997) Extreme adaptive modification in sex ratio of the Seychelles warbler's eggs. *Nature* **385**: 522–525.

Komdeur J, Pen I (2002) Adaptive sex allocation in birds: the complexities of linking theory and practice. *Philos Trans R Soc Lond B Biol Sci* **357**: 373–380.

Komen J, De Boer P, Richter C (1992a) Male sex reversal in gynogenetic XX females of common carp (*Cyprinus carpio* L.) by a recessive mutation in a sex-determining gene. *J Hered* **83**: 431–434.

Komen J, Wiegertjes G, Van Ginneken V, Eding E et al. (1992b) Gynogenesis in common carp (*Cyprinus carpio* L.). III. The effects of inbreeding on gonadal development of heterozygous and homozygous gynogenetic offspring. *Aquaculture* **104**: 51–66.

Komen J, Yamashita M, Nagahama Y (1992c) Testicular development induced by a recessive mutation during gonadal differentiation of female common carp (*Cyprinus carpio*, L.). *Dev Growth Differ* **34**: 535–544.

Kondo M, Hornung U, Nanda I, Imai S et al. (2006) Genomic organization of the sex-determining and adjacent regions of the sex chromosomes of medaka. *Genome Res* **16**: 815–826.

Kondo M, Nagao E, Mitani H, Shima A (2001) Differences in recombination frequencies during female and male meioses of the sex chromosomes of the medaka, *Oryzias latipes*. *Genet Res* **78**: 23–30.

Kondo M, Nanda I, Hornung U, Asakawa S et al. (2003) Absence of the candidate male sex-determining gene *dmrt1b(Y)* of medaka from other fish species. *Curr Biol* **13**: 416–420.

Kondo M, Nanda I, Hornung U, Schmid M et al. (2004) Evolutionary origin of the medaka Y chromosome. *Curr Biol* **14**: 1664–1669.

Kondrashov AS, Crow JF (1991) Haploidy or diploidy: which is better? *Nature* **351**: 314–315.

Koopman P, Gubbay J, Vivian N, Goodfellow P et al. (1991) Male development of chromosomally female mice transgenic for *Sry*. *Nature* **351**: 117–121.

Kopp A (2012) Dmrt genes in the development and evolution of sexual dimorphism. *Trends Genet* **28**: 175–184.

Korol AB, Iliadi KG (1994) Increased recombination frequencies resulting from directional selection for geotaxis in *Drosophila*. *Heredity* **72 (Pt 1)**: 64–68.

Korpelainen H (1998) Labile sex expression in plants. *Biol Rev* **73**: 157–180.

Korpelainen H (1994) Growth, sex determination and reproduction of *Dryopteris filix-mas* (L.) Schott gametophytes under varying nutritional conditions. *Bot J Linn Soc* **114**: 357–366.

Kozielska MA (2008) *Evolutionary dynamics of sex determination. Mechanistic theory and empirical investigations. PhD thesis*. University of Groningen.

Kozielska MA, Feldmeyer B, Pen I, Weissing FJ et al. (2008) Are autosomal sex-determining factors of the housefly (*Musca domestica*) spreading north? *Genet Res (Camb)* **90**: 157–165.

Kozielska M, Pen I, Beukeboom LW, Weissing FJ (2006) Sex ratio selection and multi-factorial sex determination in the housefly: a dynamic model. *J Evol Biol* **19**: 879–888.

Kozielska M, Weissing FJ, Beukeboom LW, Pen I (2010) Segregation distortion and the evolution of sex-determining mechanisms. *Heredity* **104**: 100–112.

Kraak SB, Pen I (2002) Sex-determining mechanisms in vertebrates. In: Hardy ICW. ed. *Sex ratios: Concepts and research methods*. Cambridge University Press, Cambridge. pp. 158–177.

Kraak SBM, De Looze EMA (1993) A new hypothesis on the evolution of sex determination in vertebrates; big females ZW, big males XY. *Neth J Zool* **43**: 260–273.

Kraemer C, Schmidt ER (1993) The sex determining region of *Chironomus thummi* is associated with highly repetitive DNA and transposable elements. *Chromosoma* **102**: 553–562.

Krizek BA, Fletcher JC (2005) Molecular mechanisms of flower development: an armchair guide. *Nature Rev Gen* **6**: 688–698.

Kronforst MR, Young LG, Kapan DD, McNeely C et al. (2006) Linkage of butterfly mate preference and wing color preference cue at the genomic location of wingless. *Proc Natl Acad Sci U S A* **103**: 6575–6580.

Kroon FJ, Munday PL, Westcott DA, Hobbs JA et al. (2005) Aromatase pathway mediates sex change in each direction. *Proc Biol Sci* **272**: 1399–1405.

Kües U, Casselton LA (1993) The origin of multiple mating types in mushrooms. *J Cell Sci* **104**: 227–230.

Kuhn G, Hijri M, Sanders IR (2001) Evidence for the evolution of multiple genomes in arbuscular mycorrhizal fungi. *Nature* **414**: 745–748.

Kuijper B, Pen I (2010) The evolution of haplodiploidy by male-killing endosymbionts: importance of population structure and endosymbiont mutualisms. *J Evol Biol* **23**: 40–52.

Kulathinal RJ, Skwarek L, Morton RA, Singh RS (2003) Rapid evolution of the sex-determining gene, transformer: structural diversity and rate heterogeneity among sibling species of *Drosophila*. *Mol Biol Evol* **20**: 441–452.

Kupper C, Augustin J, Edwards S, Szekely T et al. (2012) Triploid plover female provides support for a role of the W chromosome in avian sex determination. *Biol Lett* **8**: 787–789.

Kwon JY, Haghpanah V, Kogson-Hurtado LM, McAndrew BJ et al. (2000) Masculinization of genetic female nile tilapia (*Oreochromis niloticus*) by dietary administration of an aromatase inhibitor during sexual differentiation. *J Exp Zool* **287**: 46–53.

Labarère J, Noël T (1992) Mating type switching in the tetrapolar basidiomycete *Agrocybe aegerita*. *Genetics* **131**: 307–319.

Lagomarsino IV, Conover DO (1993) Variation in environmental and genotypic sex-determining mechanisms across a latitudinal gradient in the fish, *Menidia menidia*. *Evolution* **47**: 487–494.

Lahn BT, Page DC (1999) Four evolutionary strata on the human X chromosome. *Science* **286**: 964–967.

Lahr DJ, Parfrey LW, Mitchell EA, Katz LA et al. (2011) The chastity of amoebae: re-evaluating evidence for sex in amoeboid organisms. *Proc Biol Sci* **278**: 2081–2090.

Lande R (1994) Risk of population extinction from fixation of new deleterious mutations. *Evolution* **48**: 1460–1469.

Lande R, Seehausen O, van Alphen JJ (2001) Mechanisms of rapid sympatric speciation by sex reversal and sexual selection in cichlid fish. *Genetica* **112–113**: 435–443.

Lang JW, Andrews HV (1994) Temperature-dependent sex determination in crocodilians. *J Exp Zool* **270**: 28–44.

Lange J, Skaletsky H, van Daalen SK, Embry SL *et al.* (2009) Isodicentric Y chromosomes and sex disorders as byproducts of homologous recombination that maintains palindromes. *Cell* **138**: 855–869.

Lawrence C, Ebersole J, Kesseli R (2008) Rapid growth and out-crossing promote female development in zebrafish (*Danio rerio*). *Environ Biol Fishes* **81**: 239–246.

Leder EH, Cano JM, Leinonen T, O'Hara RB et al. (2010) Female-biased expression on the X chromosome as a key step in sex chromosome evolution in threespine sticklebacks. *Mol Biol Evol* **27**: 1495–1503.

Lee N, Bakkeren G, Wong K, Sherwood JE et al. (1999) The mating-type and pathogenicity locus of the fungus *Ustilago hordei* spans a 500-kb region. *Proc Natl Acad Sci U S A* **96**: 15026–15031.

Lee SC, Corradi N, Doan S, Dietrich FS et al. (2010a) Evolution of the *sex*-related locus and genomic features shared in microsporidia and fungi. *PLoS One* **5**: e10539.

Lee SC, Ni M, Li W, Shertz C et al. (2010b) The evolution of sex: a perspective from the fungal kingdom. *Microbiol Mol Biol* **R74**: 298–340.

Lee B, Hulata G, Kocher TD (2004) Two unlinked loci controlling the sex of blue tilapia (*Oreochromis aureus*). *Heredity* **92**: 543–549.

Lee SC, Corradi N, Byrnes EJ, Torres-Martinez S et al. (2008) Microsporidia evolved from ancestral sexual fungi. *Curr Biol* **18**: 1675–1679.

Lehtonen J, Jennions MD, Kokko H (2012) The many costs of sex. *Trends Ecol Evol* **27**: 172–178.

Lehtonen J, Kokko H (2011) Two roads to two sexes: unifying gamete competition and gamete limitation in a single model of anisogamy evolution. *Behav Ecol Sociobiol* **65**: 445–459.

Lemmon AR, Kirkpatrick M (2006) Reinforcement and the genetics of hybrid incompatibilities. *Genetics* **173**: 1145–1155.

Lemos B, Araripe LO, Hartl DL (2008) Polymorphic Y chromosomes harbor cryptic variation with manifold functional consequences. *Science* **319**: 91–93.

Lengeler KB, Fox DS, Fraser JA, Allen A et al. (2002) Mating-type locus of *Cryptococcus neoformans*: a step in the evolution of sex chromosomes. *Eukaryot Cell* **1**: 704–718.

Lenormand T (2003) The evolution of sex dimorphism in recombination. *Genetics* **163**: 811–822.

Lenormand T, Dutheil J (2005) Recombination difference between sexes: a role for haploid selection. *PLoS Biol* **3**: e63.

Lercher MJ, Hurst LD (2003) Imprinted chromosomal regions of the human genome have unusually high recombination rates. *Genetics* **165**: 1629–1632.

Lercher MJ, Urrutia AO, Hurst LD (2003) Evidence that the human X chromosome is enriched for male-specific but not female-specific genes. *Mol Biol Evol* **20**: 1113–1116.

Levialdi Ghiron JH, Amato A, Montresor M, Kooistra WHCF (2008) Plastid inheritance in the planktonic raphid pennate diatom *Pseudo-nitzschia delicatissima* (Bacillariophyceae). *Protist* **159**: 91–98.

Lewis D (1941) Male sterility in natural populations of hermaphrodite plants the equilibrium between females and hermaphrodites to be expected with different types of inheritance. *New Phytol* **40**: 56–63.

Lewis Jr WM (1983) Interruption of synthesis as a cost of sex in small organisms. *Am Nat* **121**: 825–833.

Lewis D, Jones D (1992) The genetics of heterostyly. In: Barrett, SCH ed. *Evolution and function of heterostyly*. Springer Verlag, Berlin. pp. 129–150.

Lewis LA, McCourt RM (2004) Green algae and the origin of land plants. *Am J Bot* **91**: 1535–1556.

Li J, Phillips R, Harwood A, Koop B et al. (2011) Identification of the sex chromosomes of brown trout (*Salmo trutta*) and their comparison with the corresponding chromosomes in Atlantic salmon (*Salmo salar*) and rainbow trout (*Oncorhynchus mykiss*). *Cytogenet Genome Res* **133**: 25–33.

Li B, Zhang W, Chan G, Jancso-Radek A et al. (2001) Human sex reversal due to impaired nuclear localization of SRY. A clinical correlation. *J Biol Chem* **276**: 46480–46484.

Liew WC, Bartfai R, Lim Z, Sreenivasan R et al. (2012) Polygenic sex determination system in zebrafish. *PLoS One* **7**: e34397.

Lifschytz E, Lindsley DL (1972) The role of X-chromosome inactivation during spermatogenesis. *Proc Natl Acad Sci U S A* **69**: 182–186.

Limousin D, Streiff R, Courtois B, Dupuy V et al. (2012) Genetic architecture of sexual selection: QTL mapping of male song and female receiver traits in an acoustic moth. *PLoS One* **7**: e44554.

Lin X, Hull CM, Heitman J (2005) Sexual reproduction between partners of the same mating type in *Cryptococcus neoformans*. *Nature* **434**: 1017–1021.

Lindholm A, Breden F (2002) Sex chromosomes and sexual selection in poeciliid fishes. *Am Nat* **160**: 214–224.

Liu Z, Mara C (2010) Regulatory mechanisms for floral homeotic gene expression. *Semin Cell Dev Biol* **21**: 80–86.

Liu Z, Moore PH, Ma H, Ackerman CM et al. (2004) A primitive Y chromosome in papaya marks incipient sex chromosome evolution. *Nature* **427**: 348–352.

Lloyd DG (1972) Breeding systems in *Cotula* L. (Compositae, Anthemideae). *New Phytol* **71**: 1181–1194.

Lu J, Wu C (2005) Weak selection revealed by the whole-genome comparison of the X chromosome and autosomes of human and chimpanzee. *Proc Natl Acad Sci U S A* **102**: 4063–4067.

Lubchenco J, Cubit J (1980) Heteromorphic life histories of certain marine algae as adaptations to variations in herbivory. *Ecology* **61**: 676–687.

Luijckx P, Fienberg H, Duneau D, Ebert D (2013) A matching-allele model explains host resistance to parasites. *Curr Biol* **23**: 1085–1088.

Lynn A, Schrump S, Cherry J, Hassold T et al. (2005) Sex, not genotype, determines recombination levels in mice. *Am J Hum Genet* **77**: 670–675.

Lyon MF (1961) Gene action in the X-chromosome of the mouse (*Mus musculus* L.). *Nature* **190**: 372–373.

Lyon M (1998) X-chromosome inactivation: a repeat hypothesis. *Cytogenet Cell Genet* **80**: 133–137.

Lyttle TW (1981) Experimental population genetics of meiotic drive systems. III. Neutralization of sex-ratio distortion in *Drosophila* through sex-chromosome aneuploidy. *Genetics* **98**: 317–334.

Mable BK (2004) 'Why polyploidy is rarer in animals than in plants': myths and mechanisms. *Biol J Linn Soc* **82**: 453–466.

Mable BK, Otto SP (1998) The evolution of life cycles with haploid and diploid phases. *Bioessays* **20**: 453–462.

MacCarthy T, Seymour RM, Pomiankowski A (2010) Differential regulation drives plasticity in sex determination gene networks. *BMC Evol Biol* **10**: 388–2148–10–388.

Macholán M (1996) Multivariate morphometric analysis of European species of the genus *Mus* (Mammalia, Muridae). *Zeitschrift Saugetierkunde* **61**: 304—319.

Macholán M, Baird SJE, Dufková P, Munclinger P et al. (2011) Assessing multilocus introgression patterns: a case study on the mouse X chromosome in central Europe. *Evolution* **65**: 1428–1446.

Macholán M, Munclinger P, Šugerková M, Dufková P et al. (2007) Genetic analysis of autosomal and X-linked markers across a mouse hybrid zone. *Evolution* **61**: 746–771.

Madhani HD, Fink GR (1998) The riddle of MAP kinase signaling specificity. *Trends Genet* **14**: 151–155.

Madsen JD, Waller DM (1983) A note on the evolution of gamete dimorphism in algae. *Am Nat* **121**: 443–447.

Magerhans A, Hörstgen-Schwark G (2010) Selection experiments to alter the sex ratio in rainbow trout (*Oncorhynchus mykiss*) by means of temperature treatment. *Aquaculture* **306**: 63–67.

Magnusson K, Lycett GJ, Mendes AM, Lynd A et al. (2012) Demasculinization of the *Anopheles gambiae* X chromosome. *BMC Evol Biol* **12**: 69.

Maher B (2012) ENCODE: The human encyclopaedia. *Nature* **489**: 46–48.

Mainx F (1964) The genetics of *Megaselia scalaris* Loew (Phoridae): A new type of sex determination in Diptera. *Am Nat*: **98**: 415–430.

Makino T, McLysaght A (2008) Interacting gene clusters and the evolution of the vertebrate immune system. *Mol Biol Evol* **25**: 1855–1862.

Malik S, Pightling AW, Stefaniak LM, Schurko AM *et al.* (2008) An expanded inventory of conserved meiotic genes provides evidence for sex in *Trichomonas vaginalis*. *PLoS One* **3**: e2879.

Malmberg RL (1977) The evolution of epistasis and the advantage of recombination in populations of bacteriophage T4. *Genetics* **86**: 607–621.

Malone JH, Michalak P (2008) Physiological sex predicts hybrid sterility regardless of genotype. *Science* **319**: 59–59.

Mank JE (2009) The W, X, Y and Z of sex-chromosome dosage compensation. *Trends Genet* **25**: 226–233.

Mank JE, Avise JC (2009) Evolutionary diversity and turnover of sex determination in teleost fishes. *Sex Dev* **3**: 60–67.

Mank JE, Axelsson E, Ellegren H (2007) Fast-X on the Z: Rapid evolution of sex-linked genes in birds. *Genome Research* **17**: 618–624.

Mank JE, Ellegren H (2009) Sex bias in gene expression is not the same as dosage compensation. *Heredity* **103**: 434–434.

Mank JE, Hall DW, Kirkpatrick M, Avise JC (2006) Sex chromosomes and male ornaments: a comparative evaluation in ray-finned fishes. *Proc Biol Sci* **273**: 233–236.

Mank JE, Hultin-Rosenberg L, Zwahlen M, Ellegren H (2008) Pleiotropic constraint hampers the resolution of sexual antagonism in vertebrate gene expression. *Am Nat* **171**: 35–43.

Mank JE, Vicoso B, Berlin S, Charlesworth B (2010) Effective population size and the Faster-X effect: empirical results and their interpretation. *Evolution* **64**: 663–674.

Marais GA, Campos PR, Gordo I (2010) Can intra-Y gene conversion oppose the degeneration of the human Y chromosome? A simulation study. *Genome Biol Evol* **2**: 347–357.

Marais G, Galtier N (2003) Sex chromosomes: how X-Y recombination stops. *Curr Biol* **13**: R641–643.

Marais GA, Nicolas M, Bergero R, Chambrier P et al. (2008) Evidence for degeneration of the Y chromosome in the dioecious plant *Silene latifolia*. *Curr Biol* **18**: 545–549.

Margulis L, Sagan D (1986) *Origins of sex*. Yale University Press, New Haven.

Mariani S, Sala-Bozano M, Chopelet J, Benvenuto C. (2013) Spatial and temporal patterns of size-at-sex-change in two exploited coastal fish. *Environ Biol Fish* **96**: 535–541.

Marin I, Franke A, Bashaw GJ, Baker BS (1996) The dosage compensation system of *Drosophila* is co-opted by newly evolved X chromosomes. *Nature* **383**: 160–163.

Marin I, Siegal ML, Baker BS (2000) The evolution of dosage-compensation mechanisms. *Bioessays* **22**: 1106–1114.

Martin OY, Goodacre SL (2009) Widespread infections by the bacterial endosymbiont *Cardinium* in arachnids. *J Arachnol* **37**: 106–108.

Martin A, Troadec C, Boualem A, Rajab M et al. (2009) A transposon-induced epigenetic change leads to sex determination in melon. *Nature* **461**: 1135–1138.

Martin T, Lu SW, van Tilbeurgh H, Ripoll DR et al. (2010) Tracing the origin of the fungal alpha1 domain places its ancestor in the HMG-box superfamily: implication for fungal mating-type evolution. *PLoS One* **5**: e15199.

Masly JP, Presgraves DC (2007) High-resolution genome-wide dissection of the two rules of speciation in *Drosophila*. *PLoS Biol* **5**: e243.

Matson CK, Murphy MW, Sarver AL, Griswold MD et al. (2011) DMRT1 prevents female reprogramming in the postnatal mammalian testis. *Nature* **476**: 101–104.

Matson CK, Zarkower D (2012) Sex and the singular DM domain: insights into sexual regulation, evolution and plasticity. *Nat Rev Genet* **13**: 163–174.

Matsuba C, Alho JS, Merilä J (2010) Recombination rate between sex chromosomes depends on phenotypic sex in the common frog. *Evolution* **64**: 3634–3637.

Matsubara K, Tarui H, Toriba M, Yamada K et al. (2006) Evidence for different origin of sex chromosomes in snakes, birds, and mammals and step-wise differentiation of snake sex chromosomes. *Proc Natl Acad Sci U S A* **103**: 18190–18195.

Matsuda M, Sotoyama S, Hamaguchi S, Sakaizumi M (1999) Male-specific restriction of recombination frequency in the sex chromosomes of the medaka, *Oryzias latipes*. *Genet Res* **73**: 225.

Matsuda M, Nagahama Y, Shinomiya A, Sato T et al. (2002) DMY is a Y-specific DM-domain gene required for male development in the medaka fish. *Nature* **417**: 559–563.

Maurice S, Belhassen E, Couvet D, Gouyon PH (1994) Evolution of dioecy: can nuclear-cytoplasmic interactions select for maleness? *Heredity* **73 (Pt 4)**: 346–354.

Maynard-Smith J (1978) *The Evolution of sex*. Cambridge University Press, Cambridge.

Maynard-Smith J (1982) *Evolution and the theory of games*. Cambridge University Press, Cambridge.

Maynard-Smith J, Haigh J (1974) The hitch-hiking effect of a favourable gene. *Genet Res* **23**: 23–35.

McAllister BF, McVean GA (2000) Neutral evolution of the sex-determining gene transformer in *Drosophila*. *Genetics* **154**: 1711–1720.

McClung CE (1902) The accessory chromosome—sex determinant? *Biol Bull* **3**: 43–84.

McClure B, Cruz-Garcia F, Romero C (2011) Compatibility and incompatibility in S-RNase-based systems. *Ann Bot* **108**: 647–658.

McDaniel SF, Willis JH, Shaw AJ (2007) A linkage map reveals a complex basis for segregation distortion in an interpopulation cross in the moss *Ceratodon purpureus*. *Genetics* **176**: 2489–2500.

McDermott SR, Noor MA (2010) The role of meiotic drive in hybrid male sterility. *Philos Trans R Soc Lond B Biol Sci* **365**: 1265–1272.

McGaugh SE, Janzen FJ (2011) Effective heritability of targets of sex-ratio selection under environmental sex determination. *J Evol Biol* **24**: 784–794.

McPhee CP, Robertson A (1970) The effect of suppressing crossing-over on the response to selection in *Drosophila melanogaster*. *Genet Res* **16**: 1–16.

McVean G, Hurst LD (1996) Genetic conflicts and the paradox of sex determination: three paths to the evolution of female intersexuality in a mammal. *J Theor Biol* **179**: 199–211.

Meagher TR (1988) Sex determination in plants. In: Lovett Doust J, Lovett Doust L. eds. *Plant reproductive ecology: patterns and strategies*. Oxford University Press, New York. pp. 125–138.

Meiklejohn CD, Tao Y (2010) Genetic conflict and sex chromosome evolution. *Trends Ecol Evol* **25**: 215–223.

Meiklejohn CD, Parsch J, Ranz JM, Hartl DL (2003) Rapid evolution of male-biased gene expression in *Drosophila*. *Proc Natl Acad Sci U S A* **100**: 9894–9899.

Meirmans PG, Den Nijs JC, Van Tienderen PH (2006) Male sterility in triploid dandelions: asexual females vs. asexual hermaphrodites. *Heredity* **96**: 45–52.

Meisel RP, Han MV, Hahn MW (2009) A complex suite of forces drives gene traffic from *Drosophila* X chromosomes. *Genome Biol Evol* **1**: 176–188.

Meisel RP, Malone JH, Clark AG (2012) Disentangling the relationship between sex-biased gene expression and X-linkage. *Genome Res* **22**: 1255–1265.

Menkis A, Jacobson DJ, Gustafsson T, Johannesson H (2008) The mating-type chromosome in the filamentous ascomycete *Neurospora tetrasperma* represents a model for early evolution of sex chromosomes. *PLoS Genet* **4**: e1000030.

Meyer BJ (2000) Sex in the worm: counting and compensating X-chromosome dose. *Trends Genet* **16**: 247–253.

Meyer BJ (2005) X-Chromosome dosage compensation. *WormBook*: doi/10.1895/wormbook.1.8.1, http://www.wormbook.org.

Michiels N, Kuhl A (2003) Altruistic sperm donation in a sperm-dependent parthenogenetic hermaphrodite is stabilized by reciprocal sperm exchange. *J Zool* **259**: 77–82.

Michod RE, Bernstein H, Nedelcu AM (2008) Adaptive value of sex in microbial pathogens. *Infect Genet Evol* **8**: 267–285.

Miller MG, Johnson AD (2002) White-opaque switching in *Candida albicans* is controlled by mating-type locus homeodomain proteins and allows efficient mating. *Cell* **110**: 293–302.

Miller SW, Hayward DC, Bunch TA, Miller DJ et al. (2003) A DM domain protein from a coral, *Acropora millepora*, homologous to proteins important for sex determination. *Evol Dev* **5**: 251–258.

Ming R, Yu Q, Moore PH (2007) Sex determination in papaya. *Semin Cell Dev Biol* **18**: 401–408.

Mitchell N, Nelson N, Cree A, Pledger S et al. (2006) Support for a rare pattern of temperature-dependent sex determination in archaic reptiles: evidence from two species of tuatara (*Sphenodon*). *Front Zool* **3**: 9.

Mittwoch U (1985) Erroneous theories of sex determination. *J Med Genet* **22**: 164–170.

Mittwoch U (1986) Males, females and hermaphrodites. An inaugural lecture delivered by Professor Ursula Mittwoch at University College London on 24 October 1985. *Ann Hum Genet* **50**: 103–121.

Mittwoch U (1989) Sex differentiation in mammals and tempo of growth: probabilities vs. switches. *J Theor Biol* **137**: 445–455.

Mittwoch U (1996) Sex-determining mechanisms in animals. *Trends Ecol Evol* **11**: 63–67.

Mittwoch U (2005) Sex determination in mythology and history. *Arq Bras Endocrinol Metabol* **49**: 7–13.

Miura I (2007) An evolutionary witness: the frog *Rana rugosa* underwent change of heterogametic sex from XY male to ZW female. *Sex Dev* **1**: 323–331.

Montchamp Moreau C, Ginhoux V, Atlan A (2001) The Y chromosomes of *Drosophila simulans* are highly polymorphic for their ability to suppress sex-ratio drive. *Evolution* **55**: 728–737.

Montchamp-Moreau C, Ogereau D, Chaminade N, Colard A *et al.* (2006) Organization of the sex-ratio meiotic drive region in *Drosophila simulans*. *Genetics* **174**: 1365–1371.

Moran NA, McCutcheon JP, Nakabachi A (2008) Genomics and evolution of heritable bacterial symbionts. *Annu Rev Genet* **42**: 165–190.

Morgan T (1914) No crossing over in the male of *Drosophila* of genes in the second and third pairs of chromosomes. *Biol Bull* **26**: 195–204.

Morgan T (1942) Cross-and self-fertilization in the ascidian *Styela*. *Biol Bull* **82**: 161–171.

Moritz C (1990) Patterns and processes of sex chromosome evolution in gekkonid lizards (Sauria: Reptilia). In: Olmo E., ed. *Cytogenetics of amphibians and reptiles*. Birkhäuser Verlag, Basel. pp. 205–219.

Moriyama Y, Kawano S (2010) Maternal inheritance of mitochondria: multipolarity, multiallelism and hierarchical transmission of mitochondrial DNA in the true slime mold *Physarum polycephalum*. *J Plant Res* **123**: 139–148.

Moriyama EN, Powell JR (1996) Intraspecific nuclear DNA variation in *Drosophila*. *Mol Biol Evol* **13**: 261–277.

Morran LT, Schmidt OG, Gelarden IA, Parrish RC, 2nd et al. (2011) Running with the Red Queen: host-parasite coevolution selects for biparental sex. *Science* **333**: 216–218.

Morris MR, Batra P, Ryan MJ (1992) Male-male competition and access to females in the swordtail *Xiphophorus nigrensis*. *Copeia* **1992**: 980–986.

Morrow EH, Stewart AD, Rice WR (2008) Assessing the extent of genome-wide intralocus sexual conflict via experimentally enforced gender-limited selection. *J Evol Biol* **21**: 1046–1054.

Moyle LC, Graham EB (2005) Genetics of hybrid incompatibility between *Lycopersicon esculentum* and *L. hirsutum*. *Genetics* **169**: 355–373.

Moyle LC, Muir CD, Han MV, Hahn MW (2010) The contribution of gene movement to the 'two rules of speciation'. *Evolution* **64**: 1541–1557.

Moyle LC, Nakazato T (2010) Hybrid incompatibility 'snowballs' between *Solanum* species. *Science* **329**: 1521–1523.

Mueller JL, Mahadevaiah SK, Park PJ, Warburton PE et al. (2008) The mouse X chromosome is enriched for multicopy testis genes showing postmeiotic expression. *Nat Genet* **40**: 794–799.

Muller HJ (1914) A factor for the fourth chromosome of *Drosophila*. *Science* **39**: 906.

Muller HJ (1918) Genetic variability, twin hybrids and constant hybrids, in a case of balanced lethal factors. *Genetics* **3**: 422.

Muller HJ (1925) Why polyploidy is rarer in animals than in plants. *Am Nat* **59**: 346–353.

Muller HJ (1932) Some genetic aspects of sex. *Am Nat* **66**: 118–138.

Muller HJ (1940) Bearings of the *Drosophila* work on systematics. In: Huxley JS. ed. *The new systematics*. Clarendon, Oxford. pp. 185–268.

Muller HJ (1942) Isolating mechanisms, evolution, and temperature. *Biol Symp* **6**: 71–125.

Myers S, Bowden R, Tumian A, Bontrop RE et al. (2010) Drive against hotspot motifs in primates implicates the PRDM9 gene in meiotic recombination. *Science* **327**: 876–879.

Myosho T, Otake H, Masuyama H, Matsuda M et al. (2012) Tracing the emergence of a novel sex-determining gene in medaka, *Oryzias luzonensis*. *Genetics* **191**: 163–170.

Näf U (1979) Antheridiogens and antheridial development. In: Dyer AF, ed. *The experimental biology of ferns*. Academic Press, New York. pp. 435–470.

Näf U, Nakanishi K, Endo M (1975) On the physiology and chemistry of fern antheridiogens. *Bot Rev* **41**: 315–359.

Nager RG, Monaghan P, Griffiths R, Houston DC et al. (1999) Experimental demonstration that offspring sex ratio varies with maternal condition. *Proc Natl Acad Sci U S A* **96**: 570–573.

Nakamura S, Aoyama H, van Woesik R (2003) Strict paternal transmission of mitochondrial DNA of *Chlamydomonas* species is explained by selection against maternal nucleoids. *Protoplasma* **221**: 205–210.

Nanda I, Kondo M, Hornung U, Asakawa S et al. (2002) A duplicated copy of DMRT1 in the sex-determining region of the Y chromosome of the medaka, *Oryzias latipes*. *Proc Natl Acad Sci U S A* **99**: 11778–11783.

Nasrallah JB (2005) Recognition and rejection of self in plant self-incompatibility: comparisons to animal histocompatibility. *Trends Immunol* **26**: 412–418.

Naurin S, Hansson B, Bensch S, Hasselquist D (2010) Why does dosage compensation differ between XY and ZW taxa? *Trends Genet* **26**: 15–20.

Naurin S, Hasselquist D, Bensch S, Hansson B (2012) Sex-biased gene expression on the avian Z chromosome: highly expressed genes show higher male-biased expression. *PLoS One* **7**: e46854.

Nauta MJ, Hoekstra RF (1992) Evolution of reproductive systems in filamentous ascomycetes. II. Evolution of hermaphroditism and other reproductive strategies. *Heredity* **68** (Pt 6): 537–546.

Navarro-Martin L, Vinas J, Ribas L, Diaz N et al. (2011) DNA methylation of the gonadal aromatase (cyp19a) promoter is involved in temperature-dependent sex ratio shifts in the European sea bass. *PLoS Genet* **7**: e1002447.

Naylor C, Adams J, Greenwood P (1988) Population dynamics and adaptive sexual strategies in a brackish water crustacean, *Gammarus duebeni*. *J Anim Ecol* **57**: 493–507.

Nei M (1969) Linkage modification and sex difference in recombination. *Genetics* **63**: 681–699.

Nguyen DK, Disteche CM (2006) High expression of the mammalian X chromosome in brain. *Brain Res* **1126**: 46–49.

Ni M, Feretzaki M, Sun S, Wang X *et al*. (2011) Sex in fungi. *Annu Rev Genet* **45**: 405–430.

Nicolas M, Marais G, Hykelova V, Janousek B et al. (2005) A gradual process of recombination restriction in the evolutionary history of the sex chromosomes in dioecious plants. *PLoS Biol* **3**: e4.

Nicoll M, Akerib CC, Meyer BJ (1997) X-chromosome-counting mechanisms that determine nematode sex. *Nature* **388**: 200–204.

Nielsen R, Bustamante C, Clark AG, Glanowski S et al. (2005) A scan for positively selected genes in the genomes of humans and chimpanzees. *PLoS Biol* **3**: e170.

Nielsen O, Friis T, Kjaerulff S (1996) The *Schizosaccharomyces pombe* map1 gene encodes an SRF/MCM1-related protein required for P-cell specific gene expression. *Mol Gen Genet* **253**: 387–392.

Niklas KJ, Kutschera U (2010) The evolution of the land plant life cycle. *New Phytol* **185**: 27–41.

Nishimura Y, Shikanai T, Nakamura S, Kawai-Yamada M et al. (2012) Gsp1 triggers the sexual developmental program including inheritance of chloroplast DNA and mitochondrial DNA in *Chlamydomonas reinhardtii*. *Plant Cell* **24**: 2401–2414.

Noel GR, Atibalentja N (2006) '*Candidatus Paenicardinium endonii*', an endosymbiont of the plant-parasitic nematode *Heterodera glycines* (Nemata: Tylenchida), affiliated to the phylum *Bacteroidetes*. *Int J Syst Evol Microbiol* **56**: 1697–1702.

Noor MAF, Grams KL, Bertucci LA, Reiland J (2001) Chromosomal inversions and the reproductive isolation of species. *Proc Natl Acad Sci U S A* **98**: 12084–12088.

Normark BB (2003) The evolution of alternative genetic systems in insects. *Annu Rev Entomol* **48**: 397–423.

Normark BB (2004) Haplodiploidy as an outcome of coevolution between male-killing cytoplasmic elements and their hosts. *Evolution* **58**: 790–798.

Nozaki H (2008) A new male-specific gene "OTOKOGI" in *Pleodorina starrii* (Volvocaceae, Chlorophyta) unveils the origin of male and female. *Biologia* **63**: 772–777.

Nozaki H, Mori T, Misumi O, Matsunaga S et al. (2006) Males evolved from the dominant isogametic mating type. *Curr Biol* **16**: R1018–1020.

Nur U (1989) Reproductive biology and genetics. Chromosomes, sex ratios and sex determination. In: Rosen D. ed. *Armoured scale insects, their biology, natural enemies and control. Vol A.* Elsevier Science Publishers bv, Amsterdam. pp. 179–190.

Nur U, Werren JH, Eickbush DG, Burke WD et al. (1988) A 'selfish' B chromosome that enhances its transmission by eliminating the paternal genome. *Science* **240**: 512–514.

Nygren K, Strandberg R, Gioti A, Karlsson M et al.(2012) Deciphering the relationship between mating system and the molecular evolution of the pheromone and receptor genes in *Neurospora*. *Mol Biol Evol* **29**: 3827–3842.

Nygren K, Strandberg R, Wallberg A, Nabholz B et al. (2011) A comprehensive phylogeny of *Neurospora* reveals a link between reproductive mode and molecular evolution in fungi. *Mol Phylogenet Evol* **59**: 649–663.

Ogata M, Hasegawa Y, Ohtani H, Mineyama M et al. (2008) The ZZ/ZW sex-determining mechanism originated twice and independently during evolution of the frog, *Rana rugosa*. *Heredity* **100**: 92–99.

Ohno S (1967) *Sex chromosomes and sex-linked genes.* Springer, New York.

Ojika M, Molli SD, Kanazawa H, Yajima A et al. (2011) The second *Phytophthora* mating hormone defines interspecies biosynthetic crosstalk. *Nat Chem Biol* **7**: 591–593.

Okada S, Sone T, Fujisawa M, Nakayama S et al. (2001) The Y chromosome in the liverwort *Marchantia polymorpha* has accumulated unique repeat sequences harboring a male-specific gene. *Proc Natl Acad Sci U S A* **98**: 9454–9459.

Oliver B (2007) Sex, dose, and equality. *PLoS Biol* **5**: 2778–2781.

Oliver B, Parisi M (2004) Battle of the Xs. *Bioessays* **26**: 543–548.

O'Meally D, Ezaz T, Georges A, Sarre SD et al. (2012) Are some chromosomes particularly good at sex? Insights from amniotes. *Chromosome Res* **20**: 7–19.

O'Neil MT, Belote JM (1992) Interspecific comparison of the transformer gene of *Drosophila* reveals an unusually high degree of evolutionary divergence. *Genetics* **131**: 113–128.

Oreal E, Pieau C, Mattei MG, Josso N et al. (1998) Early expression of AMH in chicken embryonic gonads precedes testicular SOX9 expression. *Dev Dyn* **212**: 522–532.

Orias E (1981) Probable somatic DNA rearrangements in mating type determination in *Tetrahymena thermophila*: A review and a model. *Dev Genet* **2**: 185–202.

Orias E, Cervantes MD, Hamilton EP (2011) *Tetrahymena thermophila*, a unicellular eukaryote with separate germline and somatic genomes. *Res Microbiol* **162**: 578–586.

Orr HA (1987) Genetics of male and female sterility in hybrids of *Drosophila pseudoobscura* and *D. persimilis*. *Genetics* **116**: 555–563.

Orr HA (1990) 'Why polyploidy is rarer in animals than in plants' revisited. *Am Nat* **6**: 759–770.

Orr HA (1993) Haldane's rule has multiple genetic causes. *Nature* **361**: 532–533.

Orr HA (1997) Haldane's rule. *Annu Rev Ecol Syst* **28**: 195–218.

Orr HA, Coyne JA (1989) The genetics of postzygotic isolation in the *Drosophila virilis* group. *Genetics* **121**: 527–537.

Orr HA, Irving S (2005) Segregation distortion in hybrids between the Bogota and USA subspecies of *Drosophila pseudoobscura*. *Genetics* **169**: 671–682.

Orr HA, Kim Y (1998) An adaptive hypothesis for the evolution of the Y chromosome. *Genetics* **150**: 1693–1698.

Orr HA, Masly JP, Phadnis N (2007) Speciation in *Drosophila*: from phenotypes to molecules. *J Hered* **98**: 103–110.

Orr HA, Otto SP (1994) Does diploidy increase the rate of adaptation? *Genetics* **136**: 1475–1480.

Orzack SH, Sohn JJ, Kallman KD, Simon A. Levin et al. (1980) Maintenance of the three sex chromosome polymorphism in the platyfish, *Xiphophorus maculatus*. *Evolution* **34**: 663 672.

Ospina-Álvarez N, Piferrer F (2008) Temperature-dependent sex determination in fish revisited: Prevalence, a single sex ratio response pattern, and possible effects of climate change. *PLoS One* **3**: e2837.

Otake H, Masuyama H, Mashima Y, Shinomiya A et al. (2010) Heritable artificial sex chromosomes in the medaka, *Oryzias latipes*. *Heredity* **105**: 247–256.

Otto SP (2003) The advantages of segregation and the evolution of sex. *Genetics* **164**: 1099–1118.

Otto SP (2009) The evolutionary enigma of sex. *Am Nat* **174**: S1–S14.

Otto SP, Goldstein DB (1992) Recombination and the evolution of diploidy. *Genetics* **131**: 745–751.

Otto SP, Jarne P (2001) Haploids—hapless or happening? *Science* **292**: 2441–2443.

Otto SP, Lenormand T (2002) Resolving the paradox of sex and recombination. *Nat Rev Genet* **3**: 252–261.

Otto SP, Nuismer SL (2004) Species interactions and the evolution of sex. *Science* **304**: 1018–1020.

Otto SP, Pannell JR, Peichel CL, Ashman TL et al. (2011) About PAR: the distinct evolutionary dynamics of the pseudoautosomal region. *Trends Genet* **27**: 358–367.

Otto SP, Whitton J (2000) Polyploid incidence and evolution. *Annu Rev Genet* **34**: 401–437.

Ozias-Akins P, van Dijk PJ (2007) Mendelian genetics of apomixis in plants. *Annu Rev Genet* **41**: 509–537.

Pala I, Naurin S, Stervander M, Hasselquist D et al. (2012) Evidence of a neo-sex chromosome in birds. *Heredity* **108**: 264–272.

Paland S, Lynch M (2006) Transitions to asexuality result in excess amino acid substitutions. *Science* **311**: 990–992.

Paldi A, Gyapay G, Jami J (1995) Imprinted chromosomal regions of the human genome display sex-specific meiotic recombination frequencies. *Curr Biol* **5**: 1030–1035.

Pane A, Salvemini M, Delli Bovi P, Polito C et al. (2002) The transformer gene in *Ceratitis capitata* provides a genetic basis for selecting and remembering the sexual fate. *Development* **129**: 3715–3725.

Pannebakker BA, Watt R, Knott SA, West SA et al. (2011) The quantitative genetic basis of sex ratio variation in *Nasonia vitripennis*: a QTL study. *J Evol Biol* **24**: 12–22.

Pannell JR, Obbard DJ, Buggs RJA (2004) Polyploidy and the sexual system: what can we learn from *Mercurialis annua*? *Biol J Linn Soc* **82**: 547–560.

Paoletti M, Seymour FA, Alcocer MJ, Kaur N et al. (2007) Mating type and the genetic basis of self-fertility in the model fungus *Aspergillus nidulans*. *Curr Biol* **17**: 1384–1389.

Parisi M, Nuttall R, Edwards P, Minor J et al. (2004) A survey of ovary-, testis-, and soma-biased gene expression in *Drosophila melanogaster* adults. *Genome Biol* **5**: R40.

Parisi M, Nuttall R, Naiman D, Bouffard G et al. (2003) Paucity of genes on the *Drosophila* X chromosome showing male-biased expression. *Science* **299**: 697–700.

Park Y, Kuroda MI (2001) Epigenetic aspects of X-chromosome dosage compensation. *Science* **293**: 1083–1085.

Parker GA (1979) Sexual selection and sexual conflict. In: Blum MS, Blum NA. eds. *Sexual selection and reproductive competition in insects*. Academic Press, Inc., New York. pp. 123–166.

Parker GA, Baker RR, Smith VG (1972) The origin and evolution of gamete dimorphism and the male-female phenomenon. *J Theor Biol* **36**: 529–553.

Partridge L, Hurst LD (1998) Sex and conflict. *Science* **281**: 2003–2008.

Parvanov ED, Petkov PM, Paigen K (2010) *Prdm9* controls activation of mammalian recombination hotspots. *Science* **327**: 835.

Patten MM, Haig D (2009) Maintenance or loss of genetic variation under sexual and parental antagonism at a sex-linked locus. *Evolution* **63**: 2888–2895.

Paul A, Kuester J, Timme A, Arnemann J (1993) The association between rank, mating effort, and reproductive success in male Barbary macaques (*Macaca sylvanus*). *Primates* **34**: 491–502.

Payer B, Lee JT (2008) X chromosome dosage compensation: how mammals keep the balance. *Annu Rev Genet* **42**: 733–772.

Payseur BA, Krenz JG, Nachman MW (2004) Differential patterns of introgression across the X chromosome in a hybrid zone between two species of house mice. *Evolution* **58**: 2064–2078.

Pease JB, Hahn MW (2012) Sex chromosomes evolved from independent ancestral linkage groups in winged insects. *Mol Biol Evol* **29**: 1645–1653.

Peck JR (1994) A ruby in the rubbish: beneficial mutations, deleterious mutations and the evolution of sex. *Genetics* **137**: 597–606.

Peichel CL, Ross JA, Matson CK, Dickson M et al. (2004) The master sex-determination locus in threespine sticklebacks is on a nascent Y chromosome. *Curr Biol* **14**: 1416–1424.

Pen I, Uller T, Feldmeyer B, Harts A et al. (2010) Climate-driven population divergence in sex-determining systems. *Nature* **468**: 436–438.

Pennisi E (2012) ENCODE project writes eulogy for junk DNA. *Science* **337**: 1159–1161.

Perlman SJ, Magnus SA, Copley CR (2010) Pervasive associations between *Cybaeus* spiders and the bacterial symbiont *Cardinium*. *J Invertebr Pathol* **103**: 150–155.

Perrin N (2012) What uses are mating types? The 'developmental switch' model. *Evolution* **66**: 947–956.

Perrin N (2009) Sex reversal: a fountain of youth for sex chromosomes? *Evolution* **63**: 3043–3049.

Perrot V, Richerd S, Valéro M (1991) Transition from haploidy to diploidy. *Nature* **351**: 315–317.

Peters AF, Scornet D, Müller DG, Kloareg B et al. (2004) Inheritance of organelles in artificial hybrids of the isogamous multicellular chromist alga *Ectocarpus siliculosus* (Phaeophyceae) *Eur J Phycol* **39**: 235–242.

Peters AF, Scornet D, Ratin M, Charrier B et al. (2008) Life-cycle-generation-specific developmental processes are modified in the immediate upright mutant of the brown alga *Ectocarpus siliculosus*. *Development* **135**: 1503–1512.

Petersen JJ (1972) Factor affecting sex ratios of a mermithid parasite of mosquitoes. *J Nematol* **4**: 83–87.

Petersen JJ (1977) Effects of host size and parasite burden on sex ratio in the mosquito parasite *Octomyomermis muspratti*. *J Nematol* **9**: 343–346.

Petit RJ, Excoffier L (2009) Gene flow and species delimitation. *Trends Ecol Evol* **24**: 386–393.

Petraitis PS (1991) The effects of sex ratio and density on the expression of gender in the polychaete *Capitella capitata*. *Evol Ecol* **5**: 393–404.

Phadke SS, Zufall RA (2009) Rapid diversification of mating systems in ciliates. *Biol J Linn Soc* **98**: 187–197.

Phillips BC, Edmands S (2012) Does the speciation clock tick more slowly in the absence of heteromorphic sex chromosomes? *Bioessays* **34**: 166–169.

Phillips RB, Konkol NR, Reed KM, Stein JD (2001) Chromosome painting supports lack of homology among sex chromosomes in *Oncorhynchus*, *Salmo*, and *Salvelinus* (Salmonidae). *Genetica* **111**: 119–123.

Pieau C (1971) Sur la proportion sexuelle chez les embryons de deux Chéloniens (*Testudo graeca* L. et *Emys orbicularis* L.) issus d'oeufs incubés artificiellement. *C R Acad Sci Paris* **272**: 3071–3074.

Pieau C (1972) Effets de la température sur le développement des glandes génitales chez les embryons de deux Chéloniens, *Emys orbicularis* L. et *Testudo graeca* L. *C R Acad Sci Paris* **274**: 719–722.

Pieau C, Girondot M, Richard-Mercier N, Desvages G et al. (1994) Temperature sensitivity of sexual differentiation of gonads in the European pond turtle: Hormonal involvement. *J Exp Zool* **270**: 86–94.

Piergentili R (2010) Multiple roles of the Y chromosome in the biology of *Drosophila melanogaster*. *ScientificWorldJournal* **10**: 1749–1767.

Pike TW, Petrie M (2003) Potential mechanisms of avian sex manipulation. *Biol Rev Camb Philos Soc* **78**: 553–574.

Plackett AR, Thomas SG, Wilson ZA, Hedden P (2011) Gibberellin control of stamen development: a fertile field. *Trends Plant Sci* **16**: 568–578.

Pöggeler S (1999) Phylogenetic relationships between mating-type sequences from homothallic and heterothallic ascomycetes. *Curr Genet* **36**: 222–231.

Pöggeler S, Risch S, Kuck U, Osiewacz HD (1997) Mating-type genes from the homothallic fungus *Sordaria macrospora* are functionally expressed in a heterothallic ascomycete. *Genetics* **147**: 567–580.

Pokorná M, Kratochvíl L (2009) Phylogeny of sex-determining mechanisms in squamate reptiles: are sex chromosomes an evolutionary trap? *Zool J Linn Soc* **156**: 168–183.

Policansky D (1982) Sex change in plants and animals. *Annu Rev Ecol Syst* **13**: 471–495.

Pomiankowski A, Nothiger R, Wilkins A (2004) The evolution of the *Drosophila* sex-determination pathway. *Genetics* **166**: 1761–1773.

Postma E, Spyrou N, Rollins LA, Brooks RC (2011) Sex-dependent selection differentially shapes genetic variation on and off the guppy Y chromosome. *Evolution* **65**: 2145–2156.

Potrzebowski L, Vinckenbosch N, Kaessmann H (2010) The emergence of new genes on the young therian X. *Trends Genet* **26**: 1–4.

Potrzebowski L, Vinckenbosch N, Marques AC, Chalmel F et al. (2008) Chromosomal gene movements reflect the recent origin and biology of therian sex chromosomes. *PLoS Biol* **6**: e80.

Preer JR, Jr (2000) Epigenetic mechanisms affecting macronuclear development in *Paramecium* and *Tetrahymena*. *J Eukaryot Microbiol* **47**: 515–524.

Premoli MC, Sella G, Berra GP (1996) Heritable variation of sex ratio in a polychaete worm. *J Evol Biol* **9**: 845–854.

Presgraves DC (2002) Patterns of postzygotic isolation in *Lepidoptera*. *Evolution* **56**: 1168–1183.

Presgraves DC (2008) Sex chromosomes and speciation in *Drosophila*. *Trends Genet* **24**: 336–343.

Presgraves DC (2010) The molecular evolutionary basis of species formation. *Nat Rev Genet* **11**: 175–180.

Presgraves DC, Orr HA (1998) Haldane's rule in taxa lacking a hemizygous X. *Science* **282**: 952–954.

Prince EG, Kirkland D, Demuth JP (2010) Hyperexpression of the X chromosome in both sexes results in extensive female bias of X-linked genes in the flour beetle. *Genome Biol Evol* **2**: 336–346.

Pryke SR (2010) Sex chromosome linkage of mate preference and color signal maintains assortative mating between interbreeding finch morphs. *Evolution* **64**: 1301–1310.

Pryke SR, Rollins LA (2012) Mothers adjust offspring sex to match the quality of the rearing environment. *Proc Biol Sci* **279**: 4051–4057.

Puhalla JE (1970) Genetic studies of the b incompatibility locus of *Ustilago maydis*. *Genet Res* **16**: 229–232.

Quinn AE, Georges A, Sarre SD, Guarino F et al. (2007) Temperature sex reversal implies sex gene dosage in a reptile. *Science* **316**: 411–411.

Quintanilla LG, Pangua E, Amigo J, Pajarón S (2005) Comparative study of the sympatric ferns *Culcita macrocarpa* and *Woodwardia radicans*: Sexual phenotype. *Flora* **200**: 187–194.

Qvarnström A, Bailey RI (2009) Speciation through evolution of sex-linked genes. *Heredity* **102**: 4–15.

Radder RS (2007) Maternally derived egg yolk steroid hormones and sex determination: review of a paradox in reptiles. *J Biosci* **32**: 1213–1220.

Radder RS, Pike DA, Quinn AE, Shine R (2009) Offspring sex in a lizard depends on egg size. *Curr Biol* **19**: 1102–1105.

Radder RS, Quinn AE, Georges A, Sarre SD et al. (2008) Genetic evidence for co-occurrence of chromosomal and thermal sex-determining systems in a lizard. *Biol Lett* **4**: 176–178.

Ramesh MA, Malik S, Logsdon Jr. JM (2005) A phylogenomic inventory of meiotic genes: Evidence for sex in *Giardia* and an early eukaryotic origin of meiosis. *Curr Biol* **15**: 185–191.

Randerson JP, Hurst LD (2001) A comparative test of a theory for the evolution of anisogamy. *Proc Biol Sci* **268**: 879–884.

Ranz JM, Castillo-Davis CI, Meiklejohn CD, Hartl DL (2003) Sex-dependent gene expression and evolution of the *Drosophila* transcriptome. *Science* **300**: 1742–1745.

Raper JR, Baxter MG, Middleton RB (1958) The genetic structure of the incompatibility factors in *Schizophyllum commune*. *Proc Natl Acad Sci U S A* **44**: 889–900.

Rawitscher-Kunkel E, Machlis L (1962) The hormonal integration of sexual reproduction in *Oedogonium*. *Am J Bot* **49**: 177–183.

Ray-Chaudhuri S, Singh L, Sharma T (1971) Evolution of sex-chromosomes and formation of W-chromatin in snakes. *Chromosoma* **33**: 239–251.

Raychoudhury R, Werren JH (2012) Host genotype changes bidirectional to unidirectional cytoplasmic incompatibility in *Nasonia longicornis*. *Heredity* **108**: 105–114.

Raymond CS, Kettlewell JR, Hirsch B, Bardwell VJ et al. (1999) Expression of Dmrt1 in the genital ridge of mouse and chicken embryos suggests a role in vertebrate sexual development. *Dev Biol* **215**: 208–220.

Raymond CS, Shamu CE, Shen MM, Seifert KJ et al. (1998) Evidence for evolutionary conservation of sex-determining genes. *Nature* **391**: 691–695.

Rea AC, Nasrallah JB (2008) Self-incompatibility systems: barriers to self-fertilization in flowering plants. *Int J Dev Biol* **52**: 627–636.

Reeve HK, Pfennig DW (2003) Genetic biases for showy males: are some genetic systems especially conducive to sexual selection? *Proc Natl Acad Sci U S A* **100**: 1089–1094.

Reinhold K (1996) Biased primary sex ratio in the bush-cricket *Poecilimon veluchianus*, an insect with sex chromosomes. *Behav Ecol Sociobiol* **39**: 189–194.

Reinhold K (1998) Sex linkage among genes controlling sexually selected traits. *Behav Ecol Sociobiol* **44**: 1–7.

Reinke V, Gil IS, Ward S, Kazmer K (2004) Genome-wide germline-enriched and sex-biased expression profiles in *Caenorhabditis elegans*. *Development* **131**: 311–323.

Reinke V, Smith HE, Nance J, Wang J et al. (2000) A global profile of germline gene expression in *C. elegans*. *Mol Cell* **6**: 605–616.

Rhen T, Schroeder A, Sakata JT, Huang V et al. (2011) Segregating variation for temperature-dependent sex determination in a lizard. *Heredity* **106**: 649–660.

Rice WR (1984) Sex chromosomes and the evolution of sexual dimorphism. *Evolution* **38**: 735–742.

Rice WR (1986) On the instability of polygenic sex determination: The effect of sex-specific selection. *Evolution* **40**: 633–639.

Rice WR (1987) The accumulation of sexually antagonistic genes as a selective agent promoting the evolution of reduced recombination between primitive sex chromosomes. *Evolution* **41**: 911–914.

Rice WR (1994) Degeneration of a nonrecombining chromosome. *Science* **263**: 230–232.

Rice WR (1996) Evolution of the Y chromosome in animals. *BioScience* **46**: 331–343.

Rice WR (1998) Male fitness increases when females are eliminated from gene pool: implications for the Y chromosome. *Proc Natl Acad Sci U S A* **95**: 6217–6221.

Rice WR, Chippindale AK (2001) Sexual recombination and the power of natural selection. *Science* **294**: 555–559.

Rice WR, Friberg U (2008) Functionally degenerate—Y not so? *Science* **319**: 42–43.

Richerd S, Couvet D, Valero M (1993) Evolution of the alternation of haploid and diploid phases in life cycles. II. Maintenance of the haplo-diplontic cycle. *J Evol Biol* **6**: 263–280.

Rigaud T (1997) Inherited microorganisms and sex determination of arthropod hosts. In: O'Neill SL, Hoffmann AA, Werren JH. eds. *Influential passengers: inherited microorganisms and arthropod reproduction*. Oxford University Press, Oxford. pp. 81–102.

Rigaud T, Juchault P, Mocquard J (1997) The evolution of sex determination in isopod crustaceans. *Bioessays* **19**: 409–416.

Rispe C, Pierre J, Simon J, Gouyon P (1998) Models of sexual and asexual coexistence in aphids based on constraints. *J Evol Biol* **11**: 685–701.

Ritchie MG, Saarikettu M, Hoikkala A (2005) Variation, but no covariance, in female preference functions and male song in a natural population of *Drosophila montana*. *Anim Behav* **70**: 849–854.

Roberts RB, Ser JR, Kocher TD (2009) Sexual conflict resolved by invasion of a novel sex determiner in Lake Malawi cichlid fishes. *Science* **326**: 998–1001.

Robertson A (1961) Inbreeding in artificial selection programmes. *Genet Res* **2**: 189–194.

Robinson WP, Lalande M (1995) Sex-specific meiotic recombination in the Prader–Willi/Angelman syndrome imprinted region. *Hum Mol Genet* **4**: 801–806.

Rodrigues N, Betto-Colliard C, Jourdan-Pineau H, Perrin N (2013) Within-population polymorphism of sex-determination systems in the common frog (*Rana temporaria*). *J Evol Biol* **26**: 1569–1577.

Rodriguez-Mari A, Yan Y, BreMiller R, Wilson C et al. (2005) Characterization and expression pattern of zebrafish anti-Mullerian hormone (amh) relative to sox9a, sox9b, and cyp19ala, during gonad development. *Gen Expr Patterns* **5**: 655–667.

Roff DA (1986) The evolution of wing dimorphism in insects. *Evolution* **40**: 1009–1020.

Roncarati A, Melotti P, Mordenti O, Gennari L (1997) Influence of stocking density of European eel (*Anguilla anguilla*, L.) elvers on sex differentiation and zootechnical performances. *J Appl Ichthyol* **13**: 131–136.

Roose-Amsaleg C, De Vallavieille-Pope C, Brygoo Y, Levis C (2002) Characterisation of a length polymorphism in the two intergenic spacers of ribosomal RNA in *Puccinia striiformis* f. sp. *tritici*, the causal agent of wheat yellow rust. *Mycol Res* **106**: 918–924.

Roosenburg WM (1996) Maternal condition and nest site choice: An alternative for the maintenance of environmental sex determination? *Am Zool* **36**: 157–168.

Rosenfeld CS, Roberts RM (2004) Maternal diet and other factors affecting offspring sex ratio: A review. *Biol Reprod* **71**: 1063–1070.

Ross MT, Grafham DV, Coffey AJ, Scherer S et al. (2005) The DNA sequence of the human X chromosome. *Nature* **434**: 325–337.

Ross JA, Urton JR, Boland J, Shapiro MD et al. (2009) Turnover of sex chromosomes in the stickleback fishes (Gasterosteidae). *PLoS Genet* **5**: e1000391.

Rosser ZH, Balaresque P, Jobling MA (2009) Gene conversion between the X chromosome and the male-specific region of the Y chromosome at a translocation hotspot. *Am J Hum Genet* **85**: 130–134.

Rowell D (1991) Chromosomal fusion and meiotic behaviour in *Delena cancerides* (Araneae: Sparassidae). II. Chiasma position and its implications for speciation. *Genome* **34**: 567–573.

Rozen S, Skaletsky H, Marszalek JD, Minx PJ et al. (2003) Abundant gene conversion between arms of palindromes in human and ape Y chromosomes. *Nature* **423**: 873–876.

Ruddat M, Kokontis J, Birch L, Garber ED et al. (1991) Interactions of *Microbotryum violaceum* (*Ustilago violacea*) with its host plant *Silene alba*. *Plant Sci* **80**: 157–165.

Rutkowska J, Badyaev AV (2008) Meiotic drive and sex determination: molecular and cytological mechanisms of sex ratio adjustment in birds. *Philos Trans R Soc Lond B Biol Sci* **363**: 1675–1686.

Ryan MJ, Pease CM, Morris MR (1992) A genetic polymorphism in the swordtail *Xiphophorus nigrensis*: testing the prediction of equal fitnesses. *Am Nat* **139**: 21–31.

Rydholm C, Dyer PS, Lutzoni F (2007) DNA sequence characterization and molecular evolution of MAT1 and MAT2 mating-type loci of the self-compatible ascomycete mold *Neosartorya fischeri*. *Eukaryot Cell* **6**: 868–874.

Ryner LC, Baker BS (1991) Regulation of doublesex pre-mRNA processing occurs by 3′-splice site activation. *Genes Dev* **5**: 2071–2085.

Saavedra C, Reyero M, Zouros E (1997) Male-dependent doubly uniparental inheritance of mitochondrial DNA and female-dependent sex-ratio in the mussel *Mytilus galloprovincialis*. *Genetics* **145**: 1073–1082.

Saccone G, Pane A, Polito LC (2002) Sex determination in flies, fruitflies and butterflies. *Genetica* **116**: 15–23.

Sackton TB, Montenegro H, Hartl DL, Lemos B (2011) Interspecific Y chromosome introgressions disrupt testis-specific gene expression and male reproductive phenotypes in *Drosophila*. *Proc Natl Acad Sci U S A* **108**: 17046–17051.

Saether SA, Saetre GP, Borge T, Wiley C et al. (2007) Sex chromosome-linked species recognition and evolution of reproductive isolation in flycatchers. *Science* **318**: 95–97.

Saetre GP, Borge T, Lindroos K, Haavie J *et al.* (2003) Sex chromosome evolution and speciation in *Ficedula* flycatchers. *Proc Biol Sci* **270**: 53–59.

Sakai S (2000) Biased sex allocation in hermaphroditic plants. *J Plant Res* **113**: 335–342.

Salvemini M, Mauro U, Lombardo F, Milano A et al. (2011) Genomic organization and splicing evolution of the doublesex gene, a *Drosophila* regulator of sexual differentiation, in the dengue and yellow fever mosquito *Aedes aegypti*. *BMC Evol Biol* **11**: 41–2148–11–41.

Salvemini M, Robertson M, Aronson B, Atkinson P et al. (2009) *Ceratitis capitata* transformer-2 gene is required to establish and maintain the autoregulation of Cctra, the master gene for female sex determination. *Int J Dev Biol* **53**: 109–120.

Salz HK, Erickson JW (2010) Sex determination in *Drosophila*: The view from the top. *Fly (Austin)* **4**: 60–70.

Sanchez L (2008) Sex-determining mechanisms in insects. *Int J Dev Biol* **52**: 837–856.

Sanchez L, Perondini ALP (1999) Sex determination in sciarid flies: a model for the control of differential X-chromosome elimination. *J Theor Biol* **197**: 247–259.

Saridaki A, Bourtzis K (2010) *Wolbachia*: more than just a bug in insects genitals. *Curr Opin Microbiol* **13**: 67–72.

Sarre SD, Georges A, Quinn A (2004) The ends of a continuum: genetic and temperature-dependent sex determination in reptiles. *Bioessays* **26**: 639–645.

Sato T, Endo T, Yamahira K, Hamaguchi S et al. (2005) Induction of female-to-male sex reversal by high temperature treatment in medaka, *Oryzias latipes*. *Zool Sci* **22**: 985–988.

Sato Y, Shinka T, Sakamoto K, Ewis AA *et al.* (2010) The male-determining gene SRY is a hybrid of DGCR8 and SOX3, and is regulated by the transcription factor CP2. *Mol Cell Biochem* **337**: 267–275.

Saumitou-Laprade P, Vernet P, Vassiliadis C, Hoareau Y et al. (2010) A self-incompatibility system explains high male frequencies in an androdioecious plant. *Science* **327**: 1648–1650.

Sawatari E, Shikina S, Takeuchi T, Yoshizaki G (2007) A novel transforming growth factor-beta superfamily member expressed in gonadal somatic cells enhances primordial germ cell and spermatogonial proliferation in rainbow trout (*Oncorhynchus mykiss*). *Dev Biol* **301**: 266–275.

Schärer L (2009) Tests of sex allocation theory in simultaneously hermaphroditic animals. *Evolution* **63**: 1377–1405.

Schartl M (2004) Sex chromosome evolution in non-mammalian vertebrates. *Curr Opin Genet Dev* **14**: 634–641.

Scheer MA, Silliker ME (2006) Mitochondrial inheritance patterns in *Didymium iridis* are not influenced by stage of mating competency. *Mycologia* **98**: 51–56.

Schempp W, Schmid M (1981) Chromosome banding in Amphibia. VI. BrdU-replication patterns in anura and demonstration of XX/XY sex chromosomes in *Rana esculenta*. *Chromosoma* **83**: 697–710.

Schilthuizen M, Giesbers MC, Beukeboom LW (2011) Haldane's rule in the 21st century. *Heredity* **107**: 95–102.

Schmid M, Nanda I, Steinlein C, Kausch K et al. (1991) Sex-determining mechanisms and sex chromosomes in Amphibia. In: Green DM, Sessions SK. eds. *Amphibian cytogenetics and evolution*. Academic Press, Inc., New York. pp. 393–430.

Schmidt R, Hediger M, Roth S, Nothiger R et al. (1997) The Y-chromosomal and autosomal male-determining M factors of *Musca domestica* are equivalent. *Genetics* **147**: 271–280.

Schmieder S, Colinet D, Poirié M (2012) Tracing back the nascence of a new sex-determination pathway to the ancestor of bees and ants. *Nat Commun* **3**: 895.

Schoenmakers S, Wassenaar E, Hoogerbrugge JW, Laven JS et al. (2009) Female meiotic sex chromosome inactivation in chicken. *PLoS Genet* **5**: e1000466.

Schrider DR, Stevens K, Cardeno CM, Langley CH et al. (2011) Genome-wide analysis of retrogene polymorphisms in *Drosophila melanogaster*. *Genome Res* **21**: 2087–2095.

Schurko AM, Neiman M, Logsdon Jr JM (2009) Signs of sex: what we know and how we know it. *Trends Ecol Evol* **24**: 208–217.

Schütt C, Nöthiger R (2000) Structure, function and evolution of sex-determining systems in Dipteran insects. *Development* **127**: 667–677.

Scudo FM (1964) Sex population genetics. *Ric Sci* **34**, II-B: 93–146.

Scudo FM (1967) *The analysis of sex determination in a multifactorial case: some criteria and data*. Tipografia popolare. Pavia.

Searles RB (1980) The strategy of the red algal life history. *Am Nat* **115**: 113–120.

Sekido R, Lovell-Badge R (2008) Sex determination involves synergistic action of SRY and SF1 on a specific Sox9 enhancer. *Nature* **453**: 930–934.

Sella G (1980) The genetic system controlling the body length at which sex change occurs in *Ophryotrocha puerilis* (Annelida, Polychaeta). *Atti della Societa toscana di scienze naturali* Serie B **87**: 197–203.

Ser JR, Roberts RB, Kocher TD (2010) Multiple interacting loci control sex determination in lake Malawi cichlid fish. *Evolution* **64**: 486–501.

Serna E, Gorab E, Ruiz MF, Goday C et al. (2004) The gene Sex-lethal of the Sciaridae family (order Diptera, suborder Nematocera) and its phylogeny in dipteran insects. *Genetics* **168**: 907–921.

Servedio MR (2000) Reinforcement and the genetics of nonrandom mating. *Evolution* **54**: 21—29.

Servedio MR, Saetre GP (2003) Speciation as a positive feedback loop between postzygotic and prezygotic barriers to gene flow. *Proc Biol Sci* **270**: 1473–1479.

Shannon RK, Holsinger KE (2007) The genetics of sex determination in stinging nettle (*Urtica dioica*). *Sex Plant Reprod* **20**: 35–43.

Sheldon BC, West SA (2004) Maternal dominance, maternal condition, and offspring sex ratio in ungulate mammals. *Am Nat* **163**: 40–54.

Shen MM, Hodgkin J (1988) mab-3, a gene required for sex-specific yolk protein expression and a male-specific lineage in *C. elegans*. *Cell* **54**: 1019–1031.

Shine R, Elphick MJ, Donnellan S (2002) Co-occurrence of multiple, supposedly incompatible modes of sex determination in a lizard population. *Ecol Lett* **5**: 486–489.

Shoemaker CM, Queen J, Crews D (2007) Response of candidate sex-determining genes to changes in temperature reveals their involvement in the molecular network underlying temperature-dependent sex determination. *Mol Endocrinol* **21**: 2750–2763.

Silliker ME, Liles JL, Monroe JA (2002) Patterns of mitochondrial inheritance in the myxogastrid *Didymium iridis*. *Mycologia* **94**: 939–946.

Simon J, Delmotte F, Rispe C, Crease T (2003) Phylogenetic relationships between parthenogens and their sexual relatives: the possible routes to parthenogenesis in animals. *Biol J Linn Soc* **79**: 151–163.

Sin HS, Barski A, Zhang F, Kartashov AV et al. (2012) RNF8 regulates active epigenetic modifications and escape gene activation from inactive sex chromosomes in post-meiotic spermatids. *Genes Dev* **26**: 2737–2748.

Sinclair AH, Berta P, Palmer MS, Hawkins JR et al. (1990) A gene from the human sex-determining region encodes a protein with homology to a conserved DNA-binding motif. *Nature* **346**: 240–244.

Sinclair A, Smith C, Western P, McClive P (2002) A comparative analysis of vertebrate sex determination. *Novartis Found Symp* **244**: 102–111.

Skaletsky H, Kuroda-Kawaguchi T, Minx PJ, Cordum HS et al. (2003) The male-specific region of the human Y chromosome is a mosaic of discrete sequence classes. *Nature* **423**: 825–837.

Skinner SW (1982) Maternally inherited sex ratio in the parasitoid wasp *Nasonia vitripennis*. *Science* **215**: 1133–1134.

Skinner SW (1985) Son-killer: a third extrachromosomal factor affecting the sex ratio in the parasitoid wasp, *Nasonia (= Mormoniella) vitripennis*. *Genetics* **109**: 745–759.

Slattery JP, Murphy WJ, O'Brien SJ (2000) Patterns of diversity among SINE elements isolated from three Y-chromosome genes in carnivores. *Mol Biol Evol* **17**: 825–829.

Slocum CJ (1980) Differential susceptibility to grazers in two phases of an intertidal alga: Advantages of heteromorphic generations. *J Exp Mar Biol Ecol* **46**: 99–110.

Smalley SL (1993) Sex-specific recombination frequencies: a consequence of imprinting? *Am J Hum Genet* **52**: 210–212.

Smith CA, McClive PJ, Western PS, Reed KJ et al. (1999) Conservation of a sex-determining gene. *Nature* **402**: 601–602.

Smith CA, Roeszler KN, Ohnesorg T, Cummins DM et al. (2009) The avian Z-linked gene DMRT1 is required for male sex determination in the chicken. *Nature* **461**: 267–271.

Smith CA, Sinclair AH (2004) Sex determination: insights from the chicken. *Bioessays* **26**: 120–132.

Sogin ML, Elwood HJ, Gunderson JH (1986) Evolutionary diversity of eukaryotic small-subunit rRNA genes. *Proc Natl Acad Sci U S A* **83**: 1383–1387.

Sonneborn TM (1937) Sex, sex inheritance and sex determination in *Paramecium aurelia*. *Proc Natl Acad Sci U S A* **23**: 378–385.

Sonneborn TM (1975) The *Paramecium aurelia* complex of fourteen sibling species. *Trans Am Microsc Soc* **94**: 155–178.

Soullier S, Hanni C, Catzeflis F, Berta P et al. (1998) Male sex determination in the spiny rat *Tokudaia osimensis* (Rodentia: Muridae) is not Sry dependent. *Mamm Genome* **9**: 590–592.

Spigler RB, Lewers KS, Main DS, Ashman TL (2008) Genetic mapping of sex determination in a wild strawberry, *Fragaria virginiana*, reveals earliest form of sex chromosome. *Heredity* **101**: 507–517.

Spit A, Hyland RH, Mellor EJ, Casselton LA (1998) A role for heterodimerization in nuclear localization of a homeodomain protein. *Proc Natl Acad Sci U S A* **95**: 6228–6233.

Stanton BC, Giles SS, Kruzel EK, Warren CL et al. (2009) Cognate site identifier analysis reveals novel binding properties of the sex inducer homeodomain proteins of *Cryptococcus neoformans*. *Mol Microbiol* **72**: 1334–1347.

Stanton BC, Giles SS, Staudt MW, Kruzel EK et al. (2010) Allelic exchange of pheromones and their receptors reprograms sexual identity in *Cryptococcus neoformans*. *PLoS Genet* **6**: e1000860.

Stearns SC (1987) The selection-arena hypothesis. *Experientia Suppl* **55**: 337–349.

Steinemann S, Steinemann M (2005) Y chromosomes: born to be destroyed. *Bioessays* **27**: 1076–1083.

Steinmann-Zwicky M, Amrein H, Nöthiger R (1990) Genetic control of sex determination in *Drosophila*. In: Wright TRF. ed *Genetic regulatory hierarchies in development*. Academic Press, Inc., New York. pp. 189–237.

Stenberg P, Saura A (2009) Cytology of asexual animals. In: Schön I, Martens K, Dijk P, eds. *Lost sex*, Springer, Dordrecht, pp. 63–71.

Stevens NM (1905) A study of the germ cells of *Aphis rosae and Aphis oenetherac*. *J Exp Zool* **2**: 313–333.

Stöck M, Horn A, Grossen C, Lindtke D et al. (2011) Ever-young sex chromosomes in European tree frogs. *PLoS Biol* **9**: e1001062.

Stone JL (2002) Molecular mechanisms underlying the breakdown of gametophytic self-incompatibility. *Q Rev Biol* **77**: 17–32.

Stone SL, Goring DR (2001) The molecular biology of self-incompatibility systems in flowering plants. *Plant Cell Tissue Organ Cult* **67**: 93–114.

Storchova R, Divina P (2006) Nonrandom representation of sex-biased genes on chicken Z chromosome. *J Mol Evol* **63**: 676–681.

Storchova R, Reif J, Nachman MW (2010) Female heterogamety and speciation: reduced introgression of the Z chromosome between two species of nightingales. *Evolution* **64**: 456–471.

Stouthamer R, Russell JE, Vavre F, Nunney L (2010) Intragenomic conflict in populations infected by parthenogenesis-inducing *Wolbachia* ends with irreversible loss of sexual reproduction. *BMC Evol Biol* **10**: 229–2148–10–229.

Strain E, Hass B, Banks JA (2001) Characterization of mutations that feminize gametophytes of the fern *Ceratopteris*. *Genetics* **159**: 1271–1281.

Straub T, Becker PB (2007) Dosage compensation: the beginning and end of generalization. *Nat Rev Genet* **8**: 47–57.

Streelman JT, Albertson RC, Kocher TD (2003) Genome mapping of the orange blotch colour pattern in cichlid fishes. *Mol Ecol* **12**: 2465–2471.

Strüssmann CA, Takashima F, Toda K (1996) Sex differentiation and hormonal feminization in pejerrey *Odontesthes bonariensis*. *Aquaculture* **139**: 31–45.

Sturgill D, Zhang Y, Parisi M, Oliver B (2007) Demasculinization of X chromosomes in the *Drosophila* genus. *Nature* **450**: 238–241.

Sugimoto TN, Fujii T, Kayukawa T, Sakamoto H et al. (2010) Expression of a doublesex homologue is altered in sexual mosaics of *Ostrinia scapulalis* moths infected with *Wolbachia*. *Insect Biochem Mol Biol* **40**: 847–854.

Sugimoto TN, Ishikawa Y (2012) A male-killing *Wolbachia* carries a feminizing factor and is associated with degradation of the sex-determining system of its host. *Biol Lett* **8**: 412–415.

Sullivan JA, Schultz RJ (1986) Genetic and environmental basis of variable sex ratios in laboratory strains of *Poeciliopsis lucida*. *Evolution* **40**: 152–158.

Sun S, Heitman J (2012) Should Y stay or should Y go: the evolution of non-recombining sex chromosomes. *Bioessays* **34**: 938–942.

Suomalainen E, Saura A, Lokki J (1987) *Cytology and evolution in parthenogenesis*. CRC Press, Boca Raton, FL.

Sutou S, Mitsui Y, Tsuchiya K (2001) Sex determination without the Y chromosome in two Japanese rodents *Tokudaia osimensis osimensis* and *Tokudaia osimensis* spp. *Mamm Genome* **12**: 17–21.

Sutton WS (1902) On the morphology of the chromosome group in *Brachystola magna*. *Biol Bull* **4**: 24–39.

Suzuki MG (2010) Sex determination: insights from the silkworm. *J Genet* **89**: 357–363.

Sweigart AL (2010) Simple Y-autosomal incompatibilities cause hybrid male sterility in reciprocal crosses between *Drosophila virilis* and *D. americana*. *Genetics* **184**: 779–787.

Taipale M, Akhtar A (2005) Chromatin mechanisms in *Drosophila* dosage compensation. *Prog Mol Subcell Biol* **38**: 123–149.

Takano H, Onoue K, Kawano S (2010) Mitochondrial fusion and inheritance of the mitochondrial genome. *J Plant Res* **123**: 131–138.

Takehana Y, Hamaguchi S, Sakaizumi M (2008) Different origins of ZZ/ZW sex chromosomes in closely related medaka fishes, *Oryzias javanicus* and *O. hubbsi*. *Chromosome Res* **16**: 801–811.

Tanaka Y (1914) Further data on the reduplication in silkworms. *Agr Tohoku Imp Univ Sapporo* **6**: 1–16.

Tanurdzic M, Banks JA (2004) Sex-determining mechanisms in land plants. *Plant Cell* **16 Suppl**: S61–71.

Tao Y, Masly JP, Araripe L, Ke Y *et al.* (2007a) A sex-ratio meiotic drive system in *Drosophila simulans*. I: An autosomal suppressor. *PLoS Biol* **5**: e292.

Tao Y, Araripe L, Kingan SB, Ke Y *et al.* (2007b) A sex-ratio meiotic drive system in *Drosophila simulans*. II: an X-linked distorter. *PLoS Biol* **5**: e293.

Tao Y, Hartl DL, Laurie CC (2001) Sex-ratio segregation distortion associated with reproductive isolation in *Drosophila*. *Proc Nat Acad Sci* **98**: 13183–13188.

Tao Y, Hartl DL (2003) Genetic dissection of hybrid incompatibilities between *Drosophila simulans* and *D. mauritiana*. III. Heterogeneous accumulation of hybrid incompatibilities, degree of dominance, and implications for Haldane's rule. *Evolution* **57**: 2580–2598.

Tarone AM, Nasser YM, Nuzhdin SV (2005) Genetic variation for expression of the sex determination pathway genes in *Drosophila melanogaster*. *Genet Res* **86**: 31–40.

Taylor DR (1990) Evolutionary consequences of cytoplasmic sex ratio distorters. *Evol Ecol* **4**: 235–248.

Taylor DR, Trimble S, McCauley DE (1999) Ecological genetics of gynodioecy in *Silene vulgaris*: relative fitness of females and hermaphrodites during the colonization process. *Evolution* **53**: 745–751.

Tease C, Hulten MA (2004) Inter-sex variation in synaptonemal complex lengths largely determine the different recombination rates in male and female germ cells. *Cytogenet Genome Res* **107**: 208–215.

Teeter KC, Payseur BA, Harris LW, Bakewell MA *et al.* (2008) Genome-wide patterns of gene flow across a house mouse hybrid zone. *Genome Res* **18**: 67–76.

Terry RS, Smith JE, Dunn AM (1998) Impact of a novel, feminising microsporidium on its crustacean host. *J Eukaryot Microbiol* **45**: 497–501.

Tessema M, Müller-Belecke A, Hörstgen-Schwark G (2006) Effect of rearing temperatures on the sex ratios of *Oreochromis niloticus* populations. *Aquaculture* **258**: 270–277.

Tomita T, Wada Y (1989) Multifactorial sex determination in natural-populations of the housefly (*Musca domestica*) in Japan. *Jpn J Genet* **64**: 373–382.

Torgerson DG, Singh RS (2003) Sex-linked mammalian sperm proteins evolve faster than autosomal ones. *Mol Biol Evol* **20**: 1705–1709.

Torgerson DG, Singh RS (2006) Enhanced adaptive evolution of sperm-expressed genes on the mammalian X chromosome. *Heredity* **96**: 39–44.

Toups MA, Pease JB, Hahn MW (2011) No excess gene movement is detected off the avian or lepidopteran Z chromosome. *Genome Biol Evol* **3**: 1381–1390.

Traut W, Willhoeft U (1990) A jumping sex determining factor in the fly *Megaselia scalaris*. *Chromosoma* **99**: 407–412.

Trickett AJ, Butlin RK (1994) Recombination suppressors and the evolution of new species. *Heredity* **73**: 339–345.

Trivers R (1972) Parental investment and sexual selection. In: Campbell B. ed. *Sexual selection and the descent of man*. Aldine Publishing Company, Chicago. pp. 136–179.

Trivers RL (1974) Parent-offspring conflict. *Am Zool* **14**: 249–264.

Trivers RL, Willard DE (1973) Natural selection of parental ability to vary the sex ratio of offspring. *Science* **179**: 90–92.

Trombetta B, Cruciani F, Underhill PA, Sellitto D et al. (2010) Footprints of X-to-Y gene conversion in recent human evolution. *Mol Biol Evol* **27**: 714–725.

Tsanev V, Vladova R, Kraptchev B, Sokolov V (2006) Apomictic reproduction of plants and possibilities for its application in agriculture. *Genet Breed* **35**: 59–71.

Tucker PK, Sage RD, Warner J, Wilson AC *et al.* (1992) Abrupt cline for sex chromosomes in a hybrid zone between two species of mice. *Evolution* **46**: 1146–1163.

Turelli M, Moyle LC (2007) Asymmetric postmating isolation: Darwin's corollary to Haldane's rule. *Genetics* **176**: 1059–1088.

Turelli M, Orr HA (2000) Dominance, epistasis and the genetics of postzygotic isolation. *Genetics* **154**: 1663–1679.

Turelli M, Orr HA (1995) The dominance theory of Haldane's rule. *Genetics* **140**: 389–402.

Turgeon BG, Debuchy R (2007) *Cochliobolus* and *Podospora*: mechanisms of sex determination and the evolution of reproductive lifestyle. In: Heitman J, Kronstad JW, Taylor JW, Casselton LA. eds. *Sex in fungi*. ASM Press, Washington. pp. 93–121.

Turner JMA (2007) Meiotic sex chromosome inactivation. *Development* **134**: 1823–1831.

Tymon AM, Kues U, Richardson WV, Casselton LA (1992) A fungal mating type protein that regulates sexual and asexual development contains a POU-related domain. *EMBO J* **11**: 1805–1813.

Ubeda F, Normark BB (2006) Male killers and the origins of paternal genome elimination. *Theor Popul Biol* **70**: 511–526.

Uller T, Helanterä H (2011) From the origin of sex-determining factors to the evolution of sex-determining systems. *Q Rev Biol* **86**: 163–180.

Uller T, Pen I (2011) A theoretial model of the evolution of maternal effects under parent-offspring conflict. Evolution **65**: 2075–2084.

Ullerich FH (1984) Analysis of sex determination in the monogenic blowfly *Chrysomya rufifacies* by pole cell transplantation. *Mol Gen Genet* **193**: 479–487.

Umen JG (2011) Evolution of sex and mating loci: an expanded view from Volvocine algae. *Curr Opin Microbiol* **14**: 634–641.

Ushijima K, Nakano R, Bando M, Shigezane Y et al. (2012) Isolation of the floral morph-related genes in heterostylous flax (*Linum grandiflorum*): the genetic polymorphism and the transcriptional and post-transcriptional regulations of the S locus. *Plant J* **69**: 317–331.

Valenzuela N, Adams DC, Janzen FJ (2003) Pattern does not equal process: exactly when is sex environmentally determined? *Am Nat* **161**: 676–683.

Valenzuela N, Lance V (2004) *Temperature-dependent sex determination in vertebrates*. Smithsonian Books, New York.

Vallesi A, Ballarini P, Di Pretoro B, Alimenti C et al. (2005) Autocrine, mitogenic pheromone receptor loop of the ciliate *Euplotes raikovi*: pheromone-induced receptor internalization. *Eukaryot Cell* **4**: 1221–1227.

Vamosi JC, Vamosi SM (2005) Present day risk of extinction may exacerbate the lower species richness of dioecious clades. *Divers Distrib* **11**: 25–32.

Van der Sluijs I, Seehausen O, Van Dooren TJ, van Alphen JJ (2010) No evidence for a genetic association between female mating preference and male secondary sexual trait in a Lake Victoria cichlid fish. *Curr Zool* **56**: 57–64.

Van Dooren TJ, Leimar O (2003) The evolution of environmental and genetic sex determination in fluctuating environments. *Evolution* **57**: 2667–2677.

Van Doorn GS, Kirkpatrick M (2007) Turnover of sex chromosomes induced by sexual conflict. *Nature* **449**: 909–912.

Van Doorn GS, Kirkpatrick M (2010) Transitions between male and female heterogamety caused by sex-antagonistic selection. *Genetics* **186**: 629–645.

Van Wilgenburg E, Driessen G, Beukeboom LW (2006) Single locus complementary sex determination in Hymenoptera: an 'unintelligent' design? *Front Zool* **3**: 1.

Vandeputte M, Dupont-Nivet M, Chavanne H, Chatain B (2007) A polygenic hypothesis for sex determination in the European sea bass *Dicentrarchus labrax*. *Genetics* **176**: 1049–1057.

Vanstechelman I, Sabbe K, Vyverman W, Vanormelingen P *et al*. (2013) Linkage mapping identifies the sex determining region as a single locus in the Pennate diatom *Seminavis robusta*. *PLoS One* **8**: e60132.

Varandas FR, Sampaio MC, Carvalho AB (1997) Heritability of sexual proportion in experimental sex-ratio populations of *Drosophila mediopunctata*. *Heredity* **79**: 104–112.

Vassiliadis C, Lepart J, Saumitou-Laprade P, Vernet P (2000) Self-incompatibility and male fertilization success in *Phillyrea angustifolia* (Oleaceae). *Int J Plant Sci* **161**: 393–402.

Vekemans X, Schierup MH, Christiansen FB (1998) Mate availability and fecundity selection in multi-allelic self-incompatibility systems in plants. *Evolution*: 19–29.

Veltsos P, Keller I, Nichols RA (2008) The inexorable spread of a newly arisen neo-Y chromosome. *PLoS Genet* **4**: e1000082.

Verhulst EC, Beukeboom LW, Van de Zande L (2010a) Maternal control of haplodiploid sex determination in the wasp *Nasonia*. *Science* **328**: 620–623.

Verhulst EC, Van de Zande L, Beukeboom LW (2010b) Insect sex determination: it all evolves around transformer. *Curr Opin Genet Dev* **20**: 376–383.

Verhulst EC, Lynch JA, Bopp D, Beukeboom LW *et al.* (2013) A new component of the *Nasonia* sex determining cascade is maternally silenced and regulates transformer expression. *PLoS One* **8**: e63618.

Veyrunes F, Chevret P, Catalan J, Castiglia R et al. (2010) A novel sex determination system in a close relative of the house mouse. *Proc Biol Sci* **277**: 1049–1056.

Vibranovski MD, Lopes HF, Karr TL, Long M (2009) Stage-specific expression profiling of *Drosophila spermatogenesis* suggests that meiotic sex chromosome inactivation drives genomic relocation of testis-expressed genes. *PLoS Gen* **5**: e1000731.

Vicoso B, Bachtrog D (2013) Reversal of an ancient sex chromosome to an autosome in *Drosophila*. *Nature* **499**: 332–335.

Vicoso B, Bachtrog D (2009) Progress and prospects toward our understanding of the evolution of dosage compensation. *Chromosome Res* **17**: 585–602.

Vicoso B, Charlesworth B (2006) Evolution on the X chromosome: unusual patterns and processes. *Nat Rev Genet* **7**: 645–653.

Vicoso B, Charlesworth B (2009) Effective population size and the faster-X effect: an extended model. *Evolution* **63**: 2413–2426.

Vicoso B, Kaiser VB, Bachtrog D (2013) Sex-biased gene expression at homomorphic sex chromosomes in emus and its implication for sex chromosome evolution. *Proc Natl Acad Sci U S A* **110**: 6453–6458.

Vinckenbosch N, Dupanloup I, Kaessmann H (2006) Evolutionary fate of retroposed gene copies in the human genome. *Proc Natl Acad Sci U S A* **103**: 3220–3225.

Volff JN, Schartl M (2001) Variability of genetic sex determination in poeciliid fishes. *Genetica* **111**: 101–110.

Voordouw MJ, Anholt BR (2002) Heritability of sex tendency in a harpacticoid copepod, *Tigriopus californicus*. *Evolution* **56**: 1754–1763.

Votintseva AA, Filatov DA (2009) Evolutionary strata in a small mating-type-specific region of the smut fungus *Microbotryum violaceum*. *Genetics* **182**: 1391–1396.

Vrijenhoek RC (1998) Animal clones and diversity. *Bioscience* **48**: 617–628.

Vuilleumier S, Lande R, van Alphen JJ, Seehausen O (2007) Invasion and fixation of sex-reversal genes. *J Evol Biol* **20**: 913–920.

Wada H, Shimada A, Fukamachi S, Naruse K et al. (1998) Sex-linked inheritance of the lf locus in the medaka fish (*Oryzias latipes*). *Zool Sci* **15**: 123–126.

Wahl R, Zahiri A, Kämper J (2010) The *Ustilago maydis* b mating type locus controls hyphal proliferation and expression of secreted virulence factors in planta. *Mol Microbiol* **75**: 208–220.

Wajnberg E (1994) Intra-population genetic variation in *Trichogramma*. In: Wajnberg E, Hassan SA. eds. *Biological control with egg parasitoids*. CAB International, Wallingford. pp. 245–273.

Wallace H, Wallace B, Badawy G (1997) Lampbrush chromosomes and chiasmata of sex-reversed crested newts. *Chromosoma* **106**: 526–533.

Wallis MC, Waters PD, Delbridge ML, Kirby PJ et al. (2007) Sex determination in platypus and echidna: autosomal location of SOX3 confirms the absence of SRY from monotremes. *Chromosome Res* **15**: 949–959.

Wallis MC, Waters PD, Graves JA (2008) Sex determination in mammals—before and after the evolution of SRY. *Cell Mol Life Sci* **65**: 3182–3195.

Wang PJ, McCarrey JR, Yang F, Page DC (2001) An abundance of X-linked genes expressed in spermatogonia. *Nature Genetics* **27**: 422–426.

Wang J, Long M, Vibranovski MD (2012) Retrogenes moved out of the Z chromosome in the silkworm. *J Mol Evol* **74**: 113–126.

Wang J, Wurm Y, Nipitwattanaphon M, Riba-Grognuz O et al. (2013) A Y-like social chromosome causes alternative colony organization in fire ants. *Nature* **493**: 664–668.

Wang R, Zhao Y (2008) Differential barrier strength and allele frequencies in hybrid zones maintained by sex-biased hybrid incompatibilities. *Heredity* **100**: 326–336.

Warner RR (1975) The adaptive significance of sequential hermaphroditism in animals. *Am Nat* **109**: 61–82.

Warner RR (1978) The evolution of hermaphroditism and unisexuality in aquatic and terrestrial vertebrates. In: Reese ES, Lighter FJ. Eds. *Contrasts in behavior*. Wiley, Chichester. pp. 77–101.

Warner DA, Shine R (2008) The adaptive significance of temperature-dependent sex determination in a reptile. *Nature* **451**: 566–568.

Waters PD, Wallis MC, Marshall Graves JA (2007) Mammalian sex—Origin and evolution of the Y chromosome and SRY. *Semin Cell Dev Biol* **18**: 389–400.

Watson ET, Demuth JP (2012) Haldane's rule in marsupials: what happens when both sexes are functionally hemizygous? *J Hered* **103**: 453–458.

Watt PJ, Aams J (1994) Adaptive variation in sex determination in a crustacean, *Gammarus duebeni*. *J Zool* **232**: 109–116.

Weeks AR, Marec F, Breeuwer JAJ (2001) A mite species that consists entirely of haploid females. *Science* **292**: 2479–2482.

Weinzierl RP, Berthold K, Beukeboom LW, Michiels NK (1998) Reduced male allocation in the parthenogenetic hermaphrodite *Dugesia polychroa*. *Evolution* **52**: 109–115.

Weismann A (1889) The significance of sexual reproduction in the theory of natural selection. *Essays upon Heredity* **1**: 255–332.

Welch DBM, Welch JLM, Meselson M (2008) Evidence for degenerate tetraploidy in bdelloid rotifers. *Proc Natl Acad Sci U S A* **105**: 5145–5149.

Werren JH (1983) Sex ratio evolution under local mate competition in a parasitic wasp. *Evolution* **37**: 116–124.

Werren JH (1987) The coevolution of autosomal and cytoplasmic sex ratio factors. *J Theor Biol* **124**: 317–334.

Werren JH (1991) The paternal-sex-ratio chromosome of *Nasonia*. *Am Nat* **137**: 392–402.

Werren JH (2011) Selfish genetic elements, genetic conflict, and evolutionary innovation. *Proc Natl Acad Sci U S A* **108 Suppl 2**: 10863–10870.

Werren JH, Baldo L, Clark ME (2008) *Wolbachia*: master manipulators of invertebrate biology. *Nat Rev Microbiol* **6**: 741–751.

Werren JH, Beukeboom LW (1998) Sex determination, sex ratios, and genetic conflict. *Annu Rev Ecol Syst* **29**: 233–261.

Werren JH, Hatcher MJ (2000) Maternal-zygotic gene conflict over sex determination: effects of inbreeding. *Genetics* **155**: 1469–1479.

Werren JH, Hatcher MJ, Godfray HC (2002) Maternal-offspring conflict leads to the evolution of dominant zygotic sex determination. *Heredity* **88**: 102–111.

Werren JH, Richards S, Desjardins CA, Niehuis O et al. (2010) Functional and evolutionary insights from the genomes of three parasitoid *Nasonia* species. *Science* **327**: 343–348.

Werren JH, Skinner SW, Huger AM (1986) Male-killing bacteria in a parasitic wasp. *Science* **231**: 990–992.

Werren JH, Taylor PD (1984) The effects of population recruitment on sex ratio selection. *Am Nat* **124**: 143–148.

Wertheim B, Beukeboom LW, Van de Zande L (2013) Polyploidy in animals: effects of gene expression on sex determination, evolution and ecology. *Cytogenet Genome Res* **140**: 256–269.

Wessels S, Hoerstgen-Schwark G (2007) Selection experiments to increase the proportion of males in Nile tilapia (*Oreochromis niloticus*) by means of temperature treatment. *Aquaculture* **272**: S80–S87.

Wessels S, Hoerstgen-Schwark G (2011) Temperature dependent sex ratios in selected lines and crosses with a YY-male in Nile tilapia (*Oreochromis niloticus*). *Aquaculture* **318**: 79–84.

West S (2009) *Sex allocation*. Princeton University Press, Princeton.

West SA, Herre EA, Sheldon BC (2000) The benefits of allocating sex. *Science* **290**: 288–290.

West SA, Lively CM, Read AF (1999) A pluralist approach to sex and recombination. *J Evol Biol* **12**: 1003–1012.

Western PS, Harry JL, Graves JA, Sinclair AH (1999) Temperature-dependent sex determination in the American alligator: AMH precedes SOX9 expression. *Dev Dyn* **216**: 411–419.

Wetzel J, Scheibner O, Burmester A, Schimek C et al. (2009) 4-dihydrotrisporin-dehydrogenase, an enzyme of the sex hormone pathway of *Mucor mucedo*: purification, cloning of the corresponding gene, and developmental expression. *Eukaryot Cell* **8**: 88–95.

Wheeler MJ, de Graaf BH, Hadjiosif N, Perry RM et al. (2009) Identification of the pollen self-incompatibility determinant in *Papaver rhoeas*. *Nature* **459**: 992–995.

White MJD (1973) *Animal Cytology and Evolution*. Cambridge University Press, Cambridge.

Whitfield LS, Lovell-Badge R, Goodfellow PN (1993) Rapid sequence evolution of the mammalian sex-determining gene SRY. *Nature* **364**: 713–715.

Wiens D, Barlow BA (1975) Permanent translocation heterozygosity and sex determination in East African mistletoes. *Science* **187**: 1208–1209.

Wild G, Taylor PD (2004) Fitness and evolutionary stability in game theoretic models of finite populations. *Proc Biol Sci* **271**: 2345–2349.

Wiley C, Ellison CK, Shaw KL (2012) Widespread genetic linkage of mating signals and preferences in the Hawaiian cricket *Laupala*. *Proc Biol Sci* **279**: 1203–1209.

Wilhelm D, Palmer S, Koopman P (2007) Sex determination and gonadal development in mammals. *Physiol Rev* **87**: 1–28.

Wilkins AS (1995) Moving up the hierarchy: a hypothesis on the evolution of a genetic sex determination pathway. *Bioessays* **17**: 71–77.

Willett CS (2008) No evidence for faster male hybrid sterility in population crosses of an intertidal copepod (*Tigriopus californicus*). *Genetica* **133**: 129–136.

Williams G (1979) The question of adaptive sex ratio in outcrossed vertebrates. *Proc R Soc Lond B* **205**: 567–580.

Williams GC (1980) Kin selection and the paradox of sexuality. In: Barlow GW, Silverberg J. eds. *Sociobiology: Beyond natural nurture?* Westview, Boulder, CO. pp. 371–384.

Wilson EB (1905) The chromosomes in relation to the determination of sex in insects. *Science* **22**: 500–502.

Wilson DS, Colwell RK (1981) Evolution of sex ratio in structured demes. *Evolution* **35**: 882–897.

Winge O (1932) The nature of sex chromosomes. *Proc 6th Int Congr Genet Ithaca, NY* **1**: 343–355.

Winge O (1934) *The experimental alteration of sex chromosomes into autosomes and vice versa, as illustrated by* Lebistes. *C R Lab Carlsberg Ser Physiol* **21**: 1–49.

Woese CR, Kandler O, Wheelis ML (1990) Towards a natural system of organisms: proposal for the domains Archaea, Bacteria, and Eucarya. *Proc Natl Acad Sci U S A* **87**: 4576–4579.

Wolf JB, Bryk J (2011) General lack of global dosage compensation in ZZ/ZW systems? Broadening the perspective with RNA-seq. *BMC Genomics* **12**: 91–2164–12–91.

Woram RA, Gharbi K, Sakamoto T, Hoyheim B et al. (2003) Comparative genome analysis of the primary sex-determining locus in salmonid fishes. *Genome Res* **13**: 272–280.

Woram RA, McGowan C, Stout JA, Gharbi K et al. (2004) A genetic linkage map for Arctic char (*Salvelinus alpinus*): evidence for higher recombination rates and segregation distortion in hybrid versus pure strain mapping parents. *Genome* **47**: 304–315.

Wu CI Davis AW (1993) Evolution of postmating reproductive isolation: the composite nature of Haldane's rule and its genetic bases. *Am Nat* **142**: 187–212.

Wu X, Haber JE (1996) A 700 bp cis-acting region controls mating-type dependent recombination along the entire left arm of yeast chromosome III. *Cell* **87**: 277–285.

Wu CI, Ting CT (2004) Genes and speciation. *Nat Rev Genet* **5**: 114–122.

Wu CI, Xu EY (2003) Sexual antagonism and X inactivation the SAXI hypothesis. *Trends Genet* **19**: 243–247.

Xiong Y, Chen X, Chen Z, Wang X et al. (2010) RNA sequencing shows no dosage compensation of the active X-chromosome. *Nat Genet* **42**: 1043–1047.

Xu J (2005) The inheritance of organelle genes and genomes: patterns and mechanisms. *Genome* **48**: 951–958.

Xu Z, Gao X, He Y, Ju J et al. (2012) Synergistic effect of SRY and its direct target, WDR5, on Sox9 expression. *PLoS One* **7**: e34327.

Xu X, Kumakura M, Kaku E, Takahashi M (2001) Odd mating-type substances may work as precursor molecules of even mating-type substances in *Paramecium caudatum*. *J Eukaryot Microbiol* **48**: 683–689.

Yamamoto T (1961) Progenies of sex-reversal females mated with sex-reversal males in the medaka, *Oryzias latipes*. *J Exp Zool* **146**: 163–179.

Yamasaki S, Fujii N, Matsuura S, Mizusawa H et al. (2001) The M locus and ethylene-controlled sex determination in andromonoecious cucumber plants. *Plant Cell Physiol* **42**: 608–619.

Yan Z, Hull CM, Sun S, Heitman J et al. (2007) The mating type-specific homeodomain genes SXI1 alpha and SXI2a coordinately control uniparental mitochondrial inheritance in *Cryptococcus neoformans*. *Curr Genet* **51**: 187–195.

Yang X, Schadt EE, Wang S, Wang H et al. (2006) Tissue-specific expression and regulation of sexually dimorphic genes in mice. *Genome Res* **16**: 995–1004.

Yano A, Guyomard R, Nicol B, Jouanno E et al. (2012) An immune-related gene evolved into the master sex-determining gene in rainbow trout, *Oncorhynchus mykiss*. *Curr Biol* **22**: 1423–1428.

Yano A, Nicol B, Jouanno E, Quillet E et al. (2013) The sexually dimorphic on the Y-chromosome gene (sdY) is a conserved male-specific Y-chromosome sequence in many salmonids. *Evol Appl* **6**: 486–496.

Yao B, Zhou L, Wang Y, Xia W et al. (2007) Differential expression and dynamic changes of SOX3 during gametogenesis and sex reversal in protogynous hermaphroditic fish. *J Exp Zool A Ecol Genet Physiol* **307**: 207–219.

Yi W, Ross J, Zarkower D (2000) *mab-3* is a direct *tra-1* target gene regulating diverse aspects of *C.elegans* male sexual development and behavior. *Development* **127**: 4469–4480.

Yoshida M, Itoh M (1974) Karyotype of the gecko, *Gekko japonicus*. *Chromosome Infor Serv* **17**: 29–31.

Yoshida K, Terai Y, Mizoiri S, Aibara M *et al.* (2011) B chromosomes have a functional effect on female sex determination in lake victoria cichlid fishes. *PLoS Genetics* **7**: e1002203.

Yoshida K, Kitano J (2012) The contribution of female meiotic drive to the evolution of neo-sex chromosomes. *Evolution* **66**: 3198–3208.

Yoshimoto S, Ikeda N, Izutsu Y, Shiba T et al. (2010) Opposite roles of DMRT1 and its W-linked paralogue, DM-W, in sexual dimorphism of *Xenopus laevis*: implications of a ZZ/ZW-type sex-determining system. *Development* **137**: 2519–2526.

Yoshimoto S, Okada E, Umemoto H, Tamura K et al. (2008) A W-linked DM-domain gene, DM-W, participates in primary ovary development in *Xenopus laevis*. *Proc Natl Acad Sci U S A* **105**: 2469–2474.

Youngman PJ, Anderson RW, Holt CE (1981) Two multiallelic mating compatibility loci separately regulate zygote formation and zygote differentiation in the myxomycete *Physarum polycephalum*. *Genetics* **97**: 513–530.

Yun SH, Berbee ML, Yoder OC, Turgeon BG (1999) Evolution of the fungal self-fertile reproductive life style from self-sterile ancestors. *Proc Natl Acad Sci U S A* **96**: 5592–5597.

Yusa Y (2007a) Nuclear sex-determining genes cause large sex-ratio variation in the apple snail *Pomacea canaliculata*. *Genetics* **175**: 179–184.

Yusa Y (2007b) Causes of variation in sex ratio and modes of sex determination in the Mollusca—an overview. *Am Malacol Bull* **23**: 89–98.

Zander RH (1985) Bryophyte sexual systems: -oicous versus -oecious. *Bryolog Beit* **3**: 46–51.

Zeyl C, Bell G (1997) The advantage of sex in evolving yeast populations. *Nature* **388**: 465–468.

Zha X, Xia Q, Duan J, Wang C et al. (2009) Dosage analysis of Z chromosome genes using microarray in silkworm, *Bombyx mori*. *Insect Biochem Mol Biol* **39**: 315–321.

Zhang YE, Vibranovski MD, Landback P, Marais GA et al. (2010) Chromosomal redistribution of male-biased genes in mammalian evolution with two bursts of gene gain on the X chromosome. *PLoS Biol* **8**: e1000494.

Zhao L, Koopman P (2012) SRY protein function in sex determination: thinking outside the box. *Chromosome Res* **20**: 153–162.

Zhao H, Lu M, Singh R, Snell W (2001) Ectopic expression of a *Chlamydomonas* mt+ -specific homeodomain protein in mt- gametes initiates zygote development without gamete fusion. *Genes Dev* **15**: 2767–2777.

Zhao D, McBride D, Nandi S, McQueen HA et al. (2010) Somatic sex identity is cell autonomous in the chicken. *Nature* **464**: 237–115.

Zhou Q, Bachtrog D (2012a) Chromosome-wide gene silencing initiates Y degeneration in *Drosophila*. *Curr Biol* **22**: 522–525.

Zhou Q, Bachtrog D (2012b) Sex-specific adaptation drives early sex chromosome evolution in *Drosophila*. *Science* **337**: 341–345.

Zhuchenko A, Korol A, Kovtyukh L (1985) Change of the crossing-over frequency in *Drosophila* during selection for resistance to temperature fluctuations. *Genetica* **67**: 73–78.

Zimmerman JK (1991) Ecological correlates of labile sex expression in the orchid *Catasetum viridiflavum*. *Ecology* **72**: 597–608.

Zuk J (1970) Structure and function of sex chromosomes in *Rumex thyrsiflorus*. *Acta Soc Bot Pol* **39**: 539–564.

Zurovcova M, Eanes WF (1999) Lack of nucleotide polymorphism in the Y-linked sperm flagellar dynein gene Dhc-Yh3 of *Drosophila melanogaster* and *D. simulans*. *Genetics* **153**: 1709–1715.

# Taxonomic Index

Regular font = book pages
Bold font = special sex determination sections
S = electronic addendum
P1 = colour plate 1
P2 = colour plate 2

# Author Index

# Subject Index

Regular font = book pages
**Bold font** = special sex determination sections
*Italic font* = glossary
S = electronic addendum
Gene names in italics, protein names in capitals

The manufacturer's authorised representative in the EU for product safety is Oxford University Press España S.A. of El Parque Empresarial San Fernando de Henares, Avenida de Castilla, 2 - 28830 Madrid (www.oup.es/en or product.safety@oup.com). OUP España S.A. also acts as importer into Spain of products made by the manufacturer.
Printed and bound by CPI Group (UK) Ltd, Croydon, CR0 4YY
14/09/2026
02216291-0019